ENGINEERING IN HISTORY

BY

RICHARD SHELTON KIRBY
SIDNEY WITHINGTON
ARTHUR BURR DARLING
AND
FREDERICK GRIDLEY KILGOUR

DOVER PUBLICATIONS, INC., New York

Published in Canada by General Publishing Company, Ltd., 30 Lesmill Road, Don Mills, Toronto, Ontario.
Published in the United Kingdom by Constable and Company, Ltd.

This Dover edition, first published in 1990, is an unabridged, unaltered republication of the work originally published by the McGraw-Hill Book Company, New York, 1956.

Manufactured in the United States of America
Dover Publications, Inc., 31 East 2nd Street, Mineola, N.Y. 11501

Library of Congress Cataloging-in-Publication Data

Engineering in history / by Richard Shelton Kirby . . . [et al.]. — Dover ed.
    p.   cm.
  Reprint. Originally published: New York : McGraw-Hill Book Co., 1956.
  Includes bibliographical references.
  ISBN 0-486-26412-2
  1. Engineering—History. I. Kirby, Richard Shelton, 1874– .
TA15.E54   1990
620'.009—dc20   90-34547
                                                          CIP

Preface

This history of engineering is intended to present the development of engineering in Western civilization from its origins into the twentieth century. Although the emphasis is on the advance of engineering through the centuries, the activities of engineers are related to other human activities because engineering has not evolved in a vacuum. It has contributed much to human societies, and in turn these societies have conditioned the progress of engineering. The history of this interaction is discussed with the history of the accumulation of knowledge on which the advance of engineering depends.

Engineering in History is a general introduction to the subject rather than a definitive history. We have chosen to emphasize within the scope of the book achievements whose impact on civilization has appeared to us most significant. The fields covered are certain branches of civil, mechanical, electrical, and metallurgical engineering. We hope the present volume will give both student and general reader a knowledge and understanding of engineering history. Technical terms have been either omitted or defined in simple words. Readers who wish to pursue the subject further will find a short general bibliography at the end of the book and selected bibliographies following most chapters. These book lists will also provide a general documentation for the several chapters. However, the writing of the book has been based in large part on original sources. Footnotes have been kept to a minimum, merely citing sources on which possibly controversial statements have been based and the origins of most quotations.

Engineers and historians have worked jointly on the book. Richard Shelton Kirby is Associate Professor Emeritus in the School of Engineering of Yale University; Sidney Withington is former Chief Electrical Engineer of the New York, New Haven and Hartford Railroad; Arthur Burr Darling is Chairman of the History Department in Phillips Academy, having previously been Associate Professor of History in Yale University; and Frederick Gridley Kilgour is Lecturer on the History of Science and Librarian of the Medical Library at Yale University. This joint authorship has been most stimulating to the authors; each with his different specialized interest has learned much from the others.

The authors acknowledge their indebtedness to many historians and engineers. Among the historians, Stanley M. Pargellis, formerly Assistant Professor of History at Yale and now Librarian of the Newberry Library, contributed much when the book began to take shape more than a decade ago. Ferris J. Stephens of Yale's Babylonian Collection, Ludlow Bull of the Metropolitan Museum of Art, Clarence W. Mendell, and Harry M. Hubbell have furnished many helpful suggestions, as have P. J. Branner, S. McK. Crosby, and P. C. Miller. Among engineers to whom the authors are grateful for their help on one or more chapters are H. F. Brown, A. G. Conrad, Theodore Crane, B. F. Dodge, C. S. Farnham, C. R. Harte, K. T. Healy, T. K. A. Hendrick, P. G. Laurson, C. T. G. Looney, S. F. Mackay, Grant Robley, and R. H. Suttie. For the style, arrangement, and correctness of the text the authors are, however, solely responsible.

For the preparation of the manuscript and for many helpful suggestions, we are most grateful to Evelena B. Strong, Eleanor G. Cooper, and to Henrietta T. Perkins who has also done invaluable work on the index. In addition, the authors are under very special obligation to the staff of the Yale University Library for its interest and cooperation through the years. The Engineering Societies' Library, the Harvard University Library, the New York Public Library, and the Library of the University of South Carolina have all given much help.

Richard Shelton Kirby
Sidney Withington
Arthur Burr Darling
Frederick Gridley Kilgour

Contents

PREFACE   v

CHAPTER ONE   Origins   1

CHAPTER TWO   Urban Society   6

CHAPTER THREE   Greek Engineering   36

CHAPTER FOUR   Imperial Civilization   56

CHAPTER FIVE   The Revolution in Power   95

CHAPTER SIX   Foundations for Industry   124

CHAPTER SEVEN   The Industrial Revolution   159

CHAPTER EIGHT   Roads, Canals, Bridges   199

CHAPTER NINE   Steam Vessels and Locomotives   246

CHAPTER TEN   Iron and Steel   291

CHAPTER ELEVEN   Electrical Engineering   327

CHAPTER TWELVE   Modern Transportation   374

CHAPTER THIRTEEN   Sanitary and Hydraulic Engineering   426

CHAPTER FOURTEEN   Construction   464

CHAPTER FIFTEEN   Reflections   495

INDEX   517

ONE

Origins

"As a people, our strength has lain in practical application of scientific principles, rather than in original discoveries. In the past, our country has made less than its proportionate contribution to the progress of basic science." So reported in 1947 the President's Scientific Research Board in its *Science and Public Policy*. The board went on to say, "Instead, we have imported our theory from abroad and concentrated on its application to concrete and immediate problems. This was true even in the case of the atomic bomb." This quotation is accurate in saying that the United States "has made less than its proportionate contribution to the progress of basic science" but is misleading in so far as it implies that Americans have made a large proportion of the primary applications of pure science to engineering. During the last hundred years, Europeans have made the majority of the fundamental engineering advances. Even in the case of applying pure science for development of the atomic bomb, as Henry DeW. Smyth pointed out in his *Atomic Energy for Military Purposes*, "the early efforts both at restricting publication and at getting government support were stimulated largely by a small group of foreign-born physicists. . . ." Since engineering is an important factor in national security and welfare, it is of the utmost importance in the formulation of national policies for the citizen to know the extent to which a nation is dependent on others for engineering advances. The best way to gain this perspective is from history.

Broadly speaking, there are two values which accrue from the study of the history of engineering—one pragmatic, the other general. In addition to providing a comprehension of what is necessary for national development, the study of the history of engineering teaches what it is to be an engineer. To the general reader such a study can give an understanding of the lessons of engineering experience, helping him to a knowledge of the complex environment man has created for himself. It increases reverence for the past by making clear that engineers can accomplish much today because they are standing on the shoulders of men who have gone before them. As George Sarton has put it, "Reverence without progressiveness may be stupid; progressiveness without reverence is wicked and foolish." [1] Finally, as does any study undertaken for its own sake, it widens the human horizon and liberates from narrow ideas.

The purpose of any historical work is to interpret the development and activity of man. The history of engineering is but one segment of the great historical narrative, but unlike some other histories it records a human activity which is cumulative and progressive. "Progressive" in this sense does not connote any judgment of values but merely indicates advances built upon and including previously existing knowledge. The history of engineering depicts a section of that central theme of history which reveals the development of civilization.

Engineering has not been defined satisfactorily in a single sentence. In 1828 the British architect Thomas Tredgold, probably the first to make the attempt, called it "the art of directing the great sources of power in nature for the use and convenience of man." It was so defined in the charter of the Institution of Civil Engineers of which Thomas Telford was the first president. Tredgold's simple and concise definition may have been measurably satisfactory for his generation when steam transportation was a not too successful novelty and when only a few scientists vaguely sensed the possibilities inherent in the mysterious electric current. The profession has, however, expanded so rapidly through the decades since that the dozens of definitions which have been framed by lexicographers, even by engineering organizations beginning in the 1880s, no longer seem adequate.

The authors of this book, apologetically and tentatively, submit that midway in this twentieth century civilian engineers by and large are busying themselves with "the art of the practical application of scientific and empirical knowledge to the design and production or accomplishment of various sorts of constructive projects, machines and materials of use or value to man." The three most important phrases in this definition are

[1] George Sarton, *The History of Science and the New Humanism*, Harvard University Press, Cambridge, Mass., 1937, p. 52.

"application of knowledge," "design and production or accomplishment," and "use or value"; these three elements must be considered together if the word engineering is to be properly understood. It is the thesis of this book that progress in engineering results from the accumulation of knowledge. This is not, however, the whole story; whatever is produced or accomplished must be of some use or value to man. Value is not necessarily measured by an economic yardstick; the ancient pyramids and not a few structures since are of slight economic worth, while their value in terms of faith and beauty has often been considerable. Professor Hardy Cross, in his delightful *Engineers and Ivory Towers*, has generalized the position of engineering: "It is customary to think of engineering as a part of a trilogy, pure science, applied science and engineering. It needs emphasis that this trilogy is only one of a triad of trilogies into which engineering fits. The first is pure science, applied science, engineering; the second is economic theory, finance and engineering; and the third is social relations, industrial relations, engineering. Many engineering problems are as closely allied to social problems as they are to pure science."

Another recent statement describes what engineers do.

"By and large engineers are paid by society to work on systems dealing with problems whose solutions are of interest to that society. These systems seem to group conveniently into (*a*) systems for material handling, including transformation of and conservation of raw and processed materials; (*b*) systems for energy handling, including its transformation, transmission, and control; and (*c*) systems for data on information handling, involving its collection, transmission, and processing.

"In carrying out this work engineers engage in various activities ranging through engineering research, design and development, construction, operation, and management." [2]

It is the endeavor of this book to discuss engineering in history, not as though engineering occurs in a historical vacuum without reference to other human activities, but as one of many social enterprises. In other words, it is intended to integrate the history of engineering into general history. To this end the presentation is oriented about eight of the great events of history which totally changed the ways of human life. These are:

[2] W. W. Harmon, J. B. Franzini, W. G. Ireson, and S. J. Kline, "Abstract Report of the Stanford University Committee on Evaluation of Engineering Education," *The Journal of Engineering Education*, vol. 44, p. 258, December, 1953.

Food-producing revolution (ca. 6000–3000 B.C.)
Appearance of urban society (ca. 3000–2000 B.C.)
Birth of Greek science (600–300 B.C.)
Revolution in power (Middle Ages)
Rise of modern science (seventeenth century)
Steam and the Industrial Revolution (eighteenth century)
Electricity and the beginnings of applied science (nineteenth century)
Age of automatic control (twentieth century)

The fundamental changes here named stimulated engineering developments, which in turn accelerated the rate of historical revolution.

Before 6000 B.C. man's chief occupation was gathering his food. He hunted animals in the fields and forests, fished in lakes and streams, and picked edible plants wherever he could find them. He had no domesticated plants and animals to supply him with food and clothing. Many families and tribes were nomadic, following their sources of food supply wherever these sources might go, and they lived for the most part in the flimsiest of dwellings such as grass huts or tents, although in some parts of the world they eventually established more permanent homes, as the caves of Mousterians in France. The population was extremely sparse, there being few groups of dwellings that could be called villages. Engineering played no part in the existence here described.

Probably in Africa and Asia Minor some time around 6000 B.C. man started one of the most important developments in human history. He began to domesticate and cultivate plants and animals, and particularly in the valleys of the Tigris and Euphrates and of the Nile, he built houses in groups and cultivated the neighboring land. This great development which started some eight thousand years ago was still evolving in remote areas of the world during the last century. However, for the history of engineering the most important rise of a society of food producers was that which took place in the Near East approximately between 6000 and 3000 B.C.

The men who during this period settled in the Tigris-Euphrates and Nile Valleys built permanent dwellings. They also had to control and use the river water for cultivation. Their solutions to these problems of construction and irrigation were engineering ones. Structural, hydraulic, transportation, and metallurgical engineering had their beginnings in this period and helped to solve some of the problems which the new style of living generated. The food gatherers had discovered how to make and control fire, but it was the food producers who invented the wheel, the harness for oxen, the sail for boats, the plow, and the brick; they also discovered techniques for recovering copper from its ores.

Shortly after the beginning of cultivation of domestic plants and animals, the population increased very sharply. Moreover, as the effectiveness of cultivation improved, it became possible to release men from the daily business of producing food so that they could engage in other activities. A few became priests, others rulers; many became artisans, some of them developing into the first engineers. Although the new way of life produced the stimulus and opportunity which initiated engineering, there was the accompanying rise of the general practice of enslaving men and women. The continuation of the use of slaves as the principal source of power until the Middle Ages made unnecessary the development of other than human sources of power for more than three thousand years.

TWO

Urban Society

The first significant change in human living to follow the development of food production was the rise of cities, which occurred some time before 3000 B.C. Prior to that time the majority of persons lived in villages, which consisted of groups of farmers' homes. The principal difference between a village and a city was that in the village most of the inhabitants were directly engaged in the production of foods whereas very few city dwellers were so engaged. The effectiveness of food production, combined with the rise of centralized governmental administration and commerce, made it possible for many inhabitants of the larger communities which later became cities to take up activities other than farming or fishing. They became instead rulers, administrators, soldiers, priests, scribes, or craftsmen. The interaction between this new urban society and engineering was most fertile, but of equal importance was the development of knowledge and the tools of knowledge fundamental for the engineer.

It should be remembered, however, that most early knowledge was purely empirical, gained from experience and handed down from person to person. There is no evidence of the existence of any generalized or abstract geometry, nor was there any understanding of the uniformities or regularities underlying the appearances of nature. In short, there was no science as the term is understood today. The food producers transmitted, usually to apprentices by word of mouth, the knowledge they had accumulated. To assist in communicating this knowledge they later began to use elementary writing, computation, and mensuration. Writing before the third millennium B.C. consisted of complicated pictographic symbols. Extant documents are mostly accounts, contracts, and dictionaries or lists of signs that represent specific objects, and later, ideas related to the objects.

During the next few centuries the scribes simplified writing; they could now prepare new types of documents, historical texts, rituals, legal codes. The development of writing progressed rapidly as urban populations increased and prospered. Although the origin of writing had been economic, it was soon being used for all manner of purposes, including engineering. Alphabetic writing, however, was invented at least as early as the fourteenth century B.C. Computation techniques and arithmetic gradually came to be used for commercial purposes and in surveying and mensuration in the third millennium. Closely allied with the ability to compute were some very rough and highly localized standardizations of weights and measures. All of these changes had their influence on the engineering of the period.

The growth of the cities also stimulated engineering in other ways. An increase in wealth, an extension of political power, and a growth of trade accompanied the rise of cities. Before 3000 B.C. most buildings were modest homes, but after that time structural engineering was no longer merely functional; it was also architectural. Great palaces were built for princes and great temples for priests. One consequence of the rise of an organized religion with its large structures was an increase in engineering activity and knowledge. The new wealth and religious activities also led to the construction of monumental tombs, of which the pyramids are outstanding examples. Their construction advanced engineering, but it must be remembered that the motive for building them was religious.

The lower alluvial valleys of the Tigris-Euphrates and Nile are singularly lacking in most natural resources. There was clay available for bricks, but there was no stone for palaces, temples, and tombs. Stone blocks had to be transported over long distances, as did other materials, such as timber. Other requirements for communications and transportation arose from the expansion of political power and the need to maintain administrative control over large areas. The development of specialized craftsmen and manufactures in the cities meant an increase in trade, with still another demand for improved transportation. Engineers solved the transportation problems by learning how to build roads, bridges, and bigger ships which could travel over greater distances.

The cities also produced problems which the evolution of hydraulic engineering gradually solved. Open drains were constructed to remove surface water, and in a few instances, underground drains kept building foundations from sinking into mud. The need for increased efficiency in food production resulted in the construction of levees, dams, reservoirs, and canals for flood control and irrigation. City dwellers need water, and tunnels were sometimes constructed to bring water from nearby springs to pools within the city, but it was not until about 700 B.C. that stone aqueducts were first built to bring water to cities.

So it was with the urban revolution that political, economic, religious, and social factors stimulated the developments of engineering, and they have influenced it ever since. In turn the new engineering influenced politics, economics, religion, and social living by making available the means by which these activities could continue to evolve. There can be no doubt that engineering came into being to solve new problems of a new society, but once engineering was established, there arose an interaction by which it in turn influenced the evolution of society. This state of affairs continues today and will continue to exist as long as civilization remains dynamic and maintains its evolutionary course.

Building in Mesopotamia

Too little is known about the achievements of the Mesopotamians, for their cities have disappeared, concealed by the desert. No impressive monuments like the pyramids of Egypt perpetuate traditions of magnificence. Those who have described the engineering of the Babylonians and Assyrians have had to rely upon the tales of ancient historians. The famous Greek, Herodotus (ca. 484–425 B.C.), who visited Babylon in the fifth century B.C., wove into his accounts of things as he saw them the legends which were told to him, with scarcely an intimation of the difference. Strabo and Diodorus and the elder Pliny in Roman times drew heavily upon tradition as well as upon the records of their day.

Some of the tales of Herodotus have proved more accurate than was believed a generation ago. But archaeologists have yet to find the evidence which will support the statement of Diodorus that there was a vaulted brick tunnel under the Euphrates. Pliny's assertion that the walls of Babylon were 60 miles long and 200 feet high has been refuted by actual measurements on the site. Babylon was roughly some 2 miles by 3 (Figure 2.2). As for walls, the Great Wall of China was never much over 50 feet high at any place. A 200-foot wall would be as high as a present-day 20-story office building. Few of the inscriptions that have been found give accurate information. Kings are not devoted to caution

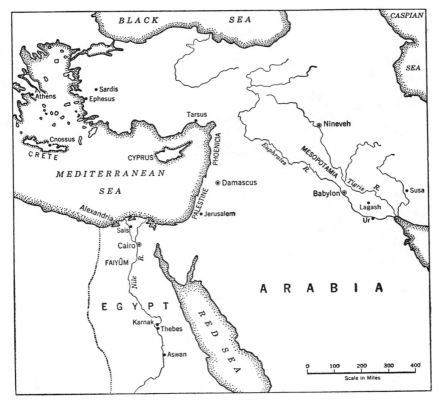

Figure 2.1. The Near East in ancient times

when writing to posterity about themselves. Nebuchadnezzar ruled in Babylon from 605 to 561 B.C., and some of his exploits, dreams, and tragic career appear in the Bible associated with the fame of Daniel, the Hebrew captive. It is particularly interesting, therefore, to learn that Nebuchadnezzar also had reconstructed the temples of Babylon, repaired its irrigation ditches, paved the streets, and built the wall about the city. But we must be allowed some doubt that he set the foundations of his wall "upon the bosom of the abyss . . . its top . . . raised mountain high." [1]

Thousands of clay tablets, which have been uncovered at the sites and are now treasured in museums the world over, have far more importance as a source of information. Most are about 3 by 4 inches, though some are as large as a good-sized book. Many date even from the time of Hammurabi. These records in cuneiform writing deal often with prob-

[1] William H. Lane, *Babylonian Problems*, John Murray, London, 1923, p. 179.

Figure 2.2
A restoration of ancient Babylon (From E. Unger, *Babylon, die Heilige Stadt,* 1931; courtesy Walter De Gruyter & Co.)

lems in practical mathematics that disclose the knowledge and the activities of the Babylonian engineers. They understood right triangles and computed areas of land, volumes of masonry, the cubic contents of an excavation for a canal. They solved many types of simple algebraic equations and applied the principles directly to their work. Instead of 10, the base for their number system was 60, which we still use for measuring angles and for telling time.

The architects of Mesopotamia may have made drawings of their proposed structures. There are statues of a petty ruler, Gudea, who lived about 2200 B.C., that show him seated with drafting materials and a plan on his lap (Figure 2.3). But all the sketches which have come to us are very simple and often crude, and although it is possible that more specific and elaborate plans on parchment or papyrus were made, they have not been preserved and no record of them exists. The most important sources of information about the engineering of Mesopotamia are the ruins of the cities themselves. Archaeologists have found at Ur, south of Babylon, enough of the ziggurat, or temple tower, to give us a very good idea of what it was in its original form. As stone was scarce, the builders used

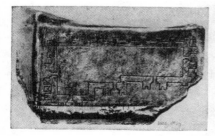

Figure 2.3
Gudea and the drawing on his lap
(From Ernest de Sarzec,
Découvertes en Chaldée, 1891)

brick—usually sun-dried. Fuel for firing was hard to get. High walls cannot be made with brick that has merely been dried in the sun; such
masonry cannot withstand heavy pressure. When the engineers approached the limit of safety—and we may assume they soon learned from
experience what that was—they started a second wall behind and above
the first, upon an interior mound of earth. There were often a third and a
fourth wall, sometimes more. In general outline the taller ziggurats must
have been quite a bit like our more recent skyscrapers with walls stepped
back.

   Like modern penthouse dwellers, the kings of Mesopotamia often
planted the series of flat roofs. The hanging gardens of Babylon, built
by Nebuchadnezzar, were one of the Seven Wonders of the World. And
it is probably from the temple tower of Marduk, a marvel of height to

Figure 2.4
Ziggurat of Ur, present state and a restoration (From C. L. Woolley, *The Ziggurat and Its Surroundings*, 1939; courtesy British Museum and the University Museum of the University of Pennsylvania)

people who lived close to the ground, that the story of the Tower of Babel has come to us in the Bible. Its actual height was probably not over 200 feet. The remains at Ur indicate that its ziggurat was a many-angled and solid pyramid, approximately 200 by 150 feet and 70 feet tall. It was made of sun-dried brick, faced with burnt brick and with some stone (Figure 2.4). There were a number of stairways by which processions of priests and worshippers could ascend to open-air shrines at the summit. This brick tower survived some two thousand years, with extensive alterations and rebuilding, for the temples of Mesopotamia were the cumulative work of many centuries. The material commonly used to bind the masonry was asphalt or bitumen; it still clings to bricks laid perhaps twenty-five centuries ago. The prize of international rivalries, it is found to this day bubbling from the ground near the site of Babylon. Evidently the early engineers never thought to use it for surfacing roads.

The palaces of later kings in Assyria and Persia were made from the same sort of materials which the ancient Sumerians in the south had used. There is still standing northeast of Nineveh a portion of Sargon II's great palace built in the eighth century B.C. Originally it was composed of three groups of buildings extending over 25 acres and containing 200 rooms. Its walls were of moist or partially baked bricks that stuck together as they dried and hardened. They were faced either with stucco or enameled brick. But in some lower parts of the walls there are great limestone monoliths, one weighing more than 20 tons, of which the architectural and engineering purposes are not apparent. The courts of the palace were elaborately paved with stone set in asphalt. There were also storm-water drains emptying into brick-lined sewers which led to main conduits covered with flat slabs of stone or with brick vaults.

The people of Mesopotamia developed two kinds of arches. The first was a corbeled, or false, arch built up with horizontal courses of brick or stone, each projecting slightly beyond the one below it until the two sides met and closed the opening at the top. The principle of the cantilever could not be carried far with such materials. The corbeled arches of the Mesopotamian engineers were therefore relatively short spans, impressive neither in width nor in height. Their second form of arch, found in gates through walls, was a true arch.[2]

Hydraulic and Sanitary Engineering

Though the Tigris and Euphrates Rivers made possible agricultural settlement upon the land, they were terrifying destroyers when uncontrolled. The story of the Deluge in the Bible appears to have come from Babylonian sources; it may be identified with a flood in the Euphrates Valley. The inhabitants of Mesopotamia must have struggled from earliest times with problems of hydraulic engineering. How were they to build levees that would keep the streams in their channels? How could dams be made to hold them back, jetties or dikes built to divert floodwaters into storage basins, and canals dug for irrigating the fields in the dry

[2] Henri Frankfort, *Iraq Excavations of the Oriental Institute, 1932–1933*, University of Chicago Press, Chicago, 1934, p. 15.

seasons? The early builders of Sumer and Babylon scored some of their most significant achievements in hydraulic engineering.

There are many legends of Marduk, the greatest of the Babylonian gods; one tells how he first conquered the dragon in the waters about the earth. More reliable are contemporaneous historical texts on clay tablets dating from about the twenty-fifth century B.C. with their frequent references either to canals that sundry kings had dug and that sometimes served as boundaries or to reservoirs that supplied certain cities. Later texts (Third Dynasty of Ur, perhaps 2000 B.C.) cite the proper names of many such canals and reservoirs and mention workmen who were stationed to repair or clean them. Letters of Hammurabi (1800 or 1750 B.C.) refer to cleaning out canals,[3] and Hammurabi's famous code has sections on the proper upkeep of irrigation ditches. On pages 76 to 91 of their *Mathematical Cuneiform Texts*, Neugebaur and Sachs refer to tablets (most of them now at Yale) which deal with the mathematics related to the business of canal construction. These tablets date from the old Babylonian period, at least as early as 1800 B.C. A recently deciphered clay tablet, dating perhaps from 1200 B.C., refers, it would seem, to a form of treadmill for raising water from an irrigation ditch. It is a legal text or document acknowledging that a man had borrowed a water-raising wheel of 17 steps, 20 feet long; it had been lost and he must replace it.[4]

What is left of the low dams and river works in Mesopotamia has been so eroded, and the courses of rivers and canals so changed, that modern engineers do not hazard specific answers to obvious questions. The mounds and ridges which remain show that the embankments and levees protecting the lands of the lower Euphrates were a hundred feet wide and hundreds of miles long; wherever feasible, there evidently were spillways to carry excessively high water into great depressions in the desert. These depressions covered as much as 650 square miles, and they could hold water to a depth of 25 feet. But how did the builders of those days make their dams and levees watertight or resistant to water wear? How were their great reservoirs opened and closed without loss of control over the water so impounded? Archaeologists have not yet discovered the answers to these questions.

There is relatively little information available about one of the

[3] *The Letters and Inscriptions of Hammurabi*, 3 vols., Luzac and Co., London, 1898–1900, vol. 3, pp. 14–19.
[4] E. Ebeling, "Akkadisch *bakrâtu* = Rad einer Bewässerungsmaschine," *Orientalia*, vol. 20, p. 14, 1951.

greatest dams of antiquity. The Arabian shore of the Red Sea, desert today, supported a thriving agricultural and commercial people for two thousand years or more, represented in the Bible by the Queen of Sheba. Their comparatively rich life in the tenth century B.C. (Solomon's day) appears to have been made possible by huge dams which stored the rains in the hills, saved the topsoil from erosion, and provided for irrigation. Only a few Western explorers, no one of whom was trained in engineering, have penetrated to the site of the largest of these, the Marib, or Yemen, Dam. According to one account, it must have been huge, for it was described as 2 miles long. 120 feet high, 500 feet wide at the base, and containing 15 million cubic yards of stone. Another story, however, has the dam no more than a quarter of that size and built of earth. It remained in use into the sixth century A.D. when it failed, probably because of neglect. Whether the Queen of Sheba's people gained their knowledge of hydraulic engineering from the Mesopotamian folk to the north beyond the desert or from the Egyptians on the other side of the Red Sea or developed it themselves is not known. If the most enthusiastic accounts are correct, they had in their Marib Dam another wonder of the world. A recent writer, George Kheirallah, tells of other Arabs who, in 1936 and 1947, examined the buried ruins of the Marib Dam. They reported that the dam was 650 meters long, that there were 5 spillway channels and 14 irrigation channels.[5] While these explorers were a bit vague as to the original height of the structure, they found remains of some exceptionally good granite masonry.

A low ridge at Jerwan above Nineveh long deceived archaeologists, who had assumed that it was either a road or a dam. Recently under excavation, it has proved to be the aqueduct (Figure 2.5) which the engineers of Sennacherib, King of Assyria, constructed in the seventh century B.C. to supply Nineveh. Its most striking feature was a bridge more than 1,000 feet long and 70 wide, rising at one point 30 feet above the valley and crossing a stream on corbeled arches. At least two million blocks of limestone went into its construction. The bed of the conduit was paved with a concrete made of lime that still has some strength. Once he had proclaimed himself king of the world, as well as of Assyria, Sennacherib made his report on this project or one like it to posterity with remarkable restraint—for a god-king. He merely said: "For a long distance, adding to it the waters of the twain Hazur River—(namely) the waters of the

5 George Kheirallah, *Arabia Reborn*, University of New Mexico Press, Albuquerque, N. Mex.. 1952. Plan opposite p. 20.

Figure 2.5 Air view of aqueduct at Jerwan, Iraq. The aqueduct runs from the upper right center to the lower left corner. (Courtesy Oriental Institute, University of Chicago, and the Royal Air Force)

River Pulpullia—(and) the waters of the town of Hanusa, the waters of the town of Gammagara, (and) the waters of the springs of the mountains to the right and left at its sides, I caused a canal to be dug to the meadows of Nineveh. Over deep-cut ravines I spanned (*lit.*, caused to step) a bridge of white stone blocks. Those waters I caused to pass over upon it." [6] Even more astonishing, he gave credit to his engineers and men. A large relief made of alabaster and placed on a wall of Sennacherib's palace in Nineveh shows how heavy statues were transported in mountainous country (Figure 2.6).

It became common practice in the Near East, wherever springs were close by, to bring water through tunnels into pools within the city walls so that there might be a supply in times of siege. Women trudged with their water jars upon their shoulders, down and up the long rock stairways of these deep pools daily, for there was no such thing then as piping into private houses. Water systems for individual dwellings are in fact even now by no means universal. A tunnel of this sort, made in the seventh century B.C., carried water from the spring called Gihon to the pool of Siloam within Jerusalem. The conduit was 6 feet high, 1,750 feet

[6] Thorkild Jacobsen and Seton Lloyd, *Sennacherib's Aqueduct at Jerwan,* University of Chicago Press, Chicago, 1935, p. 20. (Quoted with the kind permission of the University of Chicago Press.)

Figure 2.6 Transporting a heavy statue, bas relief at Nineveh (note lever) (From A. Paterson, *Assyrian Sculptures, Palace of Sinacherib,* 1912; courtesy Martinus Nijhoff)

long, shaped in a roughly reversed curve plan, and cut through solid rock. Construction began at both ends. Fortunately, each gang of laborers, groping in the darkness, heard the sound of the other's picks, and after several attempts met. The inscription relating their experience, and joy, was discovered in 1880 by boys who were bathing in the pool.

The irrigation system that, for centuries, interlaced the Mesopotamian plains along the Tigris and Euphrates Rivers has long since disappeared under the sands of the desert. Cyrus, in the sixth century B.C., may have diverted the Tigris into 30 canals and then managed to close its main channel by an earthen dam. At any rate, there are remains of a sprawling structure of earth reinforced with timbers. At some time it seems to have had a brick facing on its upper surface. The dam was named for Marduk and may well date from a thousand years before Cyrus. As late as 401 B.C. the famous Greek soldier Xenophon saw this network of canals and carefully described it in his *Anabasis;* he was constantly meeting canals which could not be crossed without bridges. Sometimes these bridges were stationary of brick and timber, sometimes of boats. These canals, he said, issued from the Tigris River, and ditches had been cut from one to another across the country. The first of these channels was large, the next were smaller, and the last mere drains.

These canals were significant engineering, but the men of Mesopo-

Figure 2.7　Harappa drain, Indus Valley (Courtesy Museum of Fine Arts, Boston)

tamia, and those who learned their canal digging from them, did not know all there was to be known about excavation. At least, so it would appear from the story of the canal which the Persian conqueror, Xerxes, on the way to Greece, had his men dig across the Isthmus of Athos in 482 B.C. He wished to move his vast army and fleet safely past the stormy headland where the previous expedition had been wrecked in the Aegean Sea. Herodotus took delight in relating how the engineer Artachaees lacked but four fingers' breadth of being 8 feet tall and had the loudest voice among them. The historian was as sensitive to facts of importance in engineering. He recorded how the excavated material was carried up from one level to another and removed from the site. He took special note that most of the laborers on the job tried to make the sides of the canal vertical, with disastrous results. The Phoenician engineers in Xerxes' army made the top of their section twice as wide as the bottom. They understood that a canal excavated in earth should have sloping banks.

The abandoned sites of Mohenjo-daro and Harappa, which lie in the Indus Valley of western India, contain excellent evidence of ancient knowledge of hydraulic engineering (Figure 2.7). It may be that their inhabitants were those, or the relatives of those, whose earlier culture influenced the Egyptians and the Sumerians. In any case, the site of Mohenjo-daro, now partially excavated, has the remains of a sewer system that was effective even according to modern standards. The authorities, as early as 2500 B.C., had provided for the householder of average means sanitary conveniences comparable to those then enjoyed elsewhere by none but royalty. Each of the larger houses, even when they lacked windows, had a well and, rather too near it, a bathroom, latrine, lavatory, cesspool, and rubbish bin. Waste water from the houses and public buildings was carried through pottery pipes into covered brick drains along every street and discharged from these into brick culverts leading into the fields or the stream. The city had a brick pool approximately 23 by 40 feet. How it was filled is a puzzle. That it stayed full, if not clean, is more certain, for its thick walls were waterproofed with clay and bitumen. Possibly it was a fish pond. More likely it was a place of ritualistic immersion and cleansing.

Roads, Bridges, and Ships

Hard-surfaced streets may have been earlier achievements in engineering than sewer systems and water supply. Men may have walked on cobblestones in Assyria four thousand years before Christ, and there were brick and limestone pavements later in Babylon. Its famous Procession Street leading to the bridge across the Euphrates had a foundation of brick covered with asphalt; the limestone flags with beveled joints were set in asphalt. The roadway was surfaced with large stones, 3 to 4 feet in length and breadth. The sidewalks were of smaller slabs of red breccia, a sort of conglomerate. Seven pier foundations of the bridge have been found in the river bed. This most ancient highway bridge of which there is a record was constructed, apparently, in the time of Nebuchadnezzar. Although its piers were made of small burnt bricks, they were substantial; each pier measured some 30 by 70 feet. They were cleverly shaped with the pointed end upstream to cut the force of the current, and the end downstream rounded to reduce the swirl which might undermine the foundations. It seems likely that two rows of closely spaced wooden stringers ran over these piers to carry the roadway, which, according to the ancient historians, was paved with stone blocks. The bridge was 400 feet long. The account by Herodotus mentions a drawspan of some sort, presumably a military precaution. He did not make clear how it was constructed.

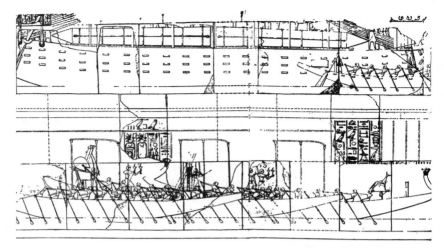

Figure 2.8 Transportation of obelisks (From E. Naville, *The Temple of Deir El Bahari*, 1908; courtesy Egypt Exploration Society)

All the early civilizations in the Near East had roads of one kind or another, designed and maintained primarily for administrative and military purposes. But the records which we have tell us very little that is specific. Assyria had a special corps of men called *ummani* to level the ground for baggage carts and to build temporary bridges. The roads which they constructed must have served for commerce and travel in times of peace as well as war. The Persian Darius had his Royal Road in the sixth century B.C. from Sardis in Asia Minor to Susa near the Persian Gulf, over 1,500 miles, and he kept his messengers going and coming upon it. This great thoroughfare provided communication through mountain passes and across wastelands. It opened the way westward to the Aegean Sea and for the historic clash of Persians with Greeks at Marathon. But, so far, archaeologists have found no physical trace of this famous highway of the past, and there is no information available about its type of engineering.

Water transport, in some areas, such as Egypt, was of greater importance than communication by road. Prior to the rise of the food-producing civilization there was little need for transporting by water any considerable amount of goods. After 3000 B.C. ships of considerable size were evolved, as their designers applied increasing knowledge of the working and joining of wood. Men propelled these vessels with oars, but even before 3000 B.C. a crude sail had been added so that the force of the wind could supplement the efforts of the oarsmen if the wind was in the right direction. These single sails were always attached to masts located forward of the point about which the ship turned, and they could not be swung very far out of their position at right angles to the ship's axis. As a result the crew could use the sail only when the wind was from astern.

The largest ships were not self-propelled but were towed by smaller vessels with their complement of oarsmen and perhaps one sail. About 2400 B.C. the Egyptians built one of these large craft that was probably about 100 feet long—that is, longer than the *Mayflower* which in 1620 carried the Pilgrims to Massachusetts and was 90 feet long. The ship of about 1600 B.C. that transported two obelisks placed end to end on her main deck was probably at least 220 feet long (Figure 2.8). It required three columns of smaller oar-propelled ships to tow it with its cargo down the Nile. Whatever their size, all ships of this period were moved primarily by human effort.

Egyptian Engineering

A settled agricultural people had been building in the fertile valley of the Nile centuries before they recorded their accomplishments in writing. The rise, culmination, and decline of Egyptian civilization, however, can be traced from 4000 B.C. The state was unified after long periods of internal struggle. Wealth and population increased with the establishment of local security. Foreign enemies either were absorbed or were conquered and reduced to slavery. Egypt passed slowly from kingdom to empire. One significant stage in that slow process was marked by the pyramids which were built in the period from approximately 2700 to 2200 B.C. The irrigation projects of the Twelfth Dynasty, from 2000 to 1800 B.C., identify a second stage of development. And the great temples of Luxor and Karnak fix attention upon a third; these monuments of the Eighteenth Dynasty were built for the most part from the sixteenth to the fourteenth century B.C. As the warlike kings of the time adopted an aggressive policy against invaders and extended Egyptian rule from Syria on the north to Libya on the west, their people enjoyed relative security, opulence, and comfort. Then came a thousand years of decay and subjection.

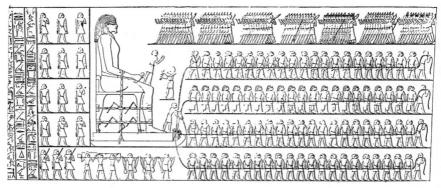

Figure 2.9 Moving an Egyptian statue, bas relief from Twelfth Dynasty tomb (From C. R. Lepsius, *Denkmaeler,* 1842)

So far as can be learned from excavations in the ruins, from inscriptions on monuments, and from records written on papyri, there was throughout the whole period comparatively little advance in the techniques of engineering. Egyptians in the third millennium B.C. were using structural methods conditioned by their environment. They were still using those methods thirty or more centuries later when their country had ceased to exist as a separate empire. Three factors determined the character of Egyptian engineering. One was the great supply of human labor. All operations were based upon unlimited use of the time and the strength of slaves (Figure 2.9) who had been captured in war or raids beyond the frontiers. Draft animals played little part in Egyptian construction. The horse, always a noble animal in the Orient, was never used for work and was indeed unknown in Egypt until about 1700 B.C.

When captive labor was not available or was inadequate, natives were summoned in relays. They seem to have had no more choice in the matter than the slaves. But evidently it is a mistake to think of Egyptian laborers as always cringing under the lash of an overseer, notwithstanding the story of Moses and the sons of Israel in the Bible. The peasant fully expected to be drafted for public tasks in that portion of the year when the climate and the river kept him from his own work in the fields. He seems to have worked rather willingly when he was not imposed upon. He took pride in the enterprise and was amenable to the discipline necessary to govern hordes of men straining at the same task. The inscriptions boast that some jobs were accomplished without accident or sickness and completed on time. The second determining factor in Egyptian engineering was the concentration of these vast armies of workmen under the absolute control of a single man and his lieutenants. Neither time nor cost made any difference to a Pharaoh once he had made up his mind to build a tomb for himself which should stand forever, or to shift the course of the Nile so that what had been desert might become fertile. Egyptian engineering was in the grand manner; its achievements were on a scale befitting one who called himself both king and god. The third factor which determined the character of Egyptian engineering was the great quantity of building stone in the ledges of the upper Nile Valley. From these quarries of limestone, sandstone, and granite came pieces weighing from 2½ to 30 tons for the largest and oldest stone structures in the world. From them also were cut obelisks weighing several hundred tons each and at least one huge block of 1,000 tons.

For this work the Egyptians used only the simplest mechanical principles and appliances. The stone was marked off to desired measurements, grooves were then cut with mallet and bronze chisel, and the pieces were finally split from bedrock by a process similar to what is known today as the plug-and-feather method. Bronze plugs were slid between thin bronze feathers or wedges and driven in until the rock split, or wooden wedges were inserted which expanded when wet and split the rock. It was in this manner that the Egyptians generally quarried limestone and sandstone. For their granite obelisks which were cut out horizontally, the Egyptians pounded the ledge with hard and tough dolerite, a coarse basalt, until they had bruised all around the edges, had cut a deep groove, and finally broken off the piece they desired. Needless to say, the process took time, strength, patience, and skill. The visitor to the quarries of Aswan today may clearly see evidences of this method, for a great stone lies where it was abandoned because of imperfections.

The metal which the Egyptians used for their wedges, saws, and cutting tools was a hard bronze. How the early peoples first obtained it is unknown. Relatively little has as yet been discovered about the technique of their metal working. Like other early peoples, the Egyptians knew almost from the beginning of their civilization and treated as more precious than gold the nickel-bearing iron which fell in meteors from the heavens. It was not much before 600 B.C. that they learned, presumably from their Hittite neighbors in Asia Minor, how to reduce the iron ores they found in the earth and put the metal effectively to work.

Besides their wedges, mallets, and chisels, the Egyptians used straight levers and the inclined plane. They had rollers made of hard acacia wood, and they obtained the wheel with the horse and chariot from Asia Minor between 1800 and 1600 B.C. It seems to be true that in Babylonia the Sumerians had wheeled chariots before the time of Sargon, probably as early as the twenty-fifth or twenty-sixth century B.C. The Egyptians seem not to have been familiar with the screw and so could not raise their heavy blocks with screw jacks. Nor did they have pulleys for their ropes. Thus lacking the greater pull that they might have obtained with block and tackle, they had need for every ounce of human strength which they could apply directly to the load.

Since each stone was shaped at the quarry, as is most often the case today, the engineer had to have in mind the general plan of his structure and the position that the stone was to occupy. The Egyptian accordingly, like his modern successor, worked from drawings, which he drew on papyrus, limestone slabs, or occasionally on wood. Unfortunately only a few of them have been preserved, but it is clear that he knew how to show details both in plan and in elevation and how to construct models to scale.

Centuries before the Greek Pythagoras demonstrated the generalized relationship among the sides of a right-angled triangle, Egypt's "rope stretchers," or surveyors, are said to have been applying the knowledge that the angle between two sides of a triangle is a right angle if the sum of their squares is equal to the square of the hypotenuse. A papyrus now in the British Museum shows that they understood also how to calculate the contents of solids and to determine the slope or amount of cutback necessary in terms of the height of a pyramid and the length of its side. Three other facts regarding their mathematical notation are of interest. They used a system based upon 10, like ours, though of course without Arabic characters. Their fractions always had the digit 1 for the numerator, except for the fraction ⅔. And in finding the area of a circle they used 3.16 for the value of pi (the ratio between the circumference and the diameter of a circle). The generally accepted approximation of this ratio today for most calculating is 3.1416.

Pyramids and Temples

It took decided skill in measuring distances and angles to transfer the plan of such a structure as a pyramid from a drawing to the site, and the engineer-architects of Egypt have left conclusive evidence that they possessed that skill. The average length of the sides of the Great Pyramid at Gizeh of Khufu, or Cheops as the Greeks called him, was 755 feet 9 inches at the base. Two of the sides varied from that figure as little as 1 inch, and two of the angles were but 3 or 4 minutes in error. This is a degree of accuracy which could hardly have been obtained with the cords commonly used at that time in surveying land. It is quite certain that rods were used, graduated in cubits, palms, and digits (an Egyptian cubit was a bit over 20 inches, a palm was 3 to 4 inches, and a digit was about ¾ inch). For the angles, it has been suggested that a huge wooden T square 8 or 10 feet long may have been employed. A right angle could have been laid off with fair accuracy by averaging the results from placing the square in position first on one of its sides and then on the other. Some device which involved sighting along or parallel to the surface of still water must have been used for leveling the base of the pyramid. How it was so accurately oriented with an axis almost on the meridian is not certainly known.

When the base course had been laid, the construction problem had barely begun. The structure must be carried upward at a uniform slope of $51°51'$ while accurately preserving the orientation of the steps of each course to avoid any twist and to maintain horizontality. All of this doubtless taxed the ingenuity and supervisory skill of the builders, as it would today. One complication not generally recognized was that the underlying ground was far from level. A ledge of rock projects up, no one knows just how far, into the body of the pyramid.

The Great Pyramid at Gizeh (Figure 2.10) dates from the twenty-seventh century B.C. With its 206 courses reaching a height of 481 feet, it is a gigantic pile of 2¼ million rough-looking but carefully squared and placed limestone blocks, averaging more than 1½ tons in weight. Originally it was faced with meticulously selected and dressed blocks of the same material, some of them weighing up to 15 tons. Egyptian masons bedded these facing stones in a thin mortar of gypsum and sand, and their best work shows extraordinarily tight, almost imperceptible joints, at least as narrow as ½₀ inch. These blocks were anchored in some way into the core—just how is unknown—making the sides of the pyramid single planes from top to bottom. In the course of the centuries since Herodotus's day practically all of this shiny white facing has found its way, piece by piece, into monumental buildings elsewhere, as far down the Nile Valley as Cairo.

The interior of the Great Pyramid is, like that of the others, mainly solid masonry. There is, however, an inclined passage, large enough for a

Figure 2.10   The Great Pyramid (From J. S. Perring, *The Pyramids of Gizeh*, 1839–1842)

man to crawl through, that leads inward and downward from the north face at nearly 26½ degrees with the horizontal. Branching from it is another leading upward into a tall but narrow "great gallery" through which one reaches the "king's chamber," nearly, but oddly enough not quite, under the apex of the pyramid. This chamber (17 by 34 feet), which Khufu trusted would hold his sarcophagus inviolate forever, is lined with granite and roofed over with corbeled granite slabs. Above the roof is a succession of five large relieving chambers obviously planned to take part of the enormous superincumbent weight of more than 300 feet of blocks, especially perhaps to guard against earthquake damage. There are also inclined shafts, some, it would seem, for ventilation, a "queen's chamber," and still another chamber deep down below the foundation.

Many scholars believe that this pyramid served in some sense as an astronomical, or astrological, observatory. Some authorities even suggest that the lower portion of the pyramid may have been used for star gazing long before the structure was finally raised over it as a tomb. But the question as to how Egyptian engineers moved heavy materials and set them in place has caused the most speculation and attracted perhaps the most attention of engineers to that ancient culture. How did they transport such overpowering blocks of stone and lift them into position without cranes or block and tackle, or even draft animals? The question why they did has not been half so intriguing to the engineer.

Prevailing opinion now is that they brought the stone down and across the Nile at high water on long barges trussed overhead from stem to stern. The barge was moored as close as possible to the quarry and steadied with sand as great gangs of men dragged the stone on wooden sleds down an inclined plane and onto the barge. Such a plane, and the roads in some cases a mile long to and from the river bank, had to be solid and smooth; a 20-ton block could not be dragged through loose sand. The surface was hardened either with stone blocks or chips from the quarries or with a mixture of wet sand, silt, and clay in proportions learned from experience. The number of men necessary in a gang was calculated mathematically by the overseer according to the weight of the block. Under any conditions, obviously, the assignment of twenty or fifty thousand men in groups of varying sizes to different tasks required organization and discipline.

There is not so much agreement about the manner in which they raised the blocks to position in the pyramids. A structure of that sort could be reproduced today within two or three years by using power-driven machinery, cranes, derricks, and a comparatively small force of men; but how the Egyptians managed it is still unknown in spite of drawings so often seen in the pictorial magazines that purport to give an exact impression. These illustrations prove the constant interest in the Egyptian pyramids on the part of those who love pictures, not the actual way in which the pyramids were built. Representations of many devices and processes have come to us from ancient times, but no authoritative illustration of the building of the pyramids has yet been discovered. The more Clarke and Engelbach studied Egyptian methods of construction, the more those experts were convinced that if any detail had to be explained by an apparatus of any degree of complexity, then the explanation was certainly wrong.

The sand-mound theory, one of the oldest, assumes that as each layer of stone was placed, it was surrounded by a bed of sand sloping downward to the surface of the ground. Up this incline were dragged the materials for the next layer. When the pyramid was completed, it was practically buried in a mountain of sand, which had then, of course, to be dug away. With the angle of repose for sand estimated at two horizontal to one vertical, the width at the bottom of such a mountain covering a pyramid 500 feet high would be more than 2,000 feet, possibly half a mile. The rocker or cradle theory, advanced more than a generation ago by the Frenchman Auguste Choisy, supposes that a stone was laid upon a solid wooden sled with a cylindrical undersurface as if cut from a tree trunk, and this was rocked by means of great levers, perhaps counterweighted. It is clear from tomb inscriptions and many small-scale models of ancient origin that the Egyptians used such a cradle sled to facilitate loading and lifting. When one side was raised, props were placed beneath it; the levers were shifted to the other side, and the operation repeated. In this way the stone blocks might have been raised stage by stage to the top of the pyramids.

The ramp theory, which now seems best supported by the available evidence, is that the Egyptians built a sloping roadway up to the level under construction. Such a ramp might have been either straight or turned about the structure. According to the straight-ramp theory, layer after layer was superimposed upon the original slope as the pyramid rose. Each, therefore, would require surfacing. The turned ramp would not seem to have presented so much difficulty. Started and maintained at a moderate grade, it could be carried on and around the pyramid several times before reaching the summit. There could be platforms, switchbacks, even hairpin turns at intervals, and a retaining wall of stone or brick at the outer edge. Archaeologists found the remains of a ramp in 1914.[7] It had been faced with sun-dried bricks varied in size from some no larger than ours to others that were as much as two men could lift. Huge granite lintels, like those carrying the roof of the king's chamber in the pyramid of Cheops (Khufu), perhaps were raised by "seesawing." Each stone weighs roughly 55 tons, and it could have been made to rest on two supports close to its center. It then could have been rocked back and forth and raised an inch or two at a time by inserting small blocks on each support alternately. A well-drilled gang of men walking from one end to the other might get the block as high as 20 feet in the course of a day's work.

The monumental buildings of Thebes, Luxor, and Karnak 300 miles or so up the Nile from Cairo and the region of the pyramids represent more than two thousand years of Egyptian construction and architecture. Though surpassed in many details since its time, the most famous of all still is the temple of Amon-Ra at Karnak (Figure 2.11). It was once, if it is no longer, the largest columnar structure in the world. Its dimensions of 338 by 1,220 feet are sufficient to contain the combined ground areas

[7] Arthur C. Mace, "Excavations at the North Pyramid of Lisht," Metropolitan Museum of Art, *Bulletin*, vol. 9, p. 220, 1914.

Figure 2.11 Great Temple of Amon at Karnak (Courtesy Yale School of the Fine Arts)

of St. Peter's at Rome, the cathedral of Milan, and of Notre Dame at Paris. The great hall of the temple, known because of its rows of columns as the Hypostyle Hall, was 329 by 170 feet. The columns stood 69 feet high in the center aisles and 42 feet along the sides. They were more than 10 feet in diameter, supporting short architraves, or crossbeams, that weighed 60 to 70 tons each. These carried the flat roof at two levels, making room for clerestory windows, apparently the first in history. Here, centuries before Christ, building principles were established for the cathedrals that in the Middle Ages rose to the adoration of God.

The darkness which the clerestory windows of Karnak did not dispel was probably enough, with the aid of incantations, to stir that fearful devotion to Amon-Ra which his priests desired. But his engineers had not been able to create the greater dread of dim and cavernous space. The tensile weakness of stone would not allow crossbeams of any length. The columns were so closely set and so thick that they took about as much room as the enclosures which they were supposed to make possible. The temple of Amon-Ra, as a housing project, can hardly be called a great success. The method of construction at Karnak was to build around the columns and along the walls falsework of brick and sand, ramps, and terraces upon which the blocks could be dragged and set in position.[8] These temporary structures were taken down by stages as the carving and painting of the interior progressed. The foundations at Karnak, however, were much less carefully designed and constructed than elsewhere. Floods have undermined them so badly that the columns have settled and most of the roof has fallen. In recent years the Egyptian government has been making an effort to right the columns and preserve what is left.

To place upright an obelisk of several hundred tons weight is an engineering feat that requires nicety of calculation and special equipment even in modern times. The length, weight, slenderness, and relative weakness under bending stress of such huge blocks of stone make the task uncertain at best. To judge from the number of these monuments in Egypt, it is obvious that its engineers solved the problem successfully, though it is not clear in just what way. Quite likely it was that described in a special study of the question by the authority Engelbach. The column of stone, strengthened by splints of timber against the bending stress due to its own length and weight, was hauled up a long ramp, tipped over a curved edge at its top, and let down upon the foundation which had been prepared in advance. From that position, perhaps three-quarters erect, it was pulled up to the vertical. Essentially the same method, though with elaborate handling machinery, has been used with more than forty obelisks which have been taken from Egypt and set up elsewhere.

[8] Earl B. Smith, *Egyptian Architecture as Cultural Expression*, Appleton-Century-Crofts, Inc., New York, 1938, p. 165.

Of these the largest outside of Egypt and the second largest in the world was transported to Rome in the time of Caligula and placed in the Circus of Nero. We have no information about the method of handling it, except that it was reportedly shipped across the Mediterranean with nearly 1,000 tons of lentils for ballast in a vessel said by Pliny to have been the largest built up to that time. It was removed in 1586 to the square before St. Peter's. As the whole of this proceeding, recorded by the director Domenico Fontana, told a great deal about the engineering equipment of the sixteenth century, we shall discuss it in detail in Chapter 6.

Three other obelisks deserve passing attention. One of the thirteenth century B.C. was removed in 1830 to 1836 from Luxor to Paris where it now stands in the Place de la Concorde. Another, famous as Cleopatra's Needle although it was quarried nearly fifteen centuries before the time of Cleopatra, journeyed from Heliopolis to Alexandria during Roman times and then in 1877 to London. By that time engineers had wire hawsers, hydraulic jacks, and an iron casing to assist in its handling. The third now stands in Central Park, New York; it was brought in 1879 to 1881 from Alexandria and removed from the ship at New York through an opening in the hull. The last 2 miles of the journey were on a specially prepared elevated track and required 112 days, progress being at the rate of only 97 feet a day. Erection at the site required another two weeks. According to John T. Johnston, the total expense was "a little over $100,000." [9]

Egyptian masons applied their simple instruments with ingenuity and care. In their best structures, as in the temples of Karnak and the facing of the Great Pyramid, the stones fitted with almost imperceptible joints even when laid dry. To ensure precise and firm bedding, they used a mortar made of gypsum with very little sand, but such precision was customary only in the outer veneer where the surface was tooled and dressed after the stone had been laid. Inside work was apt to be coarse, with more sand in the mortar. Walls of two faces were often built without connecting stones to hold them together, and sometimes the space between was filled with loose rubble. A specimen of Egyptian masonry may be seen in the Metropolitan Museum of Art, New York. The tomb of Perneb, who held high office in Memphis about 2650 B.C., was taken down piece by piece and rebuilt there in 1916. Also on exhibit are Egyptian

[9] Henry H. Gorringe, *Egyptian Obelisks,* published by the author, New York, 1882, p. 51.

masons' tools of a later period: bronze chisels, a plasterer's float, wedges, and stone plumb bobs.

In spite of remarkable achievements, especially when handling heavy materials, it cannot be said that the Egyptians made any great innovation or departure in the technique of building with stone. They seem always to have reproduced what had been made earlier in wood, mud, or brick. Their pyramids were but magnified mastabas, or rectangular superstructures, block upon block above a sunken tomb. Their temples never got beyond the post-and-lintel form, or upright and crosspiece. True arches, made of wedge-shaped mud bricks, were known in Egypt, as indeed they had been in neolithic settlements; but the Egyptians never brought them aboveground, as it were, and never adapted them to construction in stone. This significant departure in engineering was not to come until centuries later with the Etruscans.

Even so, the architects and engineers of ancient Egypt were exceptional men. They were among the first and the few, save royalty, to gain historic identity. They were appreciated in their own time and revered by succeeding generations. Imhotep, designer and builder of the first great pyramid, the Saqqara, was famous also as a physician and maker of proverbs. His career exemplified the blend of superstition and reality characteristic of Egyptian life, preoccupied with hard daily labor and thoughts of the hereafter. It was said that his plans "descended to him from heaven, to the north of Memphis." His counsel was "as if he had enquired at the oracle of God." Two thousand years later Imhotep himself was included in the awesome company of Egypt's gods.

Successors to this first great worker in stone were perhaps not so famous or adored, but without doubt they were eminent in their time. They thought no less of themselves. For one, Senmut was "great father-tutor of the king's daughter," "overseer of all the fields of Amon, Senmut triumphant," and "chief steward of the king." Possibly it was in this capacity that he was able to have his portrait behind every door in the temple of Queen Hatshepsut. Of himself, he had it inscribed, "I was the greatest of the great in the whole land." [10] Ineni, "chief of all works in Karnak," came earlier in the distinguished line of engineers, but he deserves the honor of the last place here for his own estimate of himself. "I became great beyond words," he said. ". . . I did no wrong whatever."

[10] James H. Breasted, *Ancient Records of Egypt*, 5 vols., University of Chicago Press, Chicago, 1906–1907, vol. 2, pp. 147–152.

Besides that, he was "foreman of the foremen." And still, he "never blasphemed sacred things." [11]

Irrigation

As in Mesopotamia, nature could not be relied upon in Egypt always to provide a steady and adequate supply of water. Besides, the valley of the Nile was too narrow for a large population and could not in ancient times, any more than it can now, maintain the people gathered there without supplementary irrigation of adjoining lands. The systems of to-day for the most part have been in continuous service for thousands of years. British engineers have extended and improved some, but operating side by side with their modern installations may still be found the most primitive kinds of lever-and-bucket arrangements for raising water from the Nile into irrigation ditches.

One of the notable reclamation projects in history was planned and carried through to completion by the Theban Dynasty of 2000 to 1788 B.C. This dynasty had reunited Egypt after the old regime of the pyramid builders crumbled and turned an empire into warring feudal states. The Theban kings developed an active trade with other parts of the Mediterranean and extended their influence beyond the borders of Egypt as far as Palestine. For the benefit of their people at home, they undertook to change the great oval basin in the Faiyûm desert to the west of the Lower Nile into a fertile and populated land. They threw dams across the ravines leading into the basin to impound the rains of the wet season against drought. One of these dams in a 250-foot-wide gorge had a base 143 feet thick, four times its height. It was built in three layers, the bottom of rough stones embedded in clay, the next of irregular limestone blocks, and the top of cut stone laid in steps so that the water pouring over the brim was checked in its fall and would not erode the structure. The Egyptians had also a canal from the Nile, known today as the Bahr Yusuf, or Canal of Joseph, to carry flood waters into the artificial Lake Moeris. There was developed a complex system of dikes, flood-control gates, canals, and bridges. Many buildings appeared, some of them royal palaces; the city of the province whose diety was Sobk, "crocodile god," became an important center of Egypt. According to William Willcocks, principal

[11] James Baikie, *Egyptian Antiquities in the Nile Valley*, Methuen & Co., Ltd., London, 1932, p. 587.

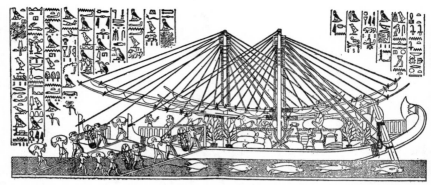

Figure 2.12 One of the ships used in Queen Hatshepsut's famous voyage to Punt (From Adolf Erman, *Aegypten,* 1887)

designer of the modern Aswan Dam, it was the management of these dikes and gates of Moeris by the rulers of Upper Egypt that reduced the flow of the Nile in Lower Egypt and caused Joseph's famines there. Recapture of the dikes by the king of Lower Egypt put an end to the hardship. Similar manipulation of local dikes and levees aided the escape of Moses and the Israelites from Pharaoh's host.

Persistent tradition has it that another Egyptian feat of engineering was a canal to connect the Nile with the Gulf of Suez, thus linking the Mediterranean with the Red Sea. It may have been begun about 1870 B.C., but according to some authorities the project was abandoned. Others think that the canal was in use during the reign of Queen Hatshepsut, who died in 1468 B.C., and that it silted in later. There is a picture of a trading expedition in her time down the Red Sea to Punt (Figure 2.12), now within the territory of Ethiopia. During Egypt's brief independence between Assyrian and Persian subjugation, the Pharaoh Neccho, born in 608 B.C., revived the famous project; in the attempt he is said by Herodotus to have sacrificed the lives of 120,000 workmen, probably to disease. During the nineteenth century the loss of life on the railroad and the French canal in Panama totaled somewhat more than 66,000. The Pharaoh stopped work on his canal either because the cost proved too great or because he feared that Egypt might be flooded from the Red Sea. Lepère, Napoleon's engineer, reported in 1803 that the Red Sea was 29 feet higher than the Mediterranean. But when the land was surveyed fifty years later for the Suez Canal, it was found that Lepère had been in error and that the two were at practically the same elevation.

Bibliography

Budge, Ernest A. T. W.: *Cleopatra's Needles and Other Egyptian Obelisks*, The Religious Tract Society, London, 1926.

Chace, Arnold B., Henry P. Manning, Raymond C. Archibald, and Ludlow S. Bull (eds.): *The Rhind Mathematical Papyrus*, 2 vols., Mathematical Association of America, Oberlin, Ohio, 1927–1929.

Clarke, Somers, and R. Engelbach: *Ancient Egyptian Masonry*, Oxford University Press, London, 1930.

Edwards, Iorwerth E. S.: *The Pyramids of Egypt*, Penguin Books, Harmondsworth, Middlesex, 1949.

Forbes, R. J.: *Notes on the History of Ancient Roads and Their Construction*, N.v. Noord-Hollandsche Uitgevers-Mij., Amsterdam, 1934.

Gorringe, Henry H.: *Egyptian Obelisks*, published by the author, New York, 1882.

Lane, William H.: *Babylonian Problems*, E. P. Dutton & Co., Inc., New York, 1923.

Legrain, Georges A.: *Les Temples de Karnak*, Vromant & Co., Brussels, 1929.

Mackay, Ernest J. H.: *Early Indus Civilizations*, 2d ed., Luzac, London, 1948.

Neugebauer, Otto: *Mathematische Keilschrift-Texte*, 3 vols., Springer Verlag OHG, Berlin, 1935–1937.

Sarton, George: *A History of Science: Ancient Science through the Golden Age of Greece*, Harvard University Press, Cambridge, Mass., 1952.

Singer, Charles J., E. J. Holmyard, and A. R. Hall (eds.): *A History of Technology*, Clarendon Press, Oxford, 1954, vol. 1.

Turner, Ralph: *The Great Cultural Traditions*, 2 vols., McGraw-Hill Book Company, Inc., New York, 1941.

Unger, Eckhard A. O.: *Babylon*, Walter de Gruyter & Co., Berlin, 1931.

THREE

Greek Engineering

To the north and west of the Suez and reaching across the entrance of the Aegean Sea lies the island of Crete. Here, more than twenty-five centuries before Christ, there developed, in touch with Egyptian civilization but distinct from it, the center of an urban culture, the Minoan, so-called after the legendary King Minos. This culture gave a unique heritage to the later Greek communities and through them to our own. Knowledge of the Aegean civilization, of which Cnossus on the island of Crete was the principal city, is remarkably recent for it all dates from 1894 when Arthur J. Evans began to uncover, literally, the archaeological evidence of the magnificent Minoan civilization. Because the discovery of this civilization is so recent, there is, of course, still much to be learned about it. Indeed, it was not until 1953 that Michael Ventris, a British architect, was the first to decipher a Minoan script.

The Minoans rose to the heights of a brilliant civilization about 2100 B.C. Their government was a monarchy with the kings living in a great palace at Cnossus. Although widely known as seafarers, Minoans were agriculturists, but they were also artisans producing remarkably beautiful vases, statuettes, ornaments, and textiles. These Cretans carried on an extensive trade by sea, and many Minoan objects have been found in Egypt, as have Egyptian products been discovered at Cnossus. Their engineering knowledge and skills were as advanced as their tools. A carpenter's house of one Minoan town yielded chisels, saws, awls, and nails.

This interesting civilization continued at its height until about 1400 B.C. when the palace and with it the political power of Cnossus were destroyed. The Minoans, however, contributed to the rising Greek civilization in many ways; they probably gave to the Greeks even their language,

36

for Ventris was able to decipher the Minoan script by working out Greek syllabic equivalents for Minoan characters.

Minoan Engineering

The terra-cotta pipes in the palace at Cnossus were tapered, presumably because they were easier to cast and remove from the mold and easier to connect. The form gave a shooting motion to the water that helped to keep the successive lengths clear of sediment. It is a skillful design of the hub and spigot which is common today. There were bathrooms, bath tubs, and sanitary facilities flushed with water that have been declared superior to any in Europe before A.D. 1800. In fact, they appear to have been more efficient than many still used in the outlying districts of any country in our own time. Minoan builders also constructed a system for collecting rain water and keeping it pure. Whether or not they understood the mathematics and chemistry involved, they certainly did observe the behavior of water and cleverly adapted their engineering to it. They built gutters descending in a series of vertical parabolic curves beside their open stairways, to let the stream down from level to level. This series of controlled cataracts kept the water from gaining too much speed over the whole descent and splashing at the bottom of the slope. Turns were banked properly to prevent water from slopping over the walkways. There were settling basins at intervals to catch sediment, and sunlight helped to cleanse the water along its steady course to the storage cisterns or washing rooms.

The Cnossus palace itself, as it appears from the ruins and from restorations which the archaeologists have made, must have been an imposing structure of squared stone, heavy, angular, and squat. Its stiffness was relieved in part by circular pillars. They, too, looked stubborn and squat. They marched in line before porches and rose beside grand stairways, but they did so with none of the graceful air which we feel in their successors, the magnificent columns of the Greek temples. Minoan Crete had an exuberant and lively spirit, but its people expressed their enthusiasm and daring in bullbaiting and in decorative art, not in architectural form.

Cretan engineers built in the manner of their Egyptian contemporaries. Cretan structures give the impression that there were on the island, as in Egypt, plenty of stone, ample time, and a great deal of manpower, presumably slave labor. Both countries contain abundant evidence of frequent mutual contact and a steady exchange of ideas and goods. Not

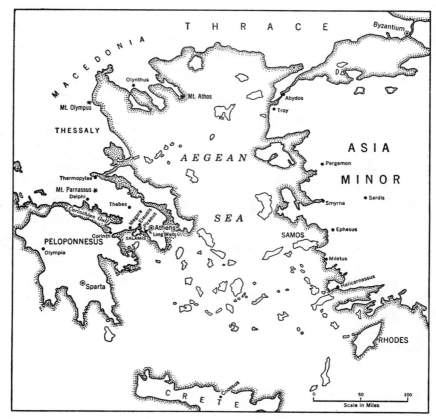

Figure 3.1 Ancient Greece

much is known about Cretan methods and tools; their techniques may have been like the Egyptians'. The Cretans made no significant departures in construction that have as yet been discovered. They, too, built with the post and lintel and made no use of even the corbeled arch in their palaces, storehouses, and dwellings. But they showed ingenuity, whether more or less than their Egyptian neighbors it is hard to say, in giving access to successive levels within their structures by stairways, in placing room after room about courts, and opening light wells into these inner mazes. For these accomplishments, they gained perhaps their greatest fame and immortality in the Greek legend of Ariadne, Theseus, and the Labyrinth. Archaeologists working over the sites have been struck by the appearance of curved walls and an oval or elliptical house. But, whatever the artistic appeal of such a design, there seems to have been no engineering purpose involved. The elements of the construction were rectilinear. In any case, the curves were horizontal. Cretan engineers did not use vertical curves in their building, not even in their viaducts (except the parabolic gutters already mentioned), so far as is known to the present time.

The remains of the great causeway across the island southward from Cnossus to its port on the Mediterranean opposite Libya show massive piers of squared masonry. First thought is that these piers were once joined by corbeled arches. No suitable stone, however, has been found on the site, and it is as likely that crossbeams of wood carried the roadway from pier to pier, just as they were used above openings in the masonry within the palace. Cretan engineers also inserted crossbeams of wood to tie together the facings of thick walls whose interiors were filled with rubble. The Minoan road which has been found several feet below a Roman pavement of a much later date is as good as the latter, if not better. Judgment, of course, should not be made according to a modern standard; it should depend upon the respective purposes for which the roadways were built. The exact purposes each hard surface was intended to serve have not yet been determined. This Roman road was merely cobblestones placed upon the ground, and the Minoan lay upon a base of rough stones embedded some two feet below the ground level. Its central path, 4½ feet wide, was made of two rows of smooth slabs set close together upon the base in a clay cement. On either side were shoulders, 3½ feet wide, of pebbles and broken pottery rammed firmly into clay, and these strips lay upon rough stones which in turn rested upon the base. There was a drain along one border, but unlike the Roman above it, the Minoan road was laid flat, with no camber or crown. Clay tablets of late Minoan times show royal horses and chariots, but no evidence of wheeled traffic on the Cretan roadways has yet been found.

Mycenaean Contributions

In the meantime a city on the mainland to the north was becoming the center of a culture less ornate but destined to link Crete and Egypt with the civilizations of Greece, the Aegean islands, and Asia Minor. Mycenae, the stronghold whence Agamemnon went forth to the Trojan War, was set well back from the sea and high in the mountains of the Peloponnesus,

Figure 3.2 Gateway, Mycenae Treasury (From H. Schlieman, *Mycenae,* 1878–1880)

perhaps for the very purpose of avoiding frequent and unsolicited contact with travelers from Troy, pirates from Phoenicia, or the sea raiders of Crete.

The builders of Mycenae, like the Egyptians, evidently handled the huge blocks of their masonry without jacks or pulleys. Their famous Lion Gate, erected before 1300 B.C., has a lintel measuring 15 by 7 by 3½ feet and weighing nearly 30 tons. It rests on uprights over 10 feet tall. Either it was dragged into position upon an earth ramp which was then carried away, as in Egypt, or the Mycenaean builders may have used levers and other tools of which they have left no record. They put stones of 120 tons each on either side of an entrance. They set another limestone lintel of 100 tons in one of their tombs.

It is for their use of the corbeled, or false, arch or vault (Figure 3.2), however, that the Mycenaeans have a place in the history of engineering. They applied the principle to underground construction so as to produce great conical or beehive tombs and storerooms in stone. And they brought such construction aboveground to be used in bridges for their roadways. These roadways were designed only for pack animals and travelers on foot and so were in many places no more than a series of steps cut into rock. But elsewhere, as they crossed ledges and worked around jutting heights, they were held by stone retaining walls and carried over intervening mountain torrents on culverts of corbeled stone. The corbeled arch had come effectively into engineering; the true arch had to wait until Etruscan and Roman times.

These early inhabitants of Greece understood the value of both water supply and drainage. The Mycenaeans did not put water under pressure and raise it to a higher level within their stronghold; they brought it in a subterranean channel to a well near the walls and ran another passage underground from the fort to the well. Northwest of Athens in the hills there was a shallow body of water in a pocket 100 square miles or so in area covering good arable land. Drainage galleries and canals from this lake (Copais), which British and French engineers rebuilt in the 1880s, were first dug four thousand years ago.

The Greeks

Sometime before 3000 B.C. the first settlers arrived in Greece where Old Stone Age man apparently did not flourish. V. Gordon Childe has demonstrated in his *Prehistoric Migrations in Europe* (H. Ashehoug & Co., Oslo, 1950) that this early Greek Sesklo culture came from the east as, indeed, did its successors. These peoples had great abilities which eventually produced a culture that ever since has amazed the world.

When the Greeks first appeared theirs was an agricultural life. As they became settled in the land, their social organization became a more complicated system of richer and poorer, nobler and humbler. Their tenure of the land and the institutions which they established upon it were nothing extraordinary, but the poetry with which Homer immortalized them is glorious. By the fifth century B.C. the rural society which Homer portrayed had become on the one hand the intricate military oligarchy of Sparta, practically devoid of poetry, and on the other the even more urban democracy of Athens where the arts attained their perfection. The difference can hardly be ascribed to contrasting attitudes toward fellow men, for both Sparta and Athens set their foundations upon slave labor. What was the dynamic force that turned the simple loyalty of tribal warriors into the complex public spirit of Athenian citizens? The question has long engrossed philosophers and historians, and it will continue to perplex them. Even to comprehend what the Greeks did rather than to attempt to analyze why they did it requires serious study, so varied and so profuse were their accomplishments in art, drama, philosophy, science, and architecture.

The Greek achievements in science during the sixth to third centuries B.C. led to dramatic development in engineering. It may again be empha-- sized that almost all early knowledge was gained as the result of experience. Previous to the sixth century B.C. there were no general theories of natural phenomena or of mathematics. Neither were there any consciously expressed or defined "laws of nature," simply because men had not come to realize that within nature there is definite regularity or order underlying its every phase. Without a consciousness of regularity the idea of natural law is impossible. Indeed, there can be no science without the concept of order in nature, for it is this very order which science seeks.

Perhaps the greatest glory of Greek science was the discovery of science itself—the discovery that there are general laws in nature which man can come to know. The Greek who is usually given the most credit for having established this new type of knowledge is Thales of Miletus, who lived about 600 B.C. Thales asked the important question, "Of what is it that all things are made?" His conclusion that all things have evolved from water was, of course, most inadequate in the light of modern knowledge. Nevertheless it was Thales who pioneered in the investigation of the problem of matter—an investigation which still continues. Indeed, the problem of matter is one of the most significant problems of modern physics yet to be solved.

It is believed that Thales invented or defined abstract geometry. The Mesopotamians and Egyptians knew how to survey plots or irregular shapes; they could also compute volumes of cylindrical objects. However, they always thought of these diverse forms as "triangular plots" or "cylindrical stones" and never in the abstract as triangles or cylinders. Although they were familiar with some of the relationships among the sides of a triangular piece of land, they had no knowledge of the general properties of triangles as such. It was Thales and a few others of his time who began to develop the general knowledge of the relations and properties of lines, angles, surfaces, and solids in the abstract without reference to any specific objects. Greek geometers developed their subject so rapidly that Euclid was able to write his classic *Elements* about 300 B.C.

Greek scientists also produced important knowledge in other fields, and their work in physics was fundamental for the future growth of science. Aristotle (384-322) was the greatest physical scientist of the period, and his works were the very foundation of the subject for two thousand years. However, there was little application of Greek natural science to engineering in antiquity. In fact, engineering contributed far more to science than science did to engineering until the latter half of the nineteenth century. In other words, fully twenty-five centuries were to elapse before this body of scientific knowledge would increase sufficiently to make it generally useful to engineers. This is not surprising when one remembers that the empirical knowledge of engineering had already been accumulating for at least twenty-five centuries when the Greeks began their scientific investigation of nature in the sixth century B.C. Although the application of natural science to engineering is relatively recent, Greek and Roman engineers began to use the new geometry as soon as it began to develop. Their principal use of geometry was in architectural engineering.

The Greeks also developed a sense of achievement disdainful of the world about them as less cultural, inferior, barbarian. They were well aware and proud of themselves, but they were realistic enough to appreciate what it was they had acquired from others and where they had learned much that they used so effectively. In the field of architecture they were skillful adapters rather than originators. Their innovations, though often significant for improving upon their Mycenaean, Minoan, or Egyptian models, expertly conserved fundamentals which had proved sound. Theirs was the contribution of the artist, a sense of proportion, an understanding of the beautiful. There was little or no experimentation with ideas or forms which were bizarre and confused. The finest Greek temples were essentially reproductions in stone of wooden and stone buildings which had preceded them in the Mycenaean age, but the difference in architectural effect is astounding.

The men who produced these marvels of beauty had no name which can be translated exactly as "engineer" according to our definition of the word. The Greek *architecton*, though primarily a technician in charge of constructing the public building, was often its designer as well. If he may be judged by what he built, he is certainly entitled to esteem as a man of ingenuity. The *architecton* worked under a contract system much as many engineers do today. When the city-state of the fifth century B.C. wished a new building, it hired private craftsmen, masons, and sculptors, who as master workmen brought along their own helpers, apprentices, and employees. These contractors took some responsibility, but they worked beside their men under the supervision of the *architecton*, who was responsible to the state. The pay of the several grades differed, of course, but not by very much. The *architecton* received about a third more than an ordinary mason.

The contract was usually inscribed on stone and placed on the site. It included specifications in some detail to guide the workmen and inform the public. Bits of such detail have been found in many locations, but none in better preservation than on the site of the naval arsenal in Piraeus, the port of Athens. This arsenal was a structure some 400 feet long, 55 wide, and nearly 40 high, built in the fourth century B.C. under the direction of Philon. His specifications, in modern type, would cover four printed pages. They outlined the general dimensions and fixed important measurements such as the thickness of walls, sizes of certain stones, the width and height of windows. Lesser details, such as we would expect on our blueprints today, were not fixed. The *architecton* and his master workmen had to carry most of that sort of information in their heads.

The Greek engineer was not the product of a specialized school of technology. In his time there was nothing like technical training in laboratory or classroom. He worked on the job, learning his craft in hard apprenticeship under men who had learned it in just that way before him. If he seems to have taken greater pleasure in the artistry of his work than perhaps the engineer of today does, it may have been because his nose was not kept in so many rolls of blueprints and pages of specifications. Perhaps, being a Greek, he was just that much more of an artist. Or possibly it was because almost everybody else shared to a degree his sense of proportion and appreciated the beauty which he was trying to express. Whatever the reason, the facts are that the Greek *architecton* was very likely to be an artist and often had high place in public regard. Ictinus and Callicrates, architects of the Parthenon, were included in the group of philosophers, artists, and statesmen whom Pericles, leader of the Athenians in their greatest period, gathered about him and with whom he delighted to talk.

Greek Engineering

The dimensions of the Parthenon were not conceived on a grand scale like that of the pyramids. Its engineers used simple mathematical principles to obtain the form which they desired, but from then on the eye of the artist took charge. Rigid application of those principles was so subtly avoided that no precisely mathematical statement of proportions can be made. The columns of the Parthenon taper ever so slightly, but enough so that the pediment does not seem to bear down upon them and make them look squat like the pillars of the Minoan palace. The broad steps which lead to the Parthenon are not horizontal; they are great vertical arcs rising to their center. The camber overcomes the optical illusion that steps actually horizontal would give of sagging at the center under the weight of the building. It may be argued that the builders did not plan it that way; they were lucky; it just happened to come out that way. If so, it was the luck of inspired artists. These refinements appear consistently in Greek architecture although there is no evidence that they had the instruments of precision necessary to obtain these results.

It is obvious in many structures that the Greek architect chose for the basis of all proportions a unit of measure which he could easily multiply. Often it was the Greek foot, 11.6 of our inches. He made the bottom drum of a column 2 or 3 units in diameter, the height of the column to its capital 10 or 12, the space between the columns 5 or 6, and so on. Thus he kept all measurements in pleasing ratios and also made it possible to carry on the work without minute specifications and plans. His masons, equipped with rods marked in unit lengths, could easily translate his proportions into stone. Such proportions would tend rapidly to become rigid and formalized according to the tastes of the most famous craftsmen. In applying theory to practice, Greek builders were held rather closely within the limits of accepted modulation. There could be very little experimenting with extreme architectural forms or speculation in strange engineering principles. During the next four hundred years their construction developed toward slighter columns and taller buildings under the disciplines of but three orders of architecture, and those were closely related. By the time of the Roman Vitruvius, first of the practicing engineers whose writings are extant, the Doric, Ionic, and Corinthian, varying only in ornamentation and modulary detail, governed the architecture and construction of the Graeco-Roman world and therefore the civilized world.

The mechanical operations of the Greek engineers are best observed in their handling of stone. They worked limestone and marble quarries in steps, making cuts around a block and splitting it with wooden wedges expanded by wetting. Bosses for levers were left at ends and sides to aid in easing the blocks down marble-paved chutes to oxcarts, wheel ruts of which may still be seen. Sometimes a column or block was encased in a drum of wood and trundled like a modern roller. Chersiphron, one of several who designed the first temple of Artemis at Ephesus, is said to have invented this method of hauling. The block was not finally shaped until it arrived at the building site where indentations were cut for lifting tongs and holes were drilled for metal cramps (rods with their ends bent at right angles to fit the holes in adjacent stones and tie them together). Channels, too, were made for the molten lead which sealed the dowels, or cramps, binding the stones together, for in their finest work the Greeks did not use mortar. A derrick and pulley, rigged on scaffolding of timbers, hoisted the block into position. The power for this lifting was supplied by the workmen themselves. Final surfacing and polishing of the stone was not done until the entire wall or column had been set.

The Greeks used not only the cutting tools, wedges and levers, inclined planes, carts, and devices that would be expected from their familiarity with Minoan and Egyptian ways, but also other tools which they appear to have employed freely for a long while. Where and when they got their derricks, their compasses, straightedges, and squares is not known; some of these instruments must have been their own invention. It is the pulley, however, that is the most interesting. Inscriptions show that the Assyrians had a kind of pulley, but whether the Greeks got it directly from the Near East, as is possible, or invented it independently has not been determined. How much they actually knew of the force exerted through the multiple pulley, or block and tackle, later credited to Archimedes, is more puzzling.

One contribution the Greeks did make to building, although it may have been suggested to them by the Minoan practice of placing wooden crossbeams in masonry. When they thought that their stone beams would not be able to withstand the weight from above, they brought the greater resistance of iron under tensile stress to the aid of the stone. They used concealed wrought-iron bars to reinforce or, as modern engineers would express it, to increase the factor of safety. The foundations of the Theban Treasury at Delphi were so strengthened with horizontal iron bars $3\frac{1}{4}$ inches wide by 4 inches thick and as long as 41 feet. The lintel of an underground doorway in the Erectheum at Athens had a groove along its lower surface in which an iron bar was sealed with lead. In the temple at Bassae, U-shaped bars within hollow marble beams largely supported the weight of the ceiling. And there was iron in the Parthenon itself, embedded to anchor cantilevers in the walls for the heavily loaded cornices.[1] Most of these reinforcements disappeared long ago in rust, so that few persons realize even now that they were used by the Greek builders. Vitruvius, though discoursing at length upon their architecture and methods of construction, had nothing to say about the Greek use of metals in this manner. The opinion prevailed that the Greeks built only with stone, mud bricks, and wood. But the grooves remain with unmistakable stains of iron rust to prove that they knew something of the problems of stress under tension as well as compression and that they found iron useful in solving them. The Greek engineers were apparently the first to make the significant advance of strengthening stonework with iron reinforcement.

They made no comparable progress in other fields of engineering. Primarily a seafaring people whose land was cut by mountains and arms of the sea into small city-states fiercely jealous of one another, they neither desired nor perhaps really needed arterial highways. In any case, they made no effort, as did the mountaineers of Peru for example, to construct roads that would link their settlements; nor did they seek to improve upon the road building of their predecessors in Cnossus and Mycenae. Short roads to shrines like Eleusis, or hard surfaces from quarry to wharf, were maintained with care. But it cannot be said that heavy traffic was expected on pavements in which grooves were cut to keep the wheels on the narrow way and which were provided with only an occasional turnout. However, cities like Corinth had some paved streets and sidewalks.

One unique project in Greek engineering did require paving on a large scale. It was the *diolkos,* or slipway, for vessels across the Isthmus of Corinth, to avoid the journey of 450 miles around the Peloponnesus. Parts of the 4 miles from shore to shore were paved to the width of some 15 feet. At either end a causeway sloped under water like a modern marine railway and made it possible to load galleys and light ships, most of them under a hundred tons, upon cradles with rollers, and then to have men or oxen haul them over the portage. Winches and pulleys may have been used on the steeper grades at the ends; Greek sailors were experienced with them for they were accustomed to beach rather than anchor their vessels. Though a canal was many times proposed, and the Roman Emperor Nero later actually began one, the task of cutting it through the

---

[1] William B. Dinsmoor, *The Architecture of Ancient Greece,* 3d ed., B. T. Batsford Ltd., London, 1950, p. 176.

solid rock was too difficult and too costly until the time of the modern ship canal at Corinth, completed in 1893.

Some centuries before Christ a new kind of ship was evolving in the Mediterranean area. The continued increase in trade presented a need for a special type of ship to carry goods. The early ships propelled by oars were long, shallow, and narrow, and not particularly suitable for carrying cargoes. Therefore engineers developed a design that was deeper and broader. Inasmuch as a large crew of oarsmen was not an economic asset, since they took up cargo space and their food occupied still more, they were dispensed with and sail was substituted as the principal source of propulsion. This Mediterranean merchantman was the first true sailing ship. However, it carried only a single square sail like its Egyptian predecessors and like them could not sail into the wind; moreover, it often was maneuvered by rowing although it did not carry a crew particularly designated as oarsmen.

Greek cities, like others in ancient times, had public supplies of water. It was relatively easy, in most cases, to dig ditches following the contour of the land from the springs in the mountains down to reservoirs in or near the city. From these reservoirs, pipes made of clay ran to fountains and basins where the people got what water they needed in their houses. There is to be seen on the island of Samos, however, a much more elaborate water-supply system of the sixth century B.C., described by Herodotus and rediscovered in 1882. Eupalinus of Megara, engineer, tunneled through a mountain of rock nearly 1,000 feet high and constructed an aqueduct ⅘ mile long and almost 6 feet square. As is still done, the tunneling was started from the opposite sides and driven into the center.[2]

The contrast between the splendid new public buildings of Athens, which had been planned, and the wretched old houses of sun-dried brick, which had been built haphazardly along crooked streets through the centuries, offended the Greek sense of appropriateness and beauty and stirred desire to do something about it. In this spirit of public benefaction, Hippodamus of Miletus, friend of Pericles, became the pioneer of modern civic planning. True it is that ancient Babylon was built somewhat according to plan with a monumental boulevard or processional street. And there had been at least one Hittite city, Zendjirli, laid out precisely in

[2] George Sarton, *Introduction to the History of Science*, 3 vols. in 5, The Williams and Wilkins Company, Baltimore, 1927–1948, vol. 1, p. 76.

Figure 3.3 Portion of Greek city of Olynthus, recently excavated (Courtesy D. M. Robinson)

circular outline about 1300 B.C. Nonetheless, the Greeks have been credited with the first real spirit of community living. Tradition holds that Hippodamus took the lead. Aristotle thought him the first *architecton* to conceive of streets and buildings as a harmonious whole. In other words, he developed a city plan the features of which were wide avenues with cross streets, forming rectangular blocks of residences, temples, theaters, gymnasiums, athletic fields, monuments, and open spaces located in proper relation to one another for the convenience of the people who were to use them.

Given the opportunity, Hippodamus reconstructed Piraeus, the port of Athens, on an elaborate scale. He laid out the Greek colonial city of Thurii in southern Italy according to an impressive rectangular plan, and it is thought that he may even have had a hand in building Olynthus (Figure 3.3), an Athenian colony on the northern shore of the Aegean. Recent excavations there have brought to light private houses grouped in blocks of ten with their walls rising from the edge of the street. As the

older Greek cities expanded, their suburbs too were sometimes constructed from plans, but general application of Hippodamian principles
had to await the new cities of the Hellenistic period.

The Hellenistic World

For its inception and in large part its character, this period has to thank
Alexander the Great more than anyone else. A pupil of Aristotle and an
enthusiast for Greek culture, Alexander took with him Greek institutions,
methods, and ways of thought wherever his armies swept over the
peoples whom the Persians once had conquered. For two centuries after
his death in 323 B.C. his successors divided his empire and fought for the
fragments, the petty states and provinces, but its cultural unity remained.
Alexander had made it a Greek world centered upon Alexandria, his own
great city in Egypt. When the Romans came in their turn to conquer
the eastern Mediterranean lands and Asia Minor as far as the borders of
India, it was a Greek civilization which put its mark upon them.

Along the banks of the Euphrates, in Babylon, on the shores of the
Black Sea, in Egypt, even upon the coasts of the western Mediterranean,
wherever men were settled in communities, lived those who spoke Greek,
disputed Greek philosophies, and built their new towns in the Greek
style. The tone of this culturally unified and prospering age was set by
its urban and well-to-do classes. Active, practical, worldly, sometimes
sophisticated to the point of decadence, they traveled extensively and
developed widening interests. They built more luxurious dwellings, larger
athletic stadia, better roads, more adequate water supplies and drainage
systems, new and more elaborate public buildings.

By its mood and its action, the Hellenistic world required hundreds of
architects, builders, and engineers. Their patrons, and the public, expected
them not merely to carry on the great traditions of classic Greece; they
were also to create structures that would impress because of their newness.
Craftsmen who could satisfy both desires were deserving of great fame.
Pytheos was one of them. He used exactly the accepted proportions for
the diameter of columns, the space between them, and their height in constructing the Mausoleum at Halicarnassus in 352 B.C., and with these proportions he combined three traditional forms, the so-called high pedestal,
or foot, of a column, the Greek temple, and the Egyptian pyramidal burial
mound. The result was a building which stirred admiration for its un-
usualness and yet satisfied established taste. Pytheos was also the first of

Figure 3.4 One of many reconstructions of the Pharos of Alexandria (From H. Thiersch, *Pharos, Antike Islam und Occident*, 1909; courtesy B. G. Teubner Verlagsgesellschaft)

many Hellenistic architect-engineers to train his apprentices in schools and to write treatises for the builders of the future. Most of these works were concerned with the mathematical proportions of the various architectural orders. None of them has survived to this day except in the writings of the Roman Vitruvius.

Other outstanding engineers of the period were Dinocrates and Sostratus. Alexander did not gratify the ambition of Dinocrates to carve Mount Athos at the head of the Aegean Sea into a gigantic seated statue of his chief, but he commissioned him to lay out the city of Alexandria at the mouth of the Nile, perhaps a more enduring monument and certainly one more worthwhile. Dinocrates gave the city two main thoroughfares, it is said, each lined with colonnades, temples, and buildings of grand proportions. There was also a long mole and causeway from the mainland to an island behind which the shipping of the Mediterranean could safely anchor.

Whether Dinocrates planned and built this harbor is not known, but Sostratus, two generations later, did construct the Pharos, its famous marble lighthouse, called one of the Seven Wonders of the World (Figure 3.4). A tradition is that it was built in three sections, the lower square, the middle octagonal, and the upper circular. Some think that an inclined walkway wound about the interior of the structure, and a coin of the period shows an entrance at the base. In Roman days, according to the historian Josephus, its beacon of burning pitch could be seen 300 stadia (35 English miles) at sea. If that report is to be accepted and the probable height of the observer from the sea assumed, and disregarding the glare from the beacon, the Pharos must have been nearly 480 feet high, nearly as tall as the highest pyramid in Egypt, and only 70 feet short of the Washington Monument. If so, probably no taller lighthouse has ever been erected.

In keeping with the Pharos, Alexandria, greatest of all Hellenistic cities, may have been equipped with the latest public utilities. What is actually known of such things as its water supply, however, is virtually nothing, but some twenty or thirty other Hellenistic communities had water systems employing tunnels. Inverted siphons appear not to have been understood by the Mycenaeans, or perhaps even by the men of Samos who tunneled their mountain. The most notable of the Hellenistic pressure systems has been uncovered at Pergamon in Asia Minor. About 200 B.C. water was carried some 35 miles in three 7-inch pipes of tile, laid side by side. They discharged into a reservoir on a hill 2 miles from the city and 100 feet higher. From there, a single 10-inch pipe, perhaps of bronze or possibly wood, reinforced at its joints by the large perforated stone blocks through which it passed, ran down across several valleys—the deepest more than 600 feet lower than the reservoir—and up over the intervening ridges to pour its stream finally into the city's fountains. For the last 2 miles, the water was thus under pressure which at the low point approached 300 pounds to the square inch, several times as much as the usual pressure in an American city. Such pressure would have been too great for tile or even lead pipe. There is doubt that bronze piping had yet been developed in ancient metallurgy, and it is believed that because of its tensile strength, wood must have been used in the inverted siphons at Pergamon.

The Hellenistic period was both end and beginning. The Greeks had brought construction in stone beyond the Minoan and Egyptian techniques to a state of architectural perfection and engineering finality. The Romans would continue building in the Greek manner even though they made departures with the arch and dome that opened new ways to the Moslems and the men of medieval Europe who followed them. But the Hellenistic Greeks did not stop with architectural perfection. Their investigations and discoveries in fields of science, which hardly stirred Roman minds, were to have incalculable influence upon engineering far beyond the Middle Ages.

Euclid (ca. 300 B.C.) organized and made original contributions to the mathematical knowledge of the time. His work constitutes for the most part the plane geometry still used, and parts of his book were still forming the basis of texts used in schools in the early twentieth century. In the field of engineering itself, others at Alexandria appear to have been the first to outline systematically certain basic principles of mechanics. Primarily speculative philosophers, they asked and answered questions but solved no problems in actual construction. None of them was a practicing engineer except Archimedes (ca. 287–212 B.C.) of Syracuse. He came to Alexandria from Sicily but had returned to his home before he worked out his well-known problems of mensuration and mechanical power. Archimedes' contributions in applied mathematics, "specific gravity," the pressures of liquids, and the principles of the action of levers are fundamental for the engineer today. The discoveries of both the compound pulley and the screw have been attributed to him, though each, of course, may have been in use some time before he made it famous. His proud boast that he could move the earth itself, if he were given a lever and a place to stand, is one of the treasured legends of history, whether or not he ever said it. Archimedes' ingenious machinery, levers, cranes, and catapults for defending Syracuse in 211 B.C. by land and sea amazed the Roman General Marcellus, whose disappointment and anger must have been genuine when he learned that, as his soldiers at last broke into the city, the great engineer and scientist had been killed.

Other Hellenistic inventions were just as significant in the field of applied power. Ctesibius of Alexandria, possibly contemporary with Archimedes, left no account of his work, and it is known only from descriptions in writings of his successor Hero, who himself may have been

Figure 3.5
Hero's aeolipile, or turbine (From *Transactions, Newcomen Society*, vol. 16; Original in Hero, *Spiritalia*, 1589; courtesy Newcomen Society)

the inventor of some of the devices. Other devices may have come from a still more shadowy figure in these centuries, Philo of Byzantium. Hero describes a hydraulic clock, hydraulic organ, fire engine, force pump, air gun, steam turbine, and most spectacular of all, an automatic theater whose puppets were made by steam or hot air to revolve and dance. These inventions were no more than mechanical toys, but their conception and the principles which they applied were significant.

The turbine ascribed to Hero was a hollow metal ball free to turn upon a horizontal axis above a cauldron and brazier (Figure 3.5). Two tubes, bent in opposite directions, protruded from the ball on opposite sides. Steam from the cauldron passed through stationary tubes on which the ball was pivoted horizontally and escaped from the bent tubes to rotate the ball. Since this motion was caused by the reaction of unbalanced forces, and not by expansion of the steam against a movable surface, Hero's apparatus was, properly speaking, a reaction turbine, but it appears to have done no useful work. Hero also tried to invent a meter for the flow of water, and he told vaguely how to find the quantity of water coming from a spring by multiplying the depth by the width of the stream, at the same time observing the velocity somehow with a sun dial. He knew the principle of the screw and described an instrument geared to a paddle wheel to indicate the speed of a ship. He wrote numerous books on stereotomy (the art and practice of stonecutting, a sort of descriptive geometry) and on measurement of areas and volumes, on pneumatics, mechanics, and surveying.

It is not easy to determine the effect which these books and experimental devices had in their own time upon practicing engineers. It is not known for certain whether the Roman engineer and writer on engineering, Vitruvius, used Hero's works. It is even possible that Hero lived later than Vitruvius. But some of Hero's ideas surely did get into use. The *dioptra* for surveying, which he described, the water screw that carries Archimedes' name, the compound pulley, the odometer for measuring distances, and perhaps the plunger pump became standard equipment for Roman engineers.

Bibliography

Baikie, James: *The Sea-kings of Crete,* 4th ed., A. & C. Black, Ltd., London, 1926.

Dinsmoor, William B.: *The Architecture of Ancient Greece,* 3d ed.,

B. T. Batsford, Ltd., London, 1950.

Evans, Arthur J.: *The Palace of Minos,* 4 vols. in 6, Macmillan & Co., Ltd., London, 1921–1935.

Feldhaus, Franz M.: *Die Technik der*

Antike und des Mittelalters, Akademische Verlagsgesellschaft Athenaion M.B.H., Wildpark-Potsdam, 1931.

Haverfield, Francis J.: *Ancient Town-planning*, Clarendon Press, Oxford, 1913.

Neuburger, Albert: *The Technical Arts and Sciences of the Ancients*, Methuen & Co., Ltd., London, 1930.

Sarton, George: *History* (see Chap. 2).

Sarton, George: *Introduction to the History of Science*, 3 vols. in 5, The Williams & Wilkins Company, Baltimore, 1927–1948, vol. 1.

Vitruvius Pollio, Marcus: *On Architecture*, 2 vols., G. P. Putnam's Sons, New York, 1931–1934.

FOUR

Imperial Civilization

Except for the short-lived empire of Alexander the Great, Greek civilization never had any centralized political control. For the most part, Greek political organization consisted of independent city-states. In classical antiquity it was the Romans who had a genius for political administration, and it was they who conquered most of the civilized world and maintained political control over it for centuries. Between 500 and 300 B.C., the period in which Greek engineering was reaching its first heights, the men of the city of Rome were conquering the Italian peninsula. Having solidified their political control of Italy, they then proceeded to foreign conquest, and during the next century and a half they conquered the Mediterranean world and established the Empire.

The political, economic, and social needs of the Empire, more particularly of its cities, presented new problems which engineers were obliged to solve. The result was an intensified stimulation of engineering which produced important advances in building construction, water supply, bridges, roads, harbors, ships, and mining. The Roman engineers were the foremost engineers of classical antiquity. It is a curious fact, however, that there never was a Roman scientist of the first rank. During the first three centuries after the Greeks had been absorbed into the Empire, Greek science continued to flourish. Indeed some of the renowned Greek scientists did their work during this period; one of them, Galen (130–201), was even active for years in the city of Rome. There was, therefore, a curious division in the Empire between the Greeks and the Romans; the former group produced outstanding scientists as well as competent engineers while the latter furnished the great engineers. The differences in cultural abilities and tradition were to have a marked effect in the Middle Ages.

Most important for the history of the West during the period covered by this chapter was the life of Jesus of Nazareth. The evolution of Christianity from the teachings of Jesus affected engineering in a way that only began to become apparent during the last centuries of the Empire. As had been the case with Mesopotamians, Egyptians, and Greeks, the main source of power for Roman engineers was the effort of human slaves. However, during the fourth and fifth centuries of our era Christian Roman slaveowners increasingly freed their slaves for a variety of reasons—some of which were economic, but also because the new Christian attitude viewed all men as being equal before God. The decreased use of slave labor was one of the many important causes leading to the Empire's dissolution; the gradual emancipation resulted in an attrition of almost the only source of power. Only a few decades before the fall of Rome did her engineers begin to make limited use of water power to replace slave labor.

The Engineers

Whatever the causes, Rome rose to supremacy over the world from Scotland to Persia. And whatever their origins, Rome's engineers added to her power and her fame. When men turn from rural to urban living, they are likely to discover interest and aptitude in themselves for devising and constructing things, in short a flair for engineering. Or they will seek out those who have it. The need becomes insistent as men gather in crowds. There is heightened appreciation of individual concern in public welfare, not necessarily a moral sense of obligation to fellow men. There is realization that one is very much involved with others. Like the Sumerians, Babylonians, Egyptians, Minoans, and Greeks, the Romans responded to the stimulus of urban living. They greatly improved the buildings, communications, and utilities of their Republic. In fact, devotion to public works remained long after republican institutions of government had become mere ornaments to a regime of emperors.

Unlike preceding periods, the time of Rome was one of few discoveries or inventions. Romans did little theorizing, but they were skillful at learning and adapting the ideas and practices of others. As they conquered and absorbed their neighbors on the north in the fourth century B.C., they appropriated Etruscan knowledge of subsurface drains and building with arches and stone blocks. They took over Greek architectural forms.

materials, tools, and methods, even the Greek engineers themselves, as they overran those neighbors in southern Italy and Sicily. In the middle of the third century B.C. virtually all that the early Romans knew about engineering came to them out of the civilizations of the eastern Mediterranean; for it is generally accepted from archaeological evidence that the Etruscans, though their records have not yet been deciphered, came to the northwestern shore of the Italian peninsula from Asia Minor and that they had close relationships with Egypt.

The engineers of Rome were developers rather than originators. They were practical men, as Frontinus boasted, expert in applying ideas, in extending and enlarging their public usefulness. To say this is not to assert that Romans thought only about the functional purposes of their work and had no appreciation of beauty. Their work was strong, solid, impressive, but it was balanced and well proportioned. It was pleasing, not wearisome, in its impressiveness. The word "grandeur" has been synonymous with Rome ever since Edgar Allan Poe wrote "To Helen." Though used until worn flat, this word still defines, as no other word can, both the architecture of the Eternal City and the range of its engineers.

One historian has declared that the cities of the Roman Empire enjoyed systems of drainage and water supply, heated houses, paved streets, meat and fish markets, public baths, and other municipal conveniences that would compare favorably with modern equipment.[1] Such generalizations can be dangerously misleading. They are relative to other factors and conditions which may not be presented to the reader at the same time. It would be a mistake to conclude from this comparison, revealing though it is, that all Romans enjoyed freely all of the facilities of their cities. The comparison nonetheless does indicate clearly the extent and the efficiency of the Romans in applying their knowledge to the problems of their daily life. And in this they were, according to our conception of the term, exceptionally able engineers. Their public baths were a notable example of civic accomplishment.

The profession of the Roman *architectus*, master technician or engineer, was greatly respected. Appius Claudius Crassus, the censor, was acclaimed down through Roman history for his aqueduct begun in 313 B.C. and his great road of the following year. Julius Caesar praised the skill of the engineers who built his bridge across the Rhine. Vitruvius

[1] Mikhail I. Rostovtsev, *The Social and Economic History of the Roman Empire*, Clarendon Press, Oxford, 1926, p. 135.

wrote for the approval of his patron Augustus a treatise on architecture and engineering that held the attention of builders far into medieval times. Agrippa, Minister to Augustus, was a noted engineer. The Emperor Claudius took a personal interest in public works. The historian Tacitus admired the genius of Nero's engineers, Severus and Celer. The Emperor Hadrian, himself an engineer, maintained a staff of experts, one of whom, Apollodorus of Damascus (ca. 98–ca. 117), built his great bridge across the Danube in record time, dwarfing the accomplishment of Caesar's men and setting an example for Charlemagne.

So keen was the interest in engineering and appreciation of its value in both military and civil affairs, that systematic training was encouraged at home and recruits for the service were sought abroad. This does not mean that institutes of technology such as we have today, or even anything like the universities of the Middle Ages, appeared in the Roman Empire. It does mean that such emperors as Trajan, Alexander Severus, Constantine, Julian, and Justinian searched for and financed the training of likely young men and that a system of apprenticeships directed by the state supplemented and improved upon the traditional method of handing down technical knowledge from father to son. Martial, sycophant to the Emperor Domitian and master of epigram, perhaps quite unconsciously drew attention to this wide interest in engineering. Certainly he paid oblique compliment to the profession with something less than his habitually light touch. "If the boy," he said, "seems to be of dull intellect, make him an auctioneer or architect." [2] In spite of the poet's sarcasm, the engineer was a man of distinction at Rome whether he were native or foreigner, patrician or plebeian, master, freedman, or slave. The American reader must caution himself here against assuming that Roman slaves were inferior because they were slaves. Roman slaves, taken captive in war or purchased in the market, were often intellectually superior to their masters and, what is more significant, were recognized as being so.

Roman engineers set their profession firmly upon economic principles. They employed exact specifications and detailed contracts and they took into account varieties of materials and types of construction best suited to particular conditions and projects. Their public buildings, aqueducts, bridges, and roads show a sense of economy and efficiency just as the legal system of Rome reveals orderliness and analytical power. As Vi-

[2] Marcus V. Martialis, *Epigrams*, 2 vols., G. P. Putnam's Sons, New York, 1919–1920, vol. 1, p. 337.

truvius expressed it, the engineers understood the "suitable disposal of supplies and the site." They knew how to make a "thrifty and wise control of expense in the works." [3]

Building

The Romans developed superior practices of their own from the techniques of masonry construction which they had obtained from the Etruscans and the Greeks. They too built with squared stone of uniform thickness, laid closely without mortar and often held in place by iron cramps. This coursed ashlar they called *opus quadratum*. Whether made of the comparatively soft tufa, the harder lava and peperino, or the very strong travertine, standard *opus quadratum* was laid in regular courses of blocks usually 2 by 2 by 4 Roman feet. It was the type of construction used for the bridges over the Tiber River and for most of the aqueducts. If necessary or desirable, Roman builders used heavier stone; some in the Colosseum are 15 feet long and weigh 5 or 6 tons. But, like the Greeks in contrast to the Egyptians, they preferred smaller pieces to large blocks. To them it was a waste of time and energy to drag large stones from the quarries and to hoist them into place if one could build as well or better with smaller materials.

Perhaps the Roman engineers knew more about masonry under compression than is generally assumed because they had no written or formal knowledge of statics, or the equilibrium of forces. The careful examination which Vitruvius gave to Greek methods indicates that the Greeks did know something of these subjects, at least empirically. He observed that the Greeks bound their walls at intervals with stones or headers running entirely through the masonry. When the Romans did not, said Vitruvius, they were in a "hurry to finish," and they filled the space between with "a lot of broken stones and mortar thrown in anyhow." When they came to building their arches and vaults, they were more careful —if they followed the expert directions of Vitruvius. "Moreover, when buildings rest upon piers," he wrote, "and arches are constructed with voussoirs and with joints directed to the centre, the end piers in the buildings are to be set out of greater width, so that they may be stronger and resist when the voussoirs, being pressed down by the weight of the

[3] Marcus Vitruvius Pollio, *On Architecture*, 2 vols., G. P. Putnam's Sons, New York, 1931–1934, vol. 1, p. 31.

walling owing to the jointing, thrust towards the centre and push out the imposts." [4] If the piers were ample, he said, they would hold together the voussoirs, or wedge-shaped stones of the arch, and give stability to the structure. Thus he explained the horizontal thrust which is brought down by the arch to the abutments.

The first Roman departure from solid work we know perhaps as random ashlar. They called it *opus incertum*. It would be a mistake to translate these words literally as "uncertain work," but the fact is that many Roman builders took less care than they should have with it despite the warnings of Vitruvius. The stones in *opus quadratum* were squared, but they were of varying sizes so that the horizontal joints were discontinuous or "broken." They were laid up irregularly in a mortar of lime and sand. This type of construction was used generally in the lofty *insulae*, or apartment houses, which often filled the city with the din of their collapsing. No other sound in the night except the cry of fire was more terrifying to the poor who crowded their upper stories. The satirist Juvenal, writing about A.D. 100, bitterly complained of the landlord's agent who propped up the old walls with the statues of a "slender flute-player" and bade "the inmates sleep at ease under a roof ready to tumble about their ears." [5] The *opus reticulatum*, or network, which for two hundred years or so replaced the *incertum* in popularity, was made of small squared stones laid with beds and joints inclining 45 degrees. Evidently this was done to please the Roman eye; it offered no structural advantages. In fact it took more time to lay and, as Vitruvius said, it was likely to crack, for "its beds and joints spread out in every direction." The Emperors Trajan and Hadrian preferred it nonetheless for warehouses, baths, villas, harbors, and other construction.

Meanwhile the Romans had been developing the use of brick as well as stone. And as they did so, they made a discovery which must be given place beside their arches and vaults among their great contributions to engineering. They were familiar with all types of brick—square, oblong, triangular, curved, using the last to build columns. They liked particularly to face their walls with triangular brick, pressing the right angles of brick into a soft backing as they laid up the walls course by course. This backing of rubble and mortar at first was not carefully proportioned.

[4] *Ibid.*, vol. 2, pp. 54–55.
[5] Decimus J. Juvenalis, *Juvenal and Persius*, G. P. Putnam's Sons, New York, 1930, p. 47.

Figure 4.1
Actual section of Eiffel-
Cologne Aqueduct
(Courtesy U.S. National
Museum)

A wet mixture of lime and sand was spread in layers, and fragments of
stone, pottery, and tile were pressed into each layer by hand. It is not
clear whether or not they used wooden forms for this work, as of course
they did when building arches and vaults.

Such mortar might easily crumble as it dried. But soon the Romans
came upon a durable cement. They added to their lime mortar almost
equal parts of a volcanic ash from the neighborhood of Vesuvius; this pozzuolana would harden even under water. They discovered similar material later in Sicily and in the Campagna. In Germany and the Low Countries they found an ancient volcanic mud, or trass, to serve the same purpose. The Romans did not know the exact chemical properties or proportions of the silica, alumina, and iron oxide in their pozzuolana. They
did know that it was very strong. The Smithsonian Institution at Washington has a section of a Roman conduit laid about A.D. 80 near Cologne
(Figure 4.1). When dug out some eighteen centuries later, it did not
crumble or crack when drilled or chipped but behaved like stone. In
fact it had become stone. With such cement as this, the Romans built
the great vaults and domes of the Pantheon, the Baths of Diocletian and
Caracalla, the law courts and banking offices which are known as the
Basilica of Constantine.

Figure 4.2 Pont du Gard, near Nîmes, France (From T. Schreiber, *Atlas of Classical Antiquities,* 1895)

Water Supply

Vitruvius in the first century B.C. extolled the cement made of pozzuolana for solidity that neither waves could break nor water dissolve. Why, then, did the Romans bother to construct of stone rather than of concrete the majestic archways that still march across the Campagna toward Rome, that span the gorges of Spain, that rise tier upon tier of the Pont du Gard (Figure 4.2), 160 feet above the river at Nîmes in France? To anyone familiar with concrete and stone masonry the answer is obvious. The elaborate "forms" needed for casting concrete compared with the absence of anything except staging for stone masonry is the explanation. The Romans understood and used inverted siphons, or conduits, which carry water across valleys, following the surface of the ground. Why did they not make high-pressure conduits of pozzuolana instead of building their famous aqueducts? There has been speculation whether the Roman engineers knew how strong their concrete was. The chances are, rather, that they realized how weak it might be, for however sturdy under compression, it could not withstand much tension. It was not strong enough to resist the bursting force of water under high pressure. Even the best of modern concrete has little tensile strength; it has to be reinforced with metal. The Romans had no such concrete.

What pressure mains they used they built of lead. The pipes were generally rolled from ¼-inch sheets on mandrels. Presumably the seams were closed by melted solder. Successive lengths were overlapped and joined in the same way. The Romans had iron but no cast-iron pipe. They knew about bronze pipe, but except in a few plumbing fixtures, they did not for some reason use it extensively, and apparently not at all in their inverted siphons. Bronze was less easy to work than lead and doubtless more costly. But the strength of bronze was obviously greater, and the matter of expense would hardly have stood in the way, especially in comparison with the cost of stone aqueducts.

Domitian, Emperor from A.D. 81 to 96, gave little thought to expense when he had an inverted siphon built to his palace on the Palatine Hill. Its lead pipe measured a foot in diameter and crossed a valley 133 feet below the reservoir fed by the Aqua Claudia. Pressure at the lowest point in the siphon thus approached 60 pounds to the square inch. The Roman record, however, seems already to have been established at Alatri, more than 50 miles east of the city. An inverted siphon there, built about 134 B.C., had a vertical drop estimated at 340 feet, resulting in a pressure of 150 pounds to the square inch. It appears that the diameter of the pipe was about 3.9 inches (10 centimeters), and the thickness of the lead sheet from which it was made was nearly 0.4 inch (10 millimeters).[6] The use of pozzuolana for conduits under water was not a problem of pressure but of convenience. Pipes, especially those of flexible lead, are much easier to handle under water than concrete. Recently many feet of lead conduit, 6 inches in diameter and larger, which had lain on the bed of the lower Rhone River near Arles some fifteen hundred years, were brought to the surface by a ship's anchor.

Vitruvius explained ways of reducing pressures in lead and clay pipes by means of frequent reservoirs, level stretches, or venters, and other retarding devices. But the Romans built arched bridges high above valleys and made deep cuts and tunnels through solid ridges to maintain the hydraulic gradient. The capacity of metal inverted or pressure siphons was more limited. The largest lead pipe which Vitruvius mentioned was approximately 100 inches in circumference, while the average dimensions of the aqueducts were about 3½ feet wide by 6 feet deep. Moreover, it was hard enough to control leakage in the channels which were not under pressure. It has been estimated that, before reaching its proper destination in the city, half of the water which entered the aqueducts

[6] Frederick W. Robins, *The Story of Water Supply*, Oxford University Press, London, 1946, p. 68.

Figure 4.3 Italian peninsula

had disappeared through leakage and theft. Perhaps the latter could have been more easily controlled in the siphons, but repairs were easier to make on the arched aqueducts. And besides, the Romans knew that there was danger in impure water, although they had no knowledge of bacteriology. Vitruvius observed poisoning of "workers in lead" and cautioned against the use of lead pipes; instead he recommended clay pipe wherever possible.

As the hydraulic engineers of Rome saw their problem, it was to obtain an ample supply of water under steady flow. This they could do with low dams, feeder canals, and settling basins at sources 1,000 feet above sea level; then it could be let down slowly through conduits underground most of the way to reservoirs in the city at an elevation of some 200 feet. The hills which sloped obliquely toward the Campagna afforded a steady decline for these aqueducts if the contours were followed, a few ridges tunneled, and ravines bridged. Thus the aqueducts of Rome came within 10 miles of the city. Then and then only were they carried upon high structures across the Campagna. These archways amounted to hardly more than an eighth of the total mileage in the water-supply system. For economy's sake, a newer aqueduct was often added to the substructure of an older one. The arches of Claudia, for example, carried the Anio Novus (Figure 4.4) across the Campagna. The Tepula and the Julia were laid upon the arcade of the Marcia.

Rome's aqueducts with their great archways were its most impressive achievement in engineering. Frontinus, water commissioner of Rome in the first century A.D., wrote proudly of his aqueducts, "Will anybody compare the idle Pyramids, or those other useless though renowned works of the Greeks with these aqueducts, with these many indispensable structures?" [7] They were copied everywhere in the Empire. Forty or fifty aspiring provincial cities had water systems on the Roman model. The aqueducts at Segovia in Spain, Athens, and Constantinople, are still in use. An outstanding example in Roman days was that which supplied Carthage. Its 7 miles of partly ruined towering arches are a feature of the African landscape, practically all there is left to mark the site of Rome's ancient foe. The engineers at Lyon, capital of Roman Gaul, used a combination of arcaded aqueduct and inverted siphon conduits. By lowering the grade more than half, they reduced both the height and the

---

[7] Sextus J. Frontinus, *The Two Books on the Water Supply of the City of Rome*, trans. by Clemens Herschel, 2d ed., Longmans, Green, and Co., New York, 1913, p. 19.

Figure 4.4 Two Roman aqueducts, Anio Novus built on Claudia (From Curt Merckel, *Die Ingenieurtechnik im Alterthum*, 1899; courtesy Julius Springer-Verlag)

length of the arched structure. And then they carried the water down one slope across the Rhone bridge and up the other in eighteen 8-inch lead pipes laid side by side. These of course were many times stronger than a single pipe of the same thickness and total capacity.

Rome itself had 11 aqueducts, ranging in length from 10 to 60 miles. The first was constructed in 312 B.C. and the last in A.D. 226, five centuries later. Four of these carried nearly three-fourths of the supply. Herschel estimated that eight in the time of Frontinus, A.D. 97, delivered 220 million gallons daily, or from 110 to 120 gallons per capita. Today in New York and modern Rome the consumption per capita is about 130 gallons. There is no accurate way of comparing the relative apportionment then and now to industrial uses, sanitation, drinking, fire fighting, and waste. Some modern cities like Paris and a few American cities have dual supplies of water. Rome had three virtually independent services. Spring waters of the Marcia, Claudia, and Virgo aqueducts were piped from the distributing reservoirs first of all to the public fountains whence the majority of Romans daily carried home their supplies for drinking and other household purposes. But it should be remembered that the Romans, like modern Europeans, did not use water as freely as Americans do in their houses. The more turbid streams of the Anio flowed into the public baths, the fulling mills, and laundries where those white or gray woolen togas that distinguished the citizenry of Rome were washed. Overflows from all the reservoirs flushed the streets and the storm sewers.

There was relatively little water for private consumption. Some of the wealthier, either by paying outright, by bribing the inspectors, or by secretly tapping the mains, contrived to have service pipes into their houses. The authorities seem never to have thought of financing the water system by sale although they leased the right to charge fees for the use of public latrines and the right to cart away night soil (human waste) for manure. Water was free in Rome, often the gift of an emperor or a wealthy citizen who had appropriated to that purpose booty taken in war. The Aqua Marcia, 55 miles long, was built in 144 B.C. from the proceeds of the victories over Corinth and Carthage at a cost estimated as 9 million dollars in the purchasing power of modern currency.

Bridges

Because stone bridges with semicircular arches and many short spans are typically Roman, there is a tendency to believe that the Roman engineers built nothing else. The fact is, they displayed skill and taste in varying their arches with the peculiarities of the site. Their seven bridges across the Tiber were of solid ashlar (squared stone) with uniformly low semicircular arches. Six are standing today, although only after extensive repairs which have amounted to rebuilding. They built a lofty viaduct more than 600 feet long across a rocky gorge at Narni on the Via Flaminia; the longest of its four arches had a span of 138 feet. They built another arch with a slightly longer span at Orense in Spain, and they raised a great bridge, perhaps their record, over the Tagus River at Alcantara, Spain, to the imposing height of 175 feet. Its six arches, the two central spans of 118 feet each, support a roadway 600 feet long. According to Lacer, the engineer who built it about A.D. 98 in the time of the Emperor Trajan, it would last forever. Though once partially destroyed, it was often repaired. It is still open to traffic.

In less traveled regions, the Romans used the "mule-back" type of bridge to save the expense of filling the approaches and to avoid the

Figure 4.5 Martorell Bridge, spanning Lobrogat River, Catalonia, Spain (From C. C. Schramm, *Historischer Schauplatz*, 1735)

construction of retaining walls. The roadway followed the slope of the arch itself, as in the case of the Martorell Bridge in Spain (Figure 4.5), attributed to Moorish builders but known to have been rebuilt on a Roman site. Where streams were shallow enough, as often in England and elsewhere, Roman engineers did not bother with bridges. They simply paved a ford with large squared stones and protected this crossing with guards consisting of horizontal beams supported by piles on either side.

The chief difficulty which the Romans had to meet and one that has not ceased to plague engineers was that of providing foundations for their bridges that could resist every vagary of the current. Sometimes they tried to overcome this difficulty by the sheer weight of their masonry. But heavy piers and short-span arches meant virtually damming the stream and thus increasing the speed of the current around the piers. As Vitruvius fully explained, it was not enough to dump loose rock or riprap about the foundations because the current was likely to burrow beneath and undermine the heaviest structure. For one reason or another Roman engineers frequently did not care to take the trouble to prepare adequate foundations, but when they did, they made a thorough job of getting below the most capricious stream even if they had to go down to bedrock. One method was to drive iron-tipped piles in close formation around the area where the pier was to stand. The machine they used to sink the piles has not been adequately described, but it must have been a simple pile

Figure 4.6 Pons Fabricius (right), built 62 B.C. and still standing, Rome (From E. Du Perac, *Topographical Study in Rome in 1581*)

driver, a heavy weight raised by tackle within a supporting frame. The material between the piles was excavated down to a firm bottom. The pier was then built under water with pozzuolana cement.

Where the water was too deep for this method, the Romans were likely to surround the area with a double-walled cofferdam of piles laced with wicker and packed watertight with clay if pozzuolana was not available. The cofferdam was then emptied by water wheels or screws, and solid dry ashlar masonry was laid up from the bedrock or from piles driven into the bottom far below the water line.[8] It is thought that the pier foundations of the Pons Aelius, now the Ponte Sant' Angelo, were built in this way for they go down more than 16 feet under the Tiber. But to judge from the number of Roman bridges that were swept away by floods, the engineers often did not use either of these painstaking and expensive procedures. The Pons Fabricius (Figure 4.6), built in 62 B.C., is still standing.

Julius Caesar's bridge (Figure 4.7) across the Rhine in 55 B.C., near the mouth of the Mosel or farther down at Bonn, is well known to school boys and girls for the troublesome Latin in which he described it. His engineers, Mamurra and others, had it finished in ten days after assembling the materials on the bank. It was ¼ mile long, built of timbers supported on piles driven in clusters 25 feet apart. How the timbers were fastened does not appear in Book IV of his *Commentaries*. Caesar was in very much of a hurry or he might have had it constructed more substantially; as a matter of fact he ordered the bridge destroyed after only eighteen days.

[8] Vitruvius, *op. cit.*, vol. 1, p. 315.

70

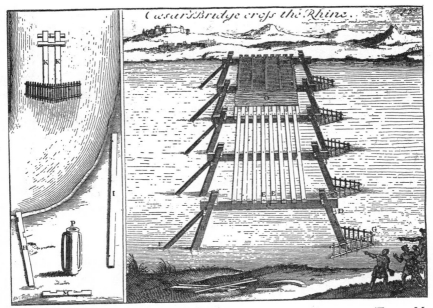

Figure 4.7 Caesar's Rhine Bridge, built and destroyed 55 B.C. (From M. Bladen, *Commentaries of Caesar*, 1719)

Later Roman bridges over the Rhine were also built of wood. No arch stone has been found in their ruins at Mainz, Treves, or Cologne. The wooden spans were set upon massive piers of stone blocks fastened with metal dowels, or cramps, sealed in the stone with lead; the piers rested upon clusters of 150 to 200 piles driven deep. Those at Cologne had actually to be blasted to get them out of the river bed; Constantine's engineers built the structure about A.D. 310.

The bridge across the Danube near the Iron Gate was the most pretentious of all Roman bridges. Or at least Trajan seems to have thought it so for he had a sketch of it cut in bas-relief (Figure 4.8) upon the famous column at Rome extolling his achievements as Emperor. It was built in A.D. 104 by the Greek Apollodorus of Damascus who, excepting Vitruvius and Frontinus, was the Empire's most noted engineer. His likeness, carved on Trajan's column, has come to us. All we know of his bridge is to be found on the column, in the meager references of contemporary historians, and from accounts of the pier foundations which may still be seen at low water.

The bridge was some 3,720 feet long and from 40 to 50 feet wide. It

Figure 4.8 Apollodorus's Danube Bridge, relief on Trajan's column (From T. Schreiber, *Atlas of Classical Antiquities*, 1895)

had 21 spans averaging 120 feet in the clear, or 177 from center to center of the piers. It seems likely that Apollodorus made a large artificial island in midstream to facilitate handling his materials. He constructed his foundations of loose rock held in cofferdams and raised the piers about 60 feet. We would expect them to have been made of cut stone, but no trace of such masonry has been found in the river. In whatever way the piers were made, it is evident they were joined by wooden arches, for that is the form shown on Trajan's column. Apollodorus did the work under pressure in two or three seasons only, a remarkable performance even if he had the whole army on that frontier at his command. How or why Apollodorus's bridge over the Danube was destroyed is not known. One account is that Trajan's successor Hadrian had it wrecked to keep the barbarians from crossing into his empire.

Roads

The highways of Rome are justly famous. Its statesmen early saw that good communications were essential to the expansion, the administration, the defense, even the daily life, of their state. They absorbed the road-building skill of their Etruscan neighbors, borrowed the ideas of their Carthaginian foes, and improved upon both. At the height of its power, Rome had great thoroughfares radiating to the farthest provinces and connecting with lateral roads and paths to constitute a well-planned system of something over 50,000 miles. This road network made possible the rapid shifting of legions from one frontier to another; it quickened and cheapened civilian travel and the shipment of goods. Together with control of the Mediterranean, after the sea power of Carthage had been broken, this system of communications made possible an empire that could not have easily been constructed without it—certainly not held together long. Without roads, the city of Rome could hardly have become or remained "The Metropolis."

Much, however, about the roads of Rome has become legend that borders on nonsense. Roman engineers naturally did not put the same effort and expenditure into the construction of a mountain path as into the Appian Way, that "queen of long roads," begun by Appius Claudius in 312 B.C. Carried south and east to Brindisi on the Adriatic Sea, it was the arterial highway to and from Greece, Egypt, the Levant, and the Orient. Just as engineers in our time classify roads with regard to the use expected of them and build accordingly, so did the Romans, with possibly even more variation. The Twelve Tables of 450 B.C. (the earliest Roman legal code) had legally distinguished roads by their width: the *semita*, or footpath, was only 1 foot wide; the *iter* for horsemen and pedestrians was but 3 feet; the *actus* for a single carriage was 4 feet; and the 2-lane *via* was about 8 feet wide. In later years these specific dimensions lost significance. The widest of the imperial highways were the costly Via Appia, or Appian Way, and its fellow to the north, the Via Flaminia. The latter carried traffic ultimately bound across the Alps or around the head of the Adriatic Sea to the Balkan regions and far distant Byzantium (Istanbul). Less important, narrower, and more cheaply built were the graveled roads in the Alps, those of cobbles packed in earth along the German frontier and surfaces hardened with cinders from neighboring foundries.

In all the important *viae*, nevertheless, whether in Italy, Africa, England, or elsewhere, certain engineering practices were standard whatever local materials may have been used. The subgrade was always drained by ditches and culverts to keep the foundations firm, the surface was sharply crowned and raised well above the adjacent terrain to shed water, and the roads were nearly always thick; modern ones are seldom more than 1½ feet thick. A few Roman *viae* were from 3 to 8 feet thick, built in four or five layers, or courses. There might be a 2-foot foundation of stone slabs laid in mortar upon the earth subgrade and leveled with sand.

Figure 4.9 Along the Via Appia in 1912

Above this, a second 9-inch layer of broken stone was mixed with mortar and tamped; then a third, also 9 inches, of fine-gravel concrete, and finally a fourth surface of hard-stone blocks each 2 or 3 feet square, 5 inches thick, and closely laid. Portions of the Via Appia (Figure 4.9) were so constructed; the U.S. Bureau of Public Roads has a model of it in Washington. There were also side paths 1½ to 8 feet wide, separated from the highway by a curb 18 inches high. Ornate mileposts, tombs, and monuments lined portions of the way. The expense of building such a road in the United States today has been estimated at $300,000 a mile.

A cheaper road in England had a 5-inch bottom layer of cobbles, an 18-inch mixture of soft limestone, sand, and earth, another 18 inches of tamped earth, and a top layer of boulders and broken stone. Still another road in London was made of 7½ feet of gravel concrete laid between stone retaining walls and surfaced with white clay tiles. A roadway across a swamp was laid upon 4-foot piles with successive courses of flint and tiles, limestone, gravel and earth, stone slabs, and fine gravel —altogether a thickness of 10 feet, although originally it may have been less. Watling Street, perhaps the best known of the Roman roads in England, reached from Dover by way of Canterbury to a crossing of the Thames near London, thence possibly to Chester, and so on north to Lancaster and beyond. It was of no single type of construction throughout; parts of its original pavement have been recently unearthed in London.

The straightness of Roman roads has been overemphasized. They appear straight on a map when really they are direct rather than obstinately straight. The Roman engineers' *groma* was not adapted to surveying long distances. In fact they probably did not align their roads by any instruments, but rather by experimental sighting along a number of straight rods, or sometimes by using smoke signals which could be seen over an intervening hill. Even so, some of the Roman roads of England do not often wander from a straight line more than ¼ or ½ mile in 20 or 30 miles. Rome's highway builders allowed much steeper grades than are used today, reaching sometimes a gradient of 1 foot in 5 or 20 per cent. Cuts, however, were often very deep. The record appears to be a gash of 117 feet into a marble cliff on the Appian Way near Terracino. These cuts had to be excavated by hand as the Romans had no blasting powder. They probably used the method of cracking the rock by heating it and then dashing it with cold water, as they did in mining. Sometimes they tunneled through the ridges. The Furlo tunnel on the Flaminian Way high up in the Apennine Mountains was 984 feet long. Where they felt the need of raising the grade regardless of length, Roman engineers did not hesitate to do so; an embankment across the Pontine marshes near Rome extended almost 18 miles.

These famous Roman roads should be judged by their adequacy in fulfilling the functions for which they were built. There is no doubt that they were highly effective in their own time, and some things can be said of them that are not so relative and have timeless significance. The Roman *viae* were built to last a long while, and they did. The original surface on stretches of the Appian Way was carrying traffic at the end of the nineteenth century. Judging from Roman accounts, we may venture to say that it was not much rougher at that time than it had been thirty years after it was laid. The poet Horace indulged his humor and left for us telling comment, also, upon a journey to Brindisi: "This distance we sluggishly cut in two, it is one for those girded higher than we; the Appian is less heavy done slowly."

Though built solidly, Roman roads lacked the smoothness of reinforced concrete and the resilience of nineteenth-century macadam. Their engineers used broken stones and perhaps road rollers on occasion; they knew asphalt and how the Babylonians had built with it, but they did not combine asphalt with broken stone. This superior type of surfacing was

not employed until centuries later. The Romans knew that their pozzuolana made a marvelous cement, but they never poured it into forms
to make slabs for their roadways. And the thought of embedding metal
in concrete for added strength apparently did not appeal to them, although the Greeks had strengthened masonry with iron bars. If compari-
sons must be made with modern roads, it is but fair to liken the more important *viae* of Rome to the pavements of granite blocks in some of our
city streets.

Lakes, Rivers, and Harbors

There were a number of engineering projects in Roman times which today would be called reclamation, river regulation, or flood control. The
Carr Dyke, 60 feet wide for some 40 miles, made it possible to drain the
fertile country around the Wash in England. Intricate series of *cloacae*
(drains) beneath Rome, some large enough for small boats like the sewers
of Paris in later times, kept the foundations of the city secure although
resting upon swampy land. Julius Caesar urged that the Pontine Marshes
and Lake Fucino be drained and farmed to augment the city's food
supply. These marshes continued to be a serious health problem until a
short time ago. Many authorities believe that malaria contributed heavily
to the enervation of the Roman people and their ultimate downfall. The
Pontine Marshes should have been drained as Caesar urged—for reasons of
which he had no knowledge. There were sporadic attempts through the
ages at the instigation of emperors, popes, and other authorities, but it
was not until Benito Mussolini took charge that the project was carried
to completion at a cost of some 35 million dollars.

Land was reclaimed in the basin of Lake Fucino at the direction of
the Emperor Claudius a hundred years after Caesar's death. The lake is
landlocked high in the Apennines 50 miles to the east of Rome. Having
no adequate outlet, its water level rose and fell with the rainfall, frequently ruining the farms around the shores. Claudius's engineer therefore planned a tunnel to pierce the mountain range and carry the surplus
water to a river 3½ miles to the west. The tunnel was driven through the
range at a depth of 1,000 feet below the summit and continued at an
average depth of 300 feet until it approached the river. The total fall in
the 3½ miles was 28 feet and the cross section of the tunnel was in area
about 100 square feet. Some 40 shafts were sunk, the deepest 400 feet, and
many were lined with brick or stone masonry or timbers. In addition,
many inclined galleries led from the surface to the shafts and to the

tunnel. The total length of the galleries and shafts was in fact at least double that of the main tunnel. The difficulties of lighting and ventilating must have been considerable. For three-quarters of the distance, the laborers faced solid rock, and for the remainder, soils that even today would be considered troublesome. They broke through the rock by chiseling and by cracking heated surfaces with water—the ancient mining practice common well into modern times. They hoisted the debris through the shafts in copper buckets bound with iron, by means of ropes and capstans turned by manpower. According to the historian Suetonius, the work required the efforts of 30,000 men for eleven years. The cost, as estimated in 1876, ran to nearly 70 million dollars. It was the longest artificial subterranean way in the world until the Mont Cenis Tunnel, twice as long, was opened in 1876.

The name of the able engineer who designed the Fucino project is unknown. His original plans were scamped by the contractor, who bore the lovely name Narcissus and was a favorite of the Emperor Claudius. He appears to have become more avaricious as the work progressed. The tunnel was not kept in line or on grade, the shafts were unevenly spaced, and the cross section was left smaller at many points than had been intended. The result of course was that the tunnel did not function as it should. Succeeding emperors had it repaired for a century or two; they even improved it, but then it fell into disuse. A new tunnel was bored in 1875, remedying the defects in the first and lowering the level of Lake Fucino considerably more than by the original plan.

The Tiber River presented quite a different problem of river control for the hydraulic engineers of Rome. They took steps in the first century B.C. to solve the problem in a manner that would be followed today. They confined the river within walls stepped back at intervals on each side so that during low water it flowed in a narrow bed, at half stage in a wider channel, and at flood in one still wider. The reasoning was that the increased current at low water would keep the channel scoured so that it would not have to be dredged. The plan worked along the upper river, but it increased silting in the lower reaches and destroyed what natural harbor there was at Ostia.

The Romans were not essentially a seafaring people like their enemies the Phoenicians of Carthage or like the Englishmen of Elizabethan times, but they established an overwhelming naval supremacy in the Mediterranean, quite as essential to their empire as were their military roads. They knew the value of sea power and the need of a port for Rome. Claudius, who ranks with Trajan among their builders, gave orders in A.D. 42 for an artificial harbor (Figure 4.10) a mile or so to the north of the Tiber's mouth. There a basin about 3,000 feet square was formed, half dug out of the shore and half by moles, or breakwaters, curving out into the sea. They must have done some dredging in the shallower places, but the evidence does not show that the Romans developed dredges of any importance. The moles, or breakwaters, were raised from loose stone or riprap thrown overboard into 15 to 18 feet of water. Claudius had an imposing lighthouse erected at the entrance to this harbor. One of the largest ships constructed up to that time, which had been used to bring an obelisk and nearly 1,000 tons of lentils from Egypt, was moored at the opening in one of the breakwaters. It was filled with a concrete made of rubble until it sank. Walls of squared ashlar were laid upon this foundation; from them the lighthouse, modeled after the Pharos of Alexandria, rose nearly 200 feet. Its beacon of burning pitch could be seen many miles at sea.

The apartment houses of Ostia indicate the density of the population

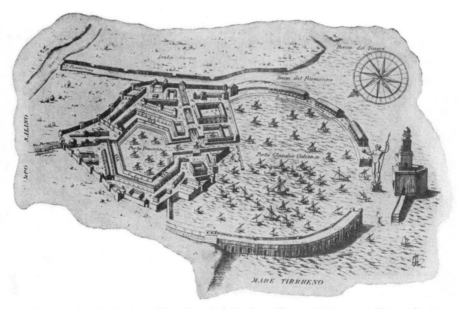

Figure 4.10. Harbors of Claudius and Trajan (From *Minutes of Proceedings, Institution of Civil Engineers*, 1845)

in this port and the busy life among the warehouses which lined the shore. In the next century the Emperor Trajan added more extensive buildings, constructed an inner basin, and connected the harbor with the Tiber by a canal. All of these bear witness to the maritime interests of the Roman people. But notwithstanding their strength on the sea and its importance to their empire, they did not rise to imperial power primarily as a seafaring people. Nor did their engineers make their greatest contributions in marine engineering.

The average Roman merchantship that used these harbors was nearly the size of the *Mayflower* with its 26-foot beam, 90-foot length, and 180 tons burden. However, some of the Roman ships which carried wheat from Egypt to Italy were larger, and there is a record of one that carried 250 tons and was 95 feet long. The ship which carried the obelisk and was later sunk to form part of the breakwater at Ostia must have been at least twice the size of the wheat ship and may have been several times larger, if Pliny's description of it was accurate.[9] Like the earlier Mediterranean merchantmen the Roman ships were primarily sailing vessels, but the Romans made one important innovation on their larger ships. In addition to the mainmast, the Romans added a second mast at the bow. This forward mast similar to a bowsprit slanted out over the water and a square sail was hung from it. This forward sail increased the speed and steadiness of the ship, but it did not permit sailing into the wind. The large Roman merchantman could make adequate headway only when the wind was considerably abaft the beam.

Measurement

The Romans did not use the tools and equipment that their structures would lead us to expect. Vitruvius declared that an engineer should have thorough knowledge of the sciences and arts, but he was exhorting his colleagues rather than describing the state of their education. They did not utilize as they might have done the mathematics which Euclid, Archimedes, and others had developed. Probably the mathematical knowledge of most did not go beyond arithmetic, and with their awkward numerical notation, even multiplying must have been vexatious. They made no use of trigonometry but based their surveying upon a rectangular system of grids. Unmistakably, they did have some empirical knowledge

[9] Secundus C. Plinius, *Natural History*, 6 vols., Henry G. Bohn, London, 1855–1857, vol. 3, pp. 419–420.

of geometrical perspective because many reliefs in bronze or marble which they kept as administrative documents show ground plans, elevations, and oblique projections or bird's-eye views of their buildings.

We have little evidence that they used conic sections to any extent— the parabola, hyperbola, or other curves such as the so-called spiral of Archimedes, whose properties and purposes were already under investigation. The Romans invariably used semicircles for their arches, never parabolic or elliptical arcs, although they did use the ellipse in horizontal construction. The great stone Colosseum, opened by the Emperor Titus in A.D. 80, has an elliptical plan, and there were earlier amphitheaters, elliptical in outline, built of wood. The ground plan of the Roman amphitheater at Pola, probably built about the time of the birth of Christ, is elliptical. A test with modern instruments showed an average deviation of no more than 6 inches from a true ellipse.[10] Such accuracy is startling enough; the wonder is that the Roman engineer could have accomplished it at all without knowing something about the properties of at least one conic section. Perhaps he did know that in an ellipse the sum of the distances from any point on the curve to the two foci is constant. Perhaps he too constructed his ellipse, as many would today, by establishing his focal points and then sliding a marker along a slack cord whose ends were fastened to these foci.

The Romans appear to have had no systematic theory regarding stresses, thrusts, and distributions of weight. There was no science of statics or the equilibrium of forces until after 1500 in the Renaissance. Roman engineers made no quantitative tests for the strength of materials under tension or compression or bending or shearing. They did not realize that the strength of a beam depends upon the shape as well as upon the area of its cross section. They did not know that by changing the shape of its cross section they might decrease the weight of a beam without reducing its strength. It is fair to presume that they would be amazed at our I beams and railroad rails. They built their huge aqueducts and bridges solidly, with caution and common sense, well within the appropriate factor of safety—or margin of ignorance, as you prefer. Some of these structures have stood through twenty centuries of comparative neglect.

At the same time, however, the Romans were deliberately raising to perilous heights apartment houses which frequently collapsed. Above the more sumptuous quarters on the ground floor, these dwellings were for the most part tenements inhabited by the poor. Possibly that explains

[10] "Roman Surveying," *Nature*, vol. 88, p. 158, Nov. 30, 1911.

their flimsy construction, as the satirists like Juvenal believed who railed at the greed of their proprietors. Perhaps they were built with an eye to quick economy, for they might have to be torn down before they fell down; Rome was growing so rapidly. In a "thrifty balancing of cost and common sense," as Vitruvius expressed it, charges for labor did not have then to be taken into account as they must on any modern engineering project. Slave labor had to be housed and fed, but so long as Roman generals won battles there was certain to be an ample supply. The consciences of Roman statesmen did not seem to have been disturbed by slavery of military prisoners.

For surveying, the engineers of Rome had the choice of two Greek instruments, the *groma* and the *dioptra*. The *groma* (Figure 4.11) was the simpler for establishing straight lines and right angles. Like the surveyor's "cross" of early America, it consisted of two arms fixed at right angles to each other with plumb lines suspended from each of the four ends,

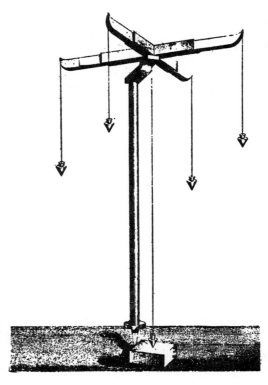

Figure 4.11
Roman groma (From Legnazzi, *Catasto Romano,* 1887)

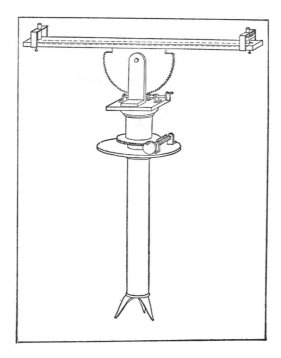

and it rested horizontally upon a staff. Sighting along either pair of these plumb lines from the opposite end of an arm, one could set a point 100 feet or so away with an error of perhaps no more than 6 inches in 100 feet if the wind were not swaying the lines. The descriptions of this instrument were so vague that it could not be visualized until 1912 when a surveyor's office was unearthed in the ruins of Pompeii and metal parts of a *groma* were found intact. The *dioptra* (Figure 4.12), ancestor of the transit and which had a primitive water level attached to it, was more accurate. Hero claimed that it could measure "the distances which separate stars" and determine the "eclipses of the sun and moon," [11] but just how, he did not say. Vitruvius had realized its limitations. Its metal bar, supported by a pedestal, or tripod, was free to swing on either a horizontal or vertical arc. Sights at each end of this bar made possible sighting on a definite point and determining the line to it.

Especially for their bridge foundations and long aqueducts, the Romans preferred another leveling instrument, the *chorobates*. It must have been

[11] Hero, *Heronis Alexandrini opera quae supersunt omnia,* 6 vols., B. G. Teubner, Leipzig, 1899–1914, vol. 3, p. 91.

more accurate in their hands than its simple construction would suggest, for it was nothing more than a 20-foot straightedge mounted upon two equal legs at right angles to it. As an illustration shows, it was leveled by plumb lines which struck vertical lines cut in the cross braces at each end between the straightedge and the legs. A groove filled with water in the top of the straightedge provided a supplementary level when the wind disturbed the plumb lines. For leveling shorter distances, they used the *libella*, whence comes our word "level." It was like a capital A with a plumb bob swinging from its apex to coincide with a vertical mark on the crossbar when it was level. They determined the rise or fall from an established elevation by using sliding targets on graduated leveling rods, essentially as is done today. Linear measurements of short distances were made with graduated poles of seasoned wood tipped with metal. For longer distances there were cords and ropes carefully stretched and coated with moistureproof substances to prevent shrinking. And there was a rolling device of known circumference, which is now called an odometer or perambulator.

The accuracy which the best Roman surveyors obtained with these instruments is surprising since they had no telescopic sights, verniers, spirit levels, delicate adjusting screws, or other equipment considered indispensable to modern instruments. Section lines in the Western part of the United States were required by law in the early nineteenth century to close within 33 feet every 6 miles, or the length of a township; actually the error of closure averaged much less. A Roman frontier boundary in Baden, Germany, has been found to vary no more than 7 feet on either side of a straight line for a distance of nearly 20 miles over rough ground.[12]

The Roman *agrimensor*, or surveyor, was duly admired in his own time. Cassiodorus, eminent historian and churchman of the sixth century, declared when requesting a surveyor to settle a boundary dispute that verged on war, "The Professors of this Science [surveying] are honoured with a more earnest attention than falls to the lot of any other philosophers. Arithmetic, geometry, astronomy, and music were explained" to listless audiences, sometimes to empty benches. It was not so in the surveyor's "forum." His "deserted fields" were soon "crowded with eager spectators," for he walked "not as other men walk." You might think him a "madman" as he pursued "the most devious paths," but he was "seeking for the traces of lost facts in rough woods and thickets." His path was "the book" from which he discoursed. "Like a mighty river," he was tak-

[12] "Roman Surveying," *Nature*, vol. 88, p. 158, Nov. 30, 1911.

ing away "the fields of one side to bestow them on the other." [13] Greek and Roman surveying was the first "applied science" in engineering and practically the only one for nearly twenty centuries.

Machines

Modern technology employs tremendous supplies of power to replace human muscle. Complicated machines are supplied with abundant energy from natural sources. Some of these machines are wasteful because they are inefficient. But Roman engineers would marvel at their power and their speed, for in comparison Roman machines were feeble and slow. These slow and relatively simple devices were operated most often by human hands or feet, seldom by animals and but rarely by water wheels. The great works already described were built with rudimentary forms of mechanical power and great expenditure of human energy. This does not mean that the Romans had no interest in effort-saving devices for the sake of more efficient construction, for they made wide use of counter-weights and levers as had the Greeks before them. It does mean that they cared no more than had the Egyptians, or any other ancient peoples, for the mass of human beings whose physical strength they were exploiting. The great powerhouse of the Roman engineer and all of his predecessors was human effort. More is known about the destructive military catapults, rams, and siege weapons of the Romans than about their building machines, but it is entirely reasonable to suppose that, wherever those weapons could be easily adapted, they were used in construction. Vitruvius described a few mechanical instruments. Among them was a derrick with three pulleys in its tackle. This *trispastos* (Figure 4.13) was operated by manpower through a capstan and a large wheel rotating on a horizontal axis, a kind of treadmill. A bas-relief has been preserved which shows a crane worked by men in a treadmill. It is not clear whether this was used to turn the crane or merely to operate the tackle; possibly both operations happened together. If so, the Romans knew the principle of the modern hoisting engine. What they lacked was mechanical power.

It was with such simple instruments as these pulleys and cranes and a great amount of manpower that the engineers of Rome lifted Egyptian obelisks and moved the 18 cubes of Parian marble, each weighing about 50 tons, from which Trajan's column was built. And in this way, too, they

[13] Magnus A. Cassiodorus, *The Letters of Cassiodorus*, Henry Froude, London, 1886, pp. 232–233.

Figure 4.13 Trispastos crane used in building a tomb (Courtesy Yale Collection of Classical Art)

raised the thousands of tons of cut stone that went into their great aqueduct bridges. The Claudian, rising at its highest point about 75 feet, required for its 7 miles in length over 560,000 tons of quarried stone, more
than 40,000 wagonloads per year through fourteen years of construction.

The Romans had three kinds of machines for raising water, the
tympanum, or drum, the cochlea, or Archimedean screw, and the force
pump. These appear always to have been operated by hand crank or foot
power. Many specimens have been found of the tympanum, particularly
in the Roman mines of Spain. They varied in dimensions, but in general
they were some 14 feet in diameter with foot treads along the rims at the
spokes, and they were equipped with open boxes or buckets to catch up
the water. They were usually made of wood but sometimes had bronze
axles. Rotating in series, one above the other, they could raise water as
far as the series could be established. Each, however, required the constant
effort of one or more men sitting above it and pushing the treads with
their feet. The drums apparently served well enough in the Roman mines.
There is no record of the mines being equipped with animal power until
the fourteenth century.

The cochlea (Figure 4.14), though known to the Greeks, is described
here because it was virtually standard equipment in Roman mines. As the
Romans adapted it, the Archimedean screw was from 10 to 12 feet long
and made of a wooden core with a strip of wood or copper wound edge-
wise and spirally around it and encased in a barrel of planks. It was pivoted
on iron beams at an angle of from 20 to 45 degrees, and the core was
turned by knobs or handholds. The cochleae, too, placed in series one
above the other could be made to raise water as high as local conditions
allowed. One cannot believe that high speed was attained or much power
involved, but widespread use does suggest that there was a measure of
satisfaction. The cochlea probably was more effective than the tym-

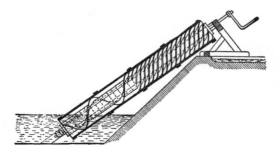

Figure 4.14
Cochlea with crank (From
T. Beck, *Beiträge zur
Geschichte des Maschinen-
baues*, 1900; courtesy of
Julius Springer-Verlag)

panum. Only one man was needed to turn the screw; the weight of the drum might require two or even three, and it may be that the flow of the cochlea was steadier. According to the authority, Oliver Davies, the cochlea made the most efficient application of power to the task of draining mines until man hit upon the idea of hitching animals to their machines in place of themselves in medieval salt mines.

In the words of Vitruvius, "we next proceed to describe the Ctesibian machine which raises water to a height." [14] We have left it to this chapter on Rome because the engineers of the Hellenistic world did nothing with it, and Vitruvius made a great deal of it. That the Roman engineers knew well the principles of the reciprocating force pump and how to construct it, we can be certain from the detailed account which Vitruvius gave of its brass levers and cylinders, oiled pistons and valves, its pipes and air pressure, its reciprocating motion. He writes, "Neither rectilinear motion without circular nor rotating movements without rectilinear can produce the raising of loads." His own sketches have not been preserved with his manuscript, but his description is so clear that there is no mistaking the construction of water-raising devices. They could easily have been connected, as he implied, with a water wheel or treadmill. How widespread or effective its use was is uncertain. Vitruvius wrote as if it were employed to raise water from low reservoirs to higher public fountains; he did not say how extensively. He said nothing further about the power which it required, and there is no other evidence from Roman times. Frontinus, who about A.D. 100 boasted of the great aqueducts and told at length of the water system of Rome, made no mention of the construction or use of such a force pump.

The Romans occasionally used water power to supplement and even to replace animals and men. They had improved the hand mill, or quern, of primitive men for grinding flour. The upper stone had become hollow and shaped like an hourglass. Its lower bell fitted over the conical base and the upper formed a hopper for the grain. Rotating levers had been inserted on either side. In the ruins of Pompeii, destroyed by Vesuvius in A.D. 79, a bas-relief shows an ass hitched to a mill of this sort. According to Pliny the Elder, who completed his *Natural History* two years before that disaster and lost his life in it, most Italians ground their flour in such a mill and by hand. But water mills had been highly developed. Vitruvius had already given a clear description of the toothed drums which meshed at right angles and transmitted the power of the undershot water wheels

[14] Vitruvius, *op. cit.*, vol. 2, p. 311.

Figure 4.15
Flour mill and undershot
water wheel (From
Perrault, *Vitruve*, 1682)

revolving on horizontal axles to the millstones which of course rotated
on vertical axles.

Other references to water wheels in Roman literature are scanty, but
they are sufficient to prove the importance of these water-powered flour
mills (Figure 4.15) to the city of Rome. There were many on the side of
the Janiculum to catch the overflow from the reservoir of the Aqua
Traiana as it rushed down the steep slope of the hill. These doubtless were
undershot wheels for they could be placed easily in the stream, but there
is evidence that the Romans knew and used an overshot wheel, set under
a millrace and designed to turn with the weight as well as the impulse of
falling water. The masonry of such a Roman water wheel may still be
seen in the Athenian Agora. It was some 10 feet in diameter and its millrace or flume was 4½ feet above it and was probably constructed between
A.D. 457 and 474.

The most famous water wheels in all Roman history, however, were
the undershot wheels which Belisarius, great commander for the Emperor Justinian, put into the Tiber River during the defense of Rome in
A.D. 537, early in the Middle Ages. The Goths had cut the aqueducts on
the far side of the Tiber and thus stopped the city's flour mills. Belisarius
had rows of hawsers stretched across the river and barges moored to
them in pairs. A water wheel was hung between each pair and connected

with millstones on the barges. Thus, it is said, Belisarius harnessed the current of the river and kept the people in the city supplied with flour throughout the siege.

Mining and Metallurgy

The Romans were as greedy in their search for metals and virtually as primitive in their methods as any other people of ancient times, but in 50 B.C. their senate restricted mining in Italy. Historians have been at a loss ever since to explain exactly why. Pliny the Elder declared that the country abounded in metals. The senate had taken this action, he said, in order to spare Italy's resources.[15] Some later historians, disagreeing with Pliny, have suggested complicated military and political reasons for the senate's action.

The Romans often left conquered fields in the possession of the inhabitants, subject thereafter to tribute, but mines and minerals almost never. Even when taken into the public domain and leased to newcomers, the fields still produced for the benefit of the locality as well as the Roman state. Not so with resources beneath the surface. The Republic appropriated to itself practically all minerals, those of former rulers by right of victory, those of private owners by confiscation or forced sale. And the Republic's successors, the emperors, who in their own minds at least were identical with the state, did the same—with more stress upon their personal gain and less upon public advantage.

Regardless of their rationalizing about title or profit, Roman authorities, both republican and imperial, held that the subsurface resources were a monopoly to be operated by and for the state. And this was true even though some mines, usually of the less precious metals, were allowed to remain in private hands, or later, as the Republic declined, were put at the disposal of favored patrician families who accumulated large fortunes on the side while they worked ostensibly for the state. Such arrant despoilers could not be expected to give much thought beyond simplest maintenance to improvements in the property or to safeguards for the slave labor which they exploited. Their manifest purpose was to get the metals out of the ground as quickly and as cheaply as possible. Human life as such meant little to the Romans, least of all in their mines where, like the Greeks, they put the vicious and the stupid. To be sure, they

---

[15] Secundus C. Plinius, *Natural History*, 7 vols., Harvard University Press, Cambridge, 1938–1952, vol. 2, p. 103.

used water screws and drums to clear their galleries and drifts, but these had continually to be in operation or the work as well as the workers would have been drowned out. Some of these techniques were no inno- vations by the Romans; the conquered peoples had been using them long before. The mining engineers of Rome were as ready to take over the ideas of others as its legions were to appropriate their resources.

Whether in the copper and silver mines of Carthaginian Spain, the iron mines of Tuscan Elba, the silver mines of Athenian Laurion, the tin and lead mines of Britain, or anywhere else, the Romans drove their tunnels and shafts down the outcroppings into the earth as had those whom they dispossessed. Some of their mines in Spain reached depths of over 600 feet. Rome's slaves hacked at the minerals underground with primitive wedges and hand picks in the murk of olive-oil lamps that made the air foul and deadly. Men, women, and children strained to carry out the ore and the waste upon their backs. Where currents of air provided the oxygen necessary, the rock was cracked with fire and water, as had been done from prehistoric times. The Greeks seem to have had some idea about ventilating by means of cross drifts, but there is no evidence that the Romans did anything of the sort. In Britain, they used extensively hushing, or sending a torrent of water down a slope to uncover the lode of ore, much as in American placer mining, but this practice, too, seems to have been in use before the Romans arrived. These methods were all primitive. No great advance in the technique of the mining engineers, Roman or any other, was made until far into the Middle Ages. It was not long before that time that animal power supplemented human strength in mining. Blasting powder did not appear in Western European mining until 1627.

Crude ores were reduced in Roman times by the ancient methods of washing away the earth and firing the residual oxides, chlorides, or sulfides of the metals. Silver and lead, often found together, could be easily sepa- rated from other materials because silver is very soluble in lead; the lead could then be removed because it has a much lower melting point than silver. Iron ore was smelted much as it had been by chance in primitive campfires, and wood provided the charcoal for producing Roman cast iron and steel. The use of coal apparently began with the Chinese, al- though there was mention of it in Greek literature about 300 B.C. The Romans may have used coal in Britain, but they did not in Gaul. Its dis- agreeable odor and its soot may have kept Europeans from employing it generally until well after Marco Polo had returned in A.D. 1295 to describe its use by the Chinese and praise it for superiority over wood in retaining fire.

There is no doubt that Rome had metal workers of very considerable skill, and their products are everywhere in museums to demonstrate it. But there was no contemporary literature explaining their methods and their formulas. Such matters of a technical nature passed orally from generation to generation of craftsmen. Practically all that is known has had to be deduced from examining the finished articles themselves. The Romans did magnificent work in metal, but how they did it is not known. For there appears to have been little association between those who wrote and those who worked with metals until the monk Theophilus, about A.D. 1050, prepared a treatise on the subject for the guidance of the individual craftsman. Pliny the Elder had tried to give some information in his *Natural History*, completed in A.D. 77, but his observations lacked both technical knowledge and practical experience. His theory that copper in iron was harmful persisted into recent times, although copper has now been successfully used in steel where resistance to corrosion is desirable. Vitruvius also wrote about metals, especially the use of lead for pipes, but his contribution to our knowledge of Roman metallurgy is trifling compared to his chapters on architecture and construction.

Byzantine Engineering

It is clear that engineering entered new phases of interest and practical significance as leadership in the Western world shifted from Rome to Constantinople in the sixth century A.D. The walls of the Byzantine capital (Istanbul) rising tier upon tier to the height of more than 40 feet are one of the sights of Europe. They held off barbarian invaders for a thousand years. They were some 14 feet thick in many places, made with a core of concrete. Masonry so strong in itself needed no facing, and at points it was covered merely with brick. In particularly vulnerable places, however, it had squared stone blocks bound with iron cramps. To supplement these fortifications, heavy iron chains were stretched across the Golden Horn at various times. Portions of the chain used in A.D. 1453, when at last the city yielded to the Turks, may still be seen in the church of Santa Irene in Istanbul (Constantinople).

For its water supply the hydraulic engineers of Constantinople copied the works of Rome, their most impressive structure being the aqueduct bridge named for Justinian, although it probably was constructed two centuries earlier in the time of Constantine and has since been rebuilt. They displayed their own ingenuity in their subterranean reservoirs within the city walls. The historian Procopius explained how the Emperor Justinian had a courtyard excavated to a great depth to bedrock in order to provide storage for the dry season and probably in times of siege. An ornate cistern known as that of the Thousand and One Columns lies 50 feet below the level of the street and has a capacity of more than 9 million gallons. What remains of another called the Underground Palace has 336 columns in 28 rows.

An account by Procopius, written about A.D. 560, is one of the earliest we have describing a kind of engineering which had been practiced for many centuries. Justinian had a sheltered harbor built where there had been none before. His engineers put overboard huge chests, or cribs, and sent them to the bottom under heavy loads of stone along converging lines that extended from the shore. Upon these foundations, or caissons, were erected breakwaters of rough-cut stone. The method was essentially the same that Claudius's engineers had employed at Ostia five hundred years earlier.

A dam near the Byzantine city of Daras on the Persian frontier was far more significant as a departure from established engineering practice. The type is familiar enough today in the canyons of the Western United States. Chryses of Alexandria, one of Justinian's engineers, mortised the ends of the dam at Daras into the cliffs and bent it upstream in a crescent, or arch, so that the pressure of the water should be carried horizontally along the arch to the ledges. Floodgates at upper and lower levels made it possible to release excess water. How these gates were opened and closed Procopius did not explain. He was more interested in reporting that Chryses, hearing the news of a disastrous flood at Daras, "went to his bed in distress" and "saw a vision." The result was a plan which coincided with the ideas that in the meantime the Emperor, "obviously moved by a divine inspiration," had "sketched out of his own head." And so, while the consulting engineers Anthemios and Isidoros considered "how God becomes a partner with this Emperor in all matters which will benefit the State," the plan of Justinian and Chryses "won the day." [16]

Anthemios of Tralles in Lydia and Isidoros of Miletus are, however, not to be remembered mainly for their share in this episode. They are

[16] Procopius, *Procopius,* 7 vols., The Macmillan Company, New York, 1914–1940, vol. 7, pp. 117, 119.

Figure 4.16 Santa Sophia, structurally analyzed, Istanbul (From Choisy, *L'Art de bâtir chez les Byzantins*, 1883)

to be honored for the reconstruction of the famous church of Santa Sophia (Figure 4.16), properly Hagia Sophia, or Divine Wisdom, which they began in 532 at the order of Justinian. Like Archimedes, Anthemios was a mathematician and an experimenter in optics and mechanics. He wrote about burning lenses, and it is said that he studied the parabola. Tradition has it also that he discovered how to use gunpowder. Only fragments of his writings have come to us. Of Isidoros little more is known than that he, too, was a mathematician of note and that he was the colleague and successor to Anthemios in rebuilding Santa Sophia.

The Byzantine engineers developed an extension of the principle of the arch, that of concentrating the stresses from a more or less uniform load so they could be carried at separated points of support, leaving clear spaces between. The Byzantine construction, which went far beyond anything the Egyptians, Greeks, or Romans had used, consisted of a dome, or hollow hemisphere, supported on the corners of a square tower, the diagonals of which were equal to the diameter of the base of the dome. The portions of the dome which otherwise would overhang were cut off in planes of the sides of the tower, forming semicircular arches and leaving over each corner of the tower an inverted spherical triangle known architecturally as a pendentive, which carried the weight of the dome and of anything resting on it to the tower and so to the foundations. In some instances a vertical cylinder resting on the main dome was capped by a smaller one.

With these principles Anthemios, Isidoros, and their associates designed and built a structure long considered the world's greatest building. Founded upon bedrock, with its complicated roof held 180 feet above the pavement upon four arches of 100 feet span and columns 25 feet square, Santa Sophia has stood fourteen centuries. We cannot give precisely the composition of the mortar used in this great building and doubtless in other structures of the period. Georgios Codinos, a Greek writer on Santa Sophia in perhaps the fifteenth century, stated that the mortar was made of "broken tiles, boiled barley, lime, and chopped elm bark" and that lukewarm water was used to retard its setting. He insisted that it became as strong as iron. Perhaps he should be believed for he appears to have had access to ancient manuscripts. No one apparently has subjected the mortar in Santa Sophia to modern testing for tensile strength.

Bibliography

Davies, Oliver: *Roman Mines in Europe*, Clarendon Press, Oxford, 1935.

Frontinus, Sextus J.: *The Two Books on the Water Supply of the City of Rome*, 2d ed., Longmans, Green, and Co., New York, 1913.

Procopius: *Procopius*, 7 vols., vol. 7, *Buildings*, The Macmillan Company, New York, 1914–1940.

Sarton, George: *Introduction*, vol. 1 (see Chap. 3).

Smith, William, William Wayte, and George E. Marindin (eds.): *A Dictionary of Greek and Roman Antiquities*, 3d ed., 2 vols., J. Murray, London, 1890–1891.

Stone, Edward N.: *Roman Surveying Instruments*, University of Washington Press, Seattle, 1928.

Van Deman, Esther B.: *The Building of the Roman Aqueducts*, Carnegie Institution of Washington, Washington, D.C., 1934.

Vitruvius Pollio, Marcus: *On Architecture* (see Chap. 3).

FIVE

The Revolution in Power

Lynn White in concluding his now classic article "Technology and Invention in the Middle Ages" wrote, "The chief glory of the later Middle Ages was not its cathedrals or its epics or its scholasticism: it was the building for the first time in history of a complex civilization which rested not on the backs of sweating slaves or coolies but primarily on non-human power." Certainly the medieval power revolution is one of the most dramatic and important developments in history. Yet the centuries which followed the decline of imperial Rome have so often been called the Dark Ages in the West that many assume there was no one except the clergy concerned with improving the condition of daily life of his fellow men in those times. But there were engineers who applied their knowledge successfully to satisfy the needs of their contemporaries. They made inventions, too, of lasting importance. It is a mistake to assume, as some do, that Western civilization reverted to primitive conditions antedating even the Sumerian or Egyptian because nomads swept Western Europe as the Roman barriers weakened. Actually, the barbarians brought much new knowledge with them, and they contributed many new devices to Western civilization. The barbarians not only brought such engineering innovations as the compact house and tall wooden spire, but also introduced such things as trousers, butter, and soap, which are commonplace in the West today.

The civilization to which the barbarians contributed their knowledge and skills had inherited largely Roman traditions of the Graeco-Roman world. Certainly so far as science and technology were concerned, the Christian West was in the Roman tradition which emphasized engineering rather than pure science. It was Islam that inherited and maintained the Greek science and technology during the Middle Ages. There was almost no scientific activity in the West until the twelfth and thirteenth centuries when Western scholars began to translate Arabic scientific works into Latin; the Moslems had translated most of the books of Greek science into Arabic five hundred years earlier. So it was that the two different traditions that came into being with the rise of the Romans in classical antiquity were continued apart until the thirteenth century. When the Latin West started to take over the Greek scientific knowledge which the Moslems had transmitted and augmented, the two traditions began to fuse.

Of great importance for the stimulation of the development of non-human resources of power was the decline of the institution of slavery and the continuing rise of Christianity. As Lynn White has pointed out, the history of medieval technology is to some extent a history of religion. The Christian ideal of the infinite worth of man and the correlative aversion to submitting men to work which required no intelligence or judgment were among the principal factors which incited the evolution of mechanical power as a replacement for human muscles. It is no exaggeration to say that the Christian ethic conditioned in part the rise of power engineering. On the other hand, power engineering has made possible a greater realization of Christian ideals and has contributed much to the growth of human dignity and freedom in the West.

Medieval engineers did not, of course, limit themselves to power engineering. Indeed, they advanced many forms of Roman engineering, with such exceptions as concrete construction and road building. Significant advances were made in the engineering of buildings, bridges, canals, and various municipal improvements. Nevertheless, it was the new power engineering, which for the first time in human history started to free men from being the principal sources of power, that was most responsible for the change in the ways of human life which began during the Middle Ages.

Power

Water, wind, and animal power were the major sources of nonhuman power which medieval engineers developed. The new prime movers were used to operate many types of machines invented during the Middle Ages. Modern machinery derives in large part from the primitive quern for grinding, made of two millstones, one rotating upon the other and operated by hand. Vitruvius explained how gears, or cogwheels of pins and

Figure 5.1 Earliest-known illustration of the use of a crank to provide rotary motion, A.D. 850 (From E. T. Dewald, *The Illustrations of the Utrecht Psalter,* 1932; courtesy Princeton University Press)

grooves, had been devised so that the motion could be transmitted at any angle from a revolving axle. Some ingenious person invented the crank to change rotary into reciprocating motion, or the reverse, and the crank was certainly in use by about 850, for the Utrecht Psalter of that period depicts it (Figure 5.1). The water wheel was probably invented in the first century B.C., and similar devices called Norse mills are known to have been used early in northern Europe. There is evidence of such an early water mill in India. Such horizontal wheels having vertical axles could not have been effective unless put at the edge of the current or in an eddy. But when the cogwheel of pins and grooves made possible setting the wheel in a vertical plane in the stream, at right angles with the axle of the millstone, the vertical water wheel with horizontal shaft began to come into use.

The Romans introduced vertical water wheels into various sections of Europe during the last centuries of the Empire. The poet Ausonius wrote of them in the fourth century noisily grinding corn and sawing stone on the Roer, tributary of the Mosel River in Germany. They were frequently attached to barges anchored in the stream, as in the classic instance when Belisarius defended Rome in 537 against the Goths. By the eleventh century they were busy at many sorts of mechanical occupations, grinding flour, moving the bellows of forges, operating saws, fulling

Figure 5.2
Three-stage pump in mine shaft, powered by undershot wheel (From G. Agricola, *De re metallica*, 1556)

cloth, making paper, pumping water (Figure 5.2), working in salt mines, breweries, and factories.

England's Domesday Book of 1086 reported some several thousand water wheels of various types and uses in the land, serving a population estimated at two million. These wheels were often set under bridges to take advantage of the swifter current there, but their vibration racked the bridges. The Pont Notre Dame at Paris, built of wood in 1413, had to be entirely rebuilt in 1440 for this reason. The wheel was adapted also to use the tides, as at London Bridge.

It is not known when it was that someone thought to take advantage of the head, or height, of a waterfall rather than the speed of running water in a stream. The flow of power would be as steady as the head of water which could be impounded and kept pouring from the millpond

Figure 5.3
Reversible overshot wheel
for hoisting material from
a mine (From G. Agricola,
De re metallica, 1556)

upon the wheel. Chapter 4 gives the evidence of a Roman overshot wheel
in the Athenian Agora which used the power of a waterfall, but there
is nothing to show that this type became common until the fourteenth
century. Although Biringuccio describes in detail all other devices used
in metallurgy, he merely says "water wheels" and takes it for granted
that his reader is thoroughly familiar with them. In 1556, Agricola pictured
an overshot wheel (Figure 5.3). Crude drawings of it had appeared in the
fifteenth century to indicate that men were appreciating its superior
features. Watertight dams and millraces were difficult to construct and
expensive to maintain. There were strong objections besides, particularly
in England, because people did not like having public waterways obstructed by private weirs and dams. Freedom from such interference with
the common right to go and come on the rivers of England had been

Figure 5.4
Seventeenth-century wind-
mill and chain pump
(From Strada, *Kunstliche
Abriss*, 1617–1618)

stipulated in Magna Charta. Undershot wheels at the side of the stream
blocked no traffic, and often they were common property available to
anyone in the community. Nevertheless, the overshot wheel, because of
its greater power, eventually replaced the undershot.

The second great source of power to be utilized was wind. The
first windmills in Europe appeared at the end of the twelfth century.
Early ones had horizontal axles with vertical sails and were necessarily
small because the entire mill had to be rotated to bring the sails into the
wind. By the fifteenth century, engineers had evolved mills with tilted
axles and sails set to catch the wind more effectively. At the end of the
fifteenth century the so-called bonnet windmill (Figure 5.4) was invented, which made possible the construction of much larger mills because it was necessary to move only the bonnet, or turret, at the top to
place the sails in the wind. Like water wheels, the windmills operated
various types of machines, and they were particularly popular on the
plains of northwestern Europe where there are not many natural waterfalls or rapidly flowing rivers. The windmill obviated the need for constructing expensive dams for overshot water wheels.

Another important development in the use of wind power was the invention of rigs for sailing ships which enabled them to tack against the wind, thereby making the ships independent of the direction of the wind and the efforts of oarsmen. As early as the time of Christ the North Europeans had built ships with square sails which enabled them to sail into the wind to some extent. They were able to tack their ships because they stepped the masts farther aft than did the Mediterranean shipbuilders, and the crew could swing the sail into a nearly fore-and-aft position. This rig was improved during the Middle Ages, and when William the Conqueror crossed the English Channel in 1066, he sailed in square-rigged ships which did not require oars.

While this advance was being made in Northern Europe, the fore-and-aft rig in the form of the lateen sail was being introduced in the Mediterranean. The lateen sail had originated in Southwest Asia, whence the Moslems brought it after their conquests of the seventh century. Greek ships were using the lateen in the ninth century, and so did the ships of the eleventh-century Italian cities. Once the Mediterranean merchantman was equipped with the lateen it no longer required oars. After the twelfth century the fore-and-aft lateen rig began to be used on all the European coasts. In the fifteenth century the lateen sail was combined with the Northern square rig, and it was with this combination rig that the great adventurers explored the world in the last of the fifteenth and early sixteenth centuries. These new rigs increased the distance which a ship could travel by materially reducing the size of the crew, more than doubled the speed of the ship, and relieved men from the heavy drudgery of rowing.

The third new source of power was the horse. In antiquity the horse had been at best a most ineffective power producer because the yoke then employed strangled the horse if he exerted much tractive effort. Moreover, without nailed shoes he often broke his hoofs and became useless. By the tenth century, however, men had developed the horse collar, which rests on the horse's shoulders and does not choke him as he pulls. In addition, the horseshoe was invented, and the tandem harness which allows more than one pair of horses to pull a load. The horse is fast and efficient compared to the ox and proved to be a valuable source of nonhuman power in agriculture and the operation of machines.

Water wheels, windmills, sails, and horses were the original prime movers to replace man as almost the only source of power. After the invention of the bonnet windmill at the end of the fifteenth century.

there were no further important developments of power sources until the invention of the steam engine.

Buildings and Builders

From the Middle Ages until the advent of bridges for heavy loads in the early nineteenth century, the principal innovations in construction were generated by the building of monumental churches. Interesting developments in civic and domestic architecture also occurred during the Middle Ages, but the remarkable evolution of structural engineering of churches is most important. Two architectural styles matured in this period—Romanesque and Gothic; each of them presented engineering problems the solutions to which produced advances.

The special requirements of the Christian worship service dictated the construction of churches with provision for extra capacity near the altars and for apses and sacristies. Dedicated in 326, Old St. Peter's at Rome was the first important example of Christian architecture. Essentially a basilican type of church of the old Graeco-Roman style, the columned nave of St. Peter's was twice as wide as that of some Gothic cathedrals of a thousand years later but was of about the same length and height. Typically, it had a wooden-truss roof. In plan, it was T-shaped, having an open-hall transept which extended from either side of the altar, and behind the latter a semicircular apse. The invention of the domed cross-structural unit at the transept led to the style of Santa Sophia discussed in the previous chapter. In the West, however, it was the addition of the wooden-staged spire to the Christian basilica that produced the first typically medieval church. The tower was not a Roman or Eastern contribution; it came rather from Northern Europe.

The great fault of the basilican church with its wooden roof and tower was its vulnerability to fire. Apparently as a result of repeated disastrous fires that were either accident or wanton destruction, engineers began in the eighth century to build and rebuild churches with fireproof stone vaults. Since this "first Romanesque architecture" appeared in regions where there were surviving Roman vaulted buildings, it seems likely that Roman construction influenced the development of the new style. In the earliest Romanesque churches, only relatively narrow bays were vaulted, for the slender columns of the nave could not support a stone vault. However, with the subsequent substitution of rugged structural piers for columns, the entire nave was vaulted. At first the nave and aisle vaults were barrel vaults, but engineers soon introduced the groined vault, which is the intersection of two barrel vaults at right angles. In the eleventh century, the cruciform plan began to replace the T form that had been used since Old St. Peter's of the fourth century. Cupolas and domes were introduced shortly after the adoption of the cruciform plan, and the wide vaults in some of the cupola churches were among the triumphs of medieval engineering.

The developments described in the previous paragraph are for the most part Roman structural techniques. It was not until medieval engineers invented the ribbed vault in the eleventh century that they made their first significant advance over Roman methods. The vaulting ribs were built independently of the vault and furnished a support for the stone-vault web. A groined vault of wide span has a tendency to sag, and it was to prevent this sagging that medieval engineers introduced the ribs that were to prove of such importance in later Gothic structures. One of the greatest Romanesque structures was the third Cluny Abbey dedicated in 1130. It was an immense building almost 300 feet long, its ribbed vaults nearly 100 feet above the pavement. Although the rest of its structure was Romanesque, its aisle arcades had typical Gothic unit-element bays with the pointed arch rib, thin stone web, and a screen wall. However, the great vault of the nave was set on walls that were not designed to take the lateral thrust. Because of their lack of understanding of principles of statics, the engineers had to build strong lateral buttresses after part of the nave vault collapsed in 1125. Apparently they thought that the vault would bear down vertically on the inner side of the wall.

The Gothic builders in the Paris area, where some of the most famous Gothic cathedrals are located, apparently profited from the structural errors made at Cluny, although they seem not to have realized immediately that buttressing would still be necessary. The first systematic use of flying buttresses was in Notre Dame at Paris about 1180, although the first consistent Gothic structure was begun at St. Denis in 1137. The twelfth-century engineers of the Paris area realized that the unit-element bay of Cluny was an important structural advance. By the end of the century they had gained important new knowledge of statics from experience and were able to balance horizontal forces with considerable economy of materials.

A Gothic cathedral of the Paris area is essentially a building whose supporting members are piers sometimes rising to 100 feet or more. Between these piers are screen walls (most of whose area is window) supporting only their own weight and not structurally built into the piers. From the piers spring pointed arches, a structural device introduced from the East at the beginning of the twelfth century. The pointed arch is more effective statically than the semicircular arch because it conforms more closely to the line of thrust. There is therefore less internal stress in the pointed arch, and it can be built using smaller stone. For the same reason, the web between the vault-supporting ribs could be lighter if accurately cut and fitted stone were used; some vault webs were only 6 inches thick. The vaults of the Beauvais Cathedral rise to 158 feet above the pavement. The roof of a Gothic cathedral is a wooden-truss rain-and-snow shed set on top of a high structural wall which the piers support. The roof was often completed before the vaults were started to protect the interior from inclement weather.

Despite the lightening of the weight of the vaults, the piers were unable to counteract the lateral thrust from the vaults, although they were perfectly adequate for transmitting the vertical thrust to the foundations. To bring the lateral thrust down to the ground medieval engineers invented the flying buttress (Figure 5.5), which absorbed the thrust at the vault springing. Flying buttresses also helped to counter both wind load and suction on the building.

Figure 5.5
Flying buttresses, choir, St. Ouen at Rouen (From J. A. Brutails, *Précis d'archéologie du moyen-âge*, 1923; courtesy Libraire A. et J. Picard et Cie.)

Although the phrase "medieval engineer" is used in this book, the men who directed the construction of the Gothic cathedrals were neither "engineers" nor "architects" in the modern sense. Rather they were the master masons who supplied the engineering and architectural direction for the construction. Master builders were paid up to three or four times as much as artisans and were of relatively high social rank. Many of them were illiterate, and all of them had learned their profession by the apprenticeship method. Some few could undoubtedly read such manuscripts as those of Vitruvius, but a man would have to know much more than is in Vitruvius to be able to direct the construction of a Gothic cathedral.

There is evidence that medieval engineers used much paper, parchment, and wood for drawings, but few of these architectural plans and specifications are in existence today. There does, however, still exist a ground plan for the monastery at St. Gallen, Switzerland, which Charlemagne's director of public works and biographer, Eginhard, is said to have designed soon after 800, and there are later medieval drawings in the British Museum and various other European collections. The most famous notebook is in the Bibliothèque Nationale at Paris, the work of Villard de Honnecourt, an architect of the thirteenth century. Thirty-three of his sketches with explanatory notes survive on parchment. It is quite possible that Villard intended that his notebook be used as a text for the instruction of young builders and that it was not primarily a personal document. One of the sketches shows a device for sawing off piles under water (Figure 5.6).

It is most doubtful that medieval architects used models of entire buildings or even of extensive sections. They did, however, make constant use of full-scale templates, or false molds, and full-scale detailed drawings of small portions. There is abundant evidence that master masons made wooden and cloth templates which they sent to quarries to facilitate the shaping of stones, and at least two lead templates have been found. There are several full-scale drawings on the walls of the Rheims Cathedral to guide the artisans in the construction of various sections of the building. The medieval architect made so many drawings and templates that he often had a special building, a "trasour," or "tracyng house," as it was called in England; it was really his drafting office.

The principal materials in medieval structures were stone, mortar, iron, timber, and nails. Medieval engineers had an extensive knowledge of the strength and properties of the materials with which they worked. In some of the great cathedrals the nonstructural wall areas are of soft

Figure 5.6
Machine for sawing off piles under water (From H. R. Hahnloser's *Villard de Honnecourt*, 1935; courtesy Anton Schroll & Co.)

stone, but the supporting piers are of hard stone. The builders sometimes used a weather-resistant stone for outer facings and usually used several methods, such as limewashing, to prevent the decay of stone. Their mortar, however, was relatively ineffective; sometimes it required several years to set, and often it lacked proper adhesive qualities. In most Gothic cathedrals there are many wrought-iron cramps, dowels, tie rods, and an occasional eyebar chain. Wrought-iron parts were boiled in tallow to prevent rusting. For roofs, the builders used wood, and where wrought-iron nails were employed, they were boiled in tallow or tinned for preservation. Medieval building tools, and, indeed, even early nineteenth-century ones, differed little from Roman tools. However, medieval engineers used treadwheel cranes and lathes for turning wood and stone. The laborsaving wheelbarrow, a medieval invention, was widely used by the mid-thirteenth century.

Roads and Bridges

Travel and commerce declined but did not stop during the Middle Ages. Roman roads, though neglected, were still in use. Bridges were scarce, but ferries were common and fords were often paved. And there is evidence that new roads met changing conditions. They were likely not to be hard-surfaced; that was too much trouble so long as they were passable in weather suitable for travel. In certain parts of Central and Western Europe, bridges were actually torn down to give communities the natural protection of their river barriers. But people continued to go and come over the highways of Europe, and the art of bridgebuilding did not perish.

The Emperor Charlemagne, like Julius Caesar before him, built a great wooden bridge across the Rhine at Mainz about the year 800. It was ten years building, only three hours burning. He planned to replace it in stone but died the next year, and his plan was abandoned. Caesar left a detailed account, but of Charlemagne's work there is only surmise. It may have been a heavy pile bridge. Foundations under water were still observable in 1881 on the site, though there is some thought that these were the remains of an earlier Roman bridge. The accidental nature of the fire is also disputed. One explanation is that robbers destroyed the bridge while seizing merchandise there. Another is that ferrymen burned it to help their own business. A third is that the Bishop of Mainz ordered the destruction to stop the robberies and murders on the bridge, causing so many luckless souls to disappear in the river. It is hard to believe that Charlemagne would have sanctioned such a remedy after having put ten years in building the bridge or that the Bishop would have dared to proceed without the Emperor's consent. In any case it was not replaced by a stone bridge until 1862.

The Moors in Spain built stone bridges worthy of note, and there were others during the ninth and tenth centuries in Italy and in Germany. These followed generally the semicircular arches, massive piers, and rigid overhead lines of the typically Roman structure. By the eleventh century, however, bridge construction in company with church building had become more active, flexible, and varied. There now were bridges with pointed arches like that built in 1035 over the Tarn at Albi in France and the Moorish Martorell Bridge in Spain (Figure 4.5). Semicircular arches still were employed for most bridges, as in the bridge over the Danube at Regensburg, built from 1135 to 1145. But flatter segments of circles of longer radius were coming more and more into use. Clearances of the spans no longer remained uniform across the stream or rose symmetrically to the center of the bridge. Variations that seem whimsical appeared. Especially was this true of the slope of the roadway. Roman engineers had used the "muleback" type of bridge, but medieval bridgebuilders put the crest anywhere along the length of a structure, wherever the location of the channel made it desirable. Their bridges differed from one another more than from those of their Roman predecessors.

Movable spans of wood, either to let boats pass through or to stop traffic as a measure of defense, were adapted from the drawbridges over the moats of medieval castles. Some, usually counterpoised, were hinged and drawn up by chains. Others were mere platforms rolling back and forth horizontally like simple gangplanks. Still others were balanced so that they could be tipped up and down. These were the *ponts à bascule* (balanced bridges) commonly used in small wooden structures of the French countryside.

The Crusades may not have driven the Moslem from the Holy Land nor have been responsible for importing his scientific knowledge into Europe, but they did greatly stimulate travel in Europe, and a large number of stone-arch bridges appeared in direct consequence. As local authority was so often inadequate or indifferent to the need of communications with other districts, self-appointed persons and religious orders took up the task. The Frères Pontifes, or Brothers of the Bridge, made it their business to assist travelers, provide lodgings and ferries at river crossings, and build bridges. They were both a traveler's-aid society and an organization of professional engineers. Of the individual enthusiasts, Bénezet, who is said to have started as a shepherd boy and ended a saint, is the most famous. To him is attributed the construction of the bridge across the Rhone at Avignon, built from 1178 to 1186 and famed in song. Whether he was a promoter and popularizer more than an engineer is not easy at this date to determine from the contemporary evidence. He left no drawings or plans. Four of the original twenty-one arches (Figure 5.7) do remain to testify to his skill. They are approximately elliptical but more sharply curved at the crown than at the haunches or springing line. The form is so unusual that it is a question whether Bénezet—or whoever was the builder—had in mind anything other than a modified Gothic arch, one with a rounded crown instead of a peak.

There are legends galore about Bénezet and the bridge. Leaving the pasture he went into Avignon and told the inhabitants, "My Lord Jesus

Figure 5.7 Ruins of the Avignon Bridge (From P. Sébillot, *Les Travaux publics,* 1894)

Christ has sent me to this town in order that I may build a bridge across the Rhone." He received supernatural aid. He lifted a stone which thirty men could not have moved—thus convincing the Bishop and the Chief Magistrate of the town, besides obtaining some 5,000 sous from the astonished populace. In all, Bénezet seems to have accomplished 18 miracles. He thwarted the Devil who sought to kill him. The Devil, so it was believed, did succeed, however, in wrecking one of the piers. As the story goes on, Bénezet obtained indulgences for the donors of his bridge from the Pope in Rome, Pontifex Maximus.

Local enthusiasm often was inert, as it is in any time, to projects beyond local experience, especially if they call for expenditures out of the ordinary. As late as the fourteenth century *Piers Plowman* was exhorting rich merchants in England to mend the roads and rebuild the bridges. The Bishop of Durham granted "forty days indulgence to all who will draw from the treasure that God has given them" [1] to aid in repairing Botyton Bridge. But in 1439 the attitude of citizens in Orléans, France, indicated greater interest. They visited in a body the foundations of their bridge, inspected the work, and advised the masons to proceed. After posted advertisement, the contract had been let to the builders who gave surety and made the most advantageous bid.

Variety had come long since. Engineers were choosing freely from among the several arch forms (segmental, pointed, or elliptical). They worked with either brick or stone. They varied the widths and heights of piers as they saw fit. Ornamental towers, houses, and indiscriminate

[1] J. Jusserand, *English Wayfaring Life in the Middle Ages,* 3d ed., T. Fisher Unwin, London, 1925, p. 36.

superstructures made outlines against the sky that no longer showed any relation to those of the old Roman aqueducts. The Saint Esprit over the Rhone River, built in 1265 to 1309 by the order known as the Hospitaliers Pontifes, is 2,700 feet long with 26 arches, for years the longest bridge in the world. Its clear spans were from 85 to 110 feet long. Its piers vary from 20 to 50 feet in thickness. The bridge at Trezzo, Italy, raised about 1375, had for centuries the longest single arch in the world, over 230 feet. But with all the variations and adaptations, there was no essential change in bridge construction until the Italian Palladio experimented with the truss during the sixteenth century.

We do not know when the first wooden structure was built across the Thames on the site of the famous London Bridge. One was there when King Sweyn of Denmark tried to take London from Ethelred in 994. Many of Sweyn's men lost their lives in the river, according to the English account, because they did not use the bridge in their flight. It was maintained until 1176 by "the liberality of divers persons" and only in part by "taxations in divers shires." Then Peter of Colechurch, priest and engineer, began a stone bridge (Figure 5.8) which was thirty-three years in building. He had been dead four years before it was finished in 1209 by three London merchants, Mercer, Almain, and Botewrite. To complete his work, they had brought over the French engineer Brother Isembert, trained in the tradition of Bénezet. As a fitting memorial, Peter was buried in the chapel of the bridge.

In addition to the draw, there were 19 irregular, pointed arches of squared stone. The piers rested upon broad foundations pointed to withstand the current of the river. According to Stow's *Survey of London*, first published in 1598, during construction the Thames was diverted into a trench from Redriffe to Battersea, a distance of 4 miles. If this was done, it was remarkable engineering, for the Thames here is a wide stream and tidal. The Archbishop of Canterbury gave a thousand marks toward the expenses. The King supplied funds, citizens of London subscribed, and inhabitants of outlying shires made voluntary contributions. Vacant lands in London were expropriated for new buildings to provide revenues. The career of the infamous King John had redeeming features other than his acceptance of Magna Charta. It is he who has been credited with expropriating these lands. He named Isembert to succeed Peter of Colechurch as the engineer.

This bridge was the London Bridge which, according to tradition, wise men went over, only fools went under. A French historian of Eng-

Figure 5.8 London Bridges of 1209 and 1831

lish wayfaring has attributed its weakening arches in part to Henry III's beloved queen, who used the tolls for her own upkeep.[2] The neglect was disastrous, for the records tell of an extreme frost and heavy snow in the year 1282 which opened great cracks in the masonry and bore down five of the arches. Moreover, builders in successive generations gave little thought to the strength, or rather the weakness, of their materials; they piled houses, shops, towers, superstructures of all sorts, upon the bridge. On one occasion a whole row of these toppled into the stream. Besides the hazards for fools who went under, the waterway was so cramped during certain stages of the tide that for hours at a time navigation was impossible. The openings constituted no more than a third of the length of the bridge, which has been called a "pierced dam" rather than a bridge. Nevertheless, despite its peculiar character and abuse, old London Bridge was well enough designed and maintained to last for 600 years. It was not replaced until 1831.

As a feat of engineering, the medieval bridge is not to be compared with the cathedral. Where the cathedral often was huge yet graceful, the bridge was heavy. Its foundations were generally more secure than those of the cathedral, but the economy of bridgebuilding was severe. It provided no more than one-way traffic over a stream. It was narrow, even where there were safety islands and turnouts on the piers. It usually was not adapted to heavily loaded vehicles. As likely as not, it had inadequate approaches. In fact there was little danger that anyone could be caught upon it in flood times and swept away; the river often would have long since spread across the lowlands on either side and submerged the causeways. And even if it were built massively of stone, the medieval bridge was vulnerable to wind and weather.

Canals and Tunnels

Early Chinese engineers had developed canals for inland transportation. During the lifetime of Confucius in the fifth century B.C., they had begun work on the Grand Canal which the Venetian traveler Marco Polo saw nearing completion about A.D. 1290. It extends some 860 miles from Hangchow on the Yangtze River northward to Tientsin. The device which they had used for centuries to raise boats from one level to another seems to have been an inclined plane equipped with rollers and windlasses operated by manpower. The Romans had canal boats drawn by mules to

[2] *Ibid.,* p. 61.

take them across the Pontine Marshes, but they were content to do most of their traveling on water by rivers and the sea because they had ready access to the sea, like the Greeks, and relatively little need for heavy hauling inland. The Romans had constructed a few small canals in England, the Netherlands, and northern Italy. These in all probability were more for irrigation and drainage than for transport.

The Emperor Charlemagne was evidently the first in Western Europe to plan a grand canal for navigation, his idea being no less than to link the Danube with the Rhine. He came in person with all his retinue to see the work begin in the autumn of 793. It was to be as important in his imperial scheme of commerce and military control as his bridge across the Rhine. The idea was grandiose; its attempted execution was a total failure. Charlemagne and his engineers knew too little about engineering. It is doubtful that they had accurate information concerning the respective water levels of the two river systems. They were unable to cope with the terrain, and a large force of men dug for some 2,000 paces to the width of 300 feet and then quit as swampy soil, continuous rains, and finally quicksand added fear to their hardships. They were sure that some evil force, perhaps the Devil himself, had the place under his spell; during the night the sand swallowed the loads of mud which they had dug out during the day. And neighboring monks were not immune from criticism. There was muttering and writing at Salzburg "against the Lord" (certainly not God). "Prudence and counsel" would not prevail with Charlemagne although every night came "the hurly-burly din of hideous noises, roaring defiance, and exulting in the laughter of derision." [3] Whether he too was scared or just disgusted, Charlemagne finally abandoned the ditch. The present (Neckar) canal between the Rhine and the Danube was in course of construction at the outbreak of the Second World War.

Many years elapsed after Charlemagne's failure before any other work meriting attention was attempted on canals. It was not until near the close of the twelfth century that further advances of any significance were made in canal engineering. Most of the effort in the meantime had been put into restoring Roman ditches for irrigation, drainage, and water supply. Then, in southern France and northern Italy almost simultaneously came several developments. In 1171 the Count of Toulouse granted to the Bishop of Cavaillon the right to divert the waters of the Durance River

[3] Jacob I. Mombert, *A History of Charles the Great*, D. Appleton & Co., New York, 1888, p. 278.

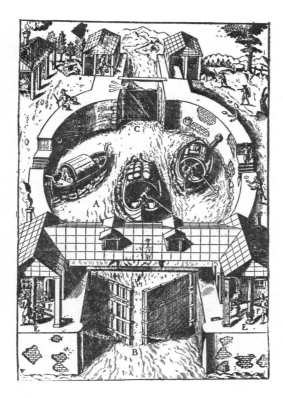

Figure 5.9
Early canal lock (From
V. Zonca, *Novo teatro di
machine et edificii,* 1656)

for flour mills. The Bishop in turn allowed the inhabitants of the district
to use the spent water on their lands. This led to construction of many
lateral channels from the main canal, and a definite system of irrigation
for the community evolved.

In Italy, engineers conceived of the Naviglio Grande to carry water
and boats from Ticino across 30 miles of plain to Milan. Construction
began in 1179 and was completed in 1258. Thus the Alpine streams of St.
Gotthard and the Simplon which were running to waste from Lake Mag-
giore to the sea were turned to make the city of Milan prosper. It was
through this canal that in 1438 the pink-and-white marble for the interior
of the Milan Cathedral was brought 60 miles from the foothills of the
Alps. To construct an effective canal took more than the simple restoration of Roman ditches. There had to be dams and other contrivances for
regulating flow and maintaining uniform depth and a fairly constant level
over long distances. What these installations were is not known. The

Italian words *sostegno,* or support, and *conca,* or chamber, describing some of the apparatus suggest barriers or gates of some kind. In any case, the modern lock was in process of development (Figure 5.9).

Medieval engineers used tunneling in their fortresses and castles, but there is no record of such work on canals until much later, in the seventeenth century. The method then was still the Roman method of using fire and water for rock excavation and picks and shovels elsewhere. Roger Bacon (ca. 1214–1294), great speculator on scientific things to come, sometimes with evidence but as often without, wrote furtively in 1248 about gunpowder: ". . . take 7 parts of saltpeter, 5 of young hazelwood (charcoal) and 5 of sulphur; . . . . and with such a mixture you will produce a bright flash and a thundering noise, if you know 'the trick.' You may find (by actual experiment) whether I am writing riddles to you or the plain truth." [4] And Albertus Magnus (ca. 1193–1280), Master of the Dominican Order at Paris, seems to have known the secret formula in 1250. Cannon date from the fourteenth century. Still there is no evidence that gunpowder was used in tunneling or excavation until well into the seventeenth century.

Dikes, Harbors, Docks

The dredges in Boston Harbor of a July morning, 1945, were sucking tons of clay and gravel from the bed of the sea to fill between two islands and enlarge the airport for the city. Their rotary cutters, electrically driven, loosened great swirls into a 30-inch pipe which carried the material under high pressure a mile to the shore. The men of the Middle Ages would have marveled at the combination of hydraulic engineering with electrical machines.

A few harbors like Ostia, port of Rome, had been abandoned because they silted in; the engineers, it is said, had not possessed the tools for clearing them. It is also to be said that most ships of the times were of so shallow draft that they could be hauled ashore rather easily; in the tideless Mediterranean there was no great demand for deep harbors and docks. But the inertia was not general. There was increasing need for access to the sea. The men of Bristol in England, not long after Londoners had turned their river aside for their bridge, dug a trench to give the Froom

[4] H. W. L. Hime, "Roger Bacon and Gunpowder," in Andrew G. Little (ed.), *Roger Bacon Essays,* Clarendon Press, Oxford, 1914, pp. 330–331.

River a larger channel from its junction with the Avon. They began work on this deep-water basin in 1239 and finished it in 1247 at a cost to the city of £5,000, a large expenditure in those days. From this greatly improved harbor with its long quay set out many of England's famous sea dogs—explorers, merchants, colonizers, and pirates.

The construction of adequate docks came with the need. Ships in ancient times, light and shallow, were beached and careened for repairs. With increasing tonnage—and the increase was rapid in the fifteenth century—the need for dry docking grew imperative. A large English warship in 1434 displaced but 400 tons; by 1474 the ship *Mary and John* of Bristol displaced 900. The early method, used in 1434 at Southampton on the south coast of England where the range of the tide is more than 13 feet, was to take the vessel inshore at high tide and surround it at low tide with a wall of timber, brush, and clay, forming a basin which could be kept reasonably dry while repairs were made. By 1496 a permanent dry dock of timber and masonry had been constructed at Portsmouth. It was not until after the introduction of steam engines some 300 years later that docks were pumped dry. An early reference to wet docks for keeping the water at a constant level in harbors where there was considerable tide range while cargoes were being loaded or unloaded appears in the diary of Samuel Pepys for April 28, 1667. By that time the lock, a necessary feature of such wet docks, had been well developed.

Netherlanders on the Continent had been driving the sea from their lands mile by mile since the tenth century, using dikes against the 20-foot tides, wind and water mills to dry the marshes, and canals to drain low-lying ground. Humble peasants, wealthy burghers, Cistercian monks, though each group worked for itself, had common interest and purpose in gaining arable soil. Disaster from storms and floods only aroused greater effort and more ambitious plans to extend and to guard the land. And in the struggle came an engineering achievement which has done more to benefit people the world over than can be readily calculated. Whether the lock first appeared in the Netherlands or in Italy or elsewhere, possibly China, is an issue for endless debate. Whoever made the discovery, it was applied successfully in the Netherlands independently of any other similar achievement elsewhere.

The Flemish city of Bruges, port of Northern Europe, was as important to the trade of the Continent in those days as Venice on the Adriatic. Bruges, however, had access to the North Sea only by way of a low canal dangerously open to storms. For its protection, a dike was built at Damme on the north, but the canal, of course, could not be shut off entirely. Someone thought of putting two gates in the dike, a sort of double sluice, so that the first could be closed behind a vessel before the second was opened and the sea let in or out as the case might be. This enclosure apparently was made of wood. Its design is not known, and there is uncertainty how the gates were hung, whether they slid up and down or swung open. But records for 1234 show that there was an *écluse à sas*, a lift lock. The contrivance was repaired in 1353, again in 1371, and replaced with one of masonry at the close of the century. If Flemish engineers were not actually the first to build locks, they did so well with dikes, locks, and harbors that their advice was in demand throughout medieval Europe.

One medieval municipal-improvement program was that undertaken by Philippe Auguste, King of France from 1180 to 1223. Philippe, twenty years old, was enveloped in a cloud of dust from the carts on the roadway below while trying to enjoy the view of the Seine from his palace window. He summoned the mayor and ordered the streets of Paris paved. This was done in 1185 with large uneven slabs of stone, some of which may still be seen in the garden of the Musée de Cluny. It was not to be compared with Roman paving. The slabs did not make for either smooth travel or quiet in nearby homes. But they did fix the level of the thoroughfares, and refuse was no longer allowed to be deposited in them. The inhabitants were required to remove the debris from in front of their houses and cart it outside the town.

Medieval communities had systems of water supply. Roman aqueducts were kept in use throughout the Middle Ages, although relatively few may have been improved. Reservoirs were provided and water pumps installed. Aldric, Bishop of Mans from 832 to 857, had an aqueduct and a vaulted reservoir built to supply two fountains. Lambert, Abbot of St. Bertin from 1095 to 1123, had a "water-elevator" connected with the abbey's mill wheels. Cistercian monks were active in this kind of hydraulic engineering. The convent of Chester in England was supplied with water in 1285 from a lead pipe 3 miles long. Canterbury Cathedral had been equipped with a water system for years. The beautiful Gothic aqueduct of Coutances (Figure 5.10) was built in 1277 and kept in repair for a long time. And yet, the usual supply of the towns and cities was drawn by buckets from well or stream and carried by hand as it had been for ages and as it is to many dwellings this day; nor was the water supply always kept free from the sewage and pollution of the neighborhood.

Slops thrown into the streets found their way down the gutters into river or well too soon for any filtering, settling, or bacterial action to purify them. Latrines, sewers, cesspools were planned, built, and cleansed periodically in medieval towns, but they were likely to be too close to water supplies. Their contents were almost certain in flood times to mingle with the streams which supplied the reservoirs and especially the family water buckets. The Romans Vitruvius and Pliny the Elder had speculated upon the relation between impure drinking water and disease. Jean Pitard, surgeon to Louis IX of France and the King's successors, anticipated in 1310 the sanitary engineering of five centuries later. Pitard dug a well within his cellar not far from the bank of the Seine, evidently sinking it below the level of the river. Regardless of flood conditions, he and his neighbors had a supply of clear and relatively pure water filtered through sand. Perhaps it had been his good fortune to hit a separate underground flow. In any case, the neighborhood used the well for the next three hundred years, but there is no record of the number who died in consequence. Municipal sanitation, carefully systematized and comprehensively developed, had to wait for the nineteenth century.

Figure 5.10 The thirteenth-century Gothic aqueduct at Coutances, Normandy (From Camille Enlart, *Manuel d'archéologie française*, 1904; courtesy A. et J. Picard et Cie.)

However, enterprising medieval real-estate men made competent plans for residential suburbs and wholly new towns. Medieval communities often had "masters of works," whom we would call city engineers. There was a board in Douai composed of two masons and two carpenters. The records of Amiens for 1292 show similar officials in the service of the aldermen; later there were *caticheurs*, or inspectors of building materials and construction. And there were road makers. The existence of these officials gives evidence of local pride in maintaining the community. The *villes neuves*, "new towns" or suburbs outside of city walls, show much better city planning. They were conceived as model communities with every convenience, in which people would be eager to settle. Conrad, Duke of Zähringen, founded Freiburg-im-Breisgau in 1120 on wasteland adjoining his castle. Each settler received a plot 50 by 100 feet for which he paid to the duke an annual rent of 1 sou. The location on the main road through the Black Forest from the Rhine to the Danube was a good one for commerce, and the town prospered. Other promoters soon followed: the Abbot Suger with his Vauchresson in France for one, Henry I of England with Newcastle on the Tyne near the Scottish border for another. There were many more from southern France to Ireland.

The *ville neuve* is significant in history for its destructive effect upon the feudal systems of lord and vassal. It had much to do with the change to the capitalistic system and industrial society. It is notable for the civic liberties, personal and economic, which the promotor conceded or was glad to offer in anticipation of profitable revenues for himself if people would come and live there. These guarantees were similar to the rights, privileges, or immunities of the old trading centers and Roman municipalities which the feudal lords had never been able to remove. But the new town is significant also for its engineering. Plans and construction varied with local situations. It was the general intention, however, to lay the streets in rectangles wherever possible and to have the ways broad or narrow according to the traffic anticipated and their relative importance as approaches to the heart of the city. Markets were placed in relation to residential zones. Service alleys ran behind the houses. Churches and public buildings stood in appropriate locations. The engineers centered their squares or plazas with regard to convenience and efficiency. There was thought, too, for the general architectural effect. Hippodamus would have approved.

The city best exemplifying the civic consciousness and municipal engineering in Europe at the close of the Middle Ages, however, was

neither an ancient trading metropolis which had survived the feudal system nor a new town which rose to destroy it. The founders of Venice
had fled from Hun and Lombard invaders to the islands and lagoons at
the head of the Adriatic Sea. By 568 they had established themselves
behind their protecting marshes and sand bars as a virtually republican
nation. Though the Venetian council came eventually under the control
of the aristocracy, efforts failed to make the doge a hereditary monarch.
The Venetians did quite well. They were able to play the Byzantine emperor against Charlemagne and to stay independent of both; by the year
1000 they had overcome the Dalmatian pirates and taken command of the
seaway to the Holy Land. For the rest of the period, well past the rise
of the Western nations—Spain, France, and England—the City of Venice,
strong in commerce with the Orient, was a power always to be watched
in the politics of Europe.

The Venetians filled in great stretches of marsh and strengthened the
shores of the *lidi,* or low islands between the estuary and the sea. Their
bridges, often made of stone, crossed from island to island on pontoons
or beams. The famous bridge of Rialto, originally designed in 1178 by
Nicolò Barattieri, though of wood, was widened after 1255 and provided
with a draw. Their first wooden dwellings, built upon platforms as were
the huts of the prehistoric lake dwellers, gave way in the fifteenth century to palaces of marble set upon close-driven piles which were completely preserved by the water. This procedure was necessary because
the subsoil of mud, peat, clay, and sand was soft to the depth of more
than 100 feet before there was a layer of hard clay 10 feet thick. Structures of any weight at all were bound to settle into such subsoil. The
problem was to have the weight so evenly distributed and balanced that
the buildings would settle uniformly. And to make the situation worse,
the whole region was subject to earthquakes and volcanic action. The
lido of Malamocco, several miles to the south, sank on Good Friday,
1102, as the sea rushed in and flames burst from the ground. The inhabitants, it is said, had time only to flee before Malamocco disappeared.

The builders of St. Mark's, the Basilica di San Marco, in Venice, began it in 830 on the foundations of an ancient Roman structure. After
destruction by fire in 976, San Marco was rebuilt during the eleventh
century in Byzantine style in the form of a Greek cross. It was given
more than 500 marble columns, a central dome 42 feet in diameter set
upon 4 piers, each 21 by 28 feet. The core of the walls of the building
is brick with a veneer of marble. The feature of St. Mark's which inter-

ests tourists today perhaps as much as its bronze horses and its pigeons is its wavy floor. This has been caused by the uneven settling of the structure, in some places 10 inches out of level. The piling beneath it is only 7 feet deep.

Undaunted, or ignorant of the subsoil, Nicolò Barattieri raised the Campanile of Venice in 1180 (Figure 5.11) to the height of nearly 200 feet above its foundations, which were less than 17 feet deep; moreover, they went down nearly vertically with little spread, and at their top they were less than 4 feet wider than the brick shaft. Fires, earthquakes, and lightning injured the Campanile, and it was frequently rebuilt. In 1517 it was raised to the height of 300 feet. But unlike the Leaning Tower of Pisa, which sank unevenly until it was 16½ feet out of plumb, the Campanile of Venice stood erect. It was examined in 1885 and found all right, a monument to the engineering skill of Nicolò Barattieri and his successors. Then on July 14, 1902, about 10 A.M., it collapsed. A cameraman happened to take a picture at the moment. Two wide vertical cracks opened. In the opinion of some, these showed that the failure was caused by old age and "fatigue" of masonry, that the foundations were not to be blamed. Precaution was taken, however, when the Campanile was rebuilt in 1905, to reinforce the foundations.

By 1328 Venice had a board of commissioners in charge of all canals, ports, quays, and works for controlling the rivers and keeping the lagoons free of silt. In time streams were diverted from their natural channels so that they emptied directly into the sea; currents were confined within millraces to make water wheels more effective. The shore of the mainland was diked so that fresh water might not mingle with the sea water, for Venetians thought that the mixture caused malaria. Prior to the sixteenth century the only water which they used came from the rain which fell on their roofs, courts, and streets. This they carried off into deep *pozzi*, or wells, after filtering through many feet of sand which was purified to some degree by repeated washing. Eventually the Republic dug the Seriola Veneta Canal from the Brenta River at Dola to Moranzani, nearest point on the mainland. Here the water was pumped into boats and ferried to the *pozzi* in the city. The total supply from both rain water and the canal, however, did not exceed 3 gallons daily for each person in a population of 183,000 in 1563.

Too often, as in other medieval cities, sewage ran in the streets and canals of Venice. But the authorities tried to keep them sanitary by laws and fines. Burials were regulated, unhealthful industries were barred from

Figure 5.11 The Campanile, Venice (Courtesy New Haven Public Library)

certain residential quarters, and hospitals were opened for the leprous and victims of other loathsome diseases. Narrow and dangerous passage-ways were lighted at public expense, and many of the streets were paved. Thus the citizens of Venice made their city an exceptional place for living as Europe came from the Middle Ages into the Renaissance, terms that have little meaning in the history of engineering, which maintained its cumulative course of development throughout both periods.

Bibliography

Adams, Thomas: *Outline of Town and City Planning: A Review of Past Efforts and Modern Aims,* Russell Sage Foundation, New York, 1935.

Briggs, Martin S.: *The Architect in History,* Clarendon Press, Oxford, 1927.

Harvey, John: *The Gothic World 1100–1600,* B. T. Batsford, Ltd., London, 1950.

Mortet, Victor: *Recueil de textes relatifs à l'histoire de l'architecture et à la condition des architectes en France, au moyen-âge, XIe–XIIe siècles,* A. Pickard et Fils, Paris, 1911.

Mortet, Victor, and Paul Deschamps: *Recueil de textes relatifs à l'histoire de l'architecture et à la condition des architectes en France, au moyen-âge, XIIe–XIIIe siècles,* Editions Auguste Picard, Paris, 1929.

Palmer, Roger L.: *English Monasteries in the Middle Ages,* Constable & Co., Ltd., London, 1930.

Porter, Arthur K.: *Medieval Architecture,* 2 vols., Baker and Taylor Co., New York, 1909.

Salzman, Louis F.: *Building in England Down to 1540,* Clarendon Press, Oxford, 1952.

Salzman, Louis F.: *English Industries of the Middle Ages,* new ed. enlarged, Clarendon Press, Oxford, 1923.

Sarton, George: *Introduction,* vols. 1–3 (see Chap. 3).

Villard de Honnecourt: *Facsimile of the Sketch-Book of Wilars de Honnecourt . . . with Commentaries and Descriptions . . . Translated and Edited, with Many Additional Articles and Notes,* J. H. and J. Parker, London, 1859.

Viollet-Le-Duc, Eugène E.: *Dictionnaire raisonné de l'architecture française du XIe au XVIe siècle,* 10 vols., B. Bance, Paris, 1858–1868.

Viollet-Le-Duc, Eugène E.: *Dictionnaire raisonné de mobilier français de l'époque carlovingienne à la Renaissance,* 6 vols., Gründ et Maguet, Paris, 1914.

White, Lynn: "Technology and Invention in the Middle Ages," *Speculum,* vol. 15, pp. 141–159, 1940.

SIX

Foundations for Industry

Though the pride of Venice accounted for resplendent engineering, Florence had by far the most famous engineer. Leonardo da Vinci, born in 1452, matured with the Renaissance as its most versatile figure—at once painter, sculptor, philosopher, scientist, practicing military and civil engineer. His mother was a peasant whom he hardly knew after infancy. His father, a well-to-do notary of Florence, gave him his name and his initial education in art under Andrea Verrochio. The youth was soon better than his teacher. His left-handedness, so complete that he wrote backwards, produced the mirror writing which is deplored today as an obstacle to the intellectual progress of children. It interfered not at all with the effectiveness of Leonardo; in fact it may have helped. Sketches along the margins of his writing indicate that his hand was free to do as he wished while his mind worked on something else. But then Leonardo da Vinci was neither in birth, upbringing, nor ability like other mortals.

The great Florentine was famous in his own day. He served the Duke of Milan for years in both military and civil capacities. It has been said that his work on the Arno River gave him high place among the engineers of all time. After François I invaded Italy, he took Leonardo back to France as an adviser. But Leonardo deserves fame in the profession rather less as practicing engineer than as prophet of engineering's future. Besides his machine guns, breech-loading cannon, tanks, a submarine, and a flying machine, Leonardo's sketches included lathes, pumps, cranes, jacks, water wheels, a canal lock, drawbridges, wheelbarrows, a diver's helmet with air hose, roller bearings, a self-propelled carriage, a double-decked city street, sprocket chains, an automatic printing press, a universal joint, a helicopter, and a wooden truss bridge. There were a great many more devices as varied and ingenious. His ideas are recorded on more than five thousand sheets of drawings and notes. But these were scattered over Europe in private collections and libraries beyond the reach of practicing engineers and were not published for centuries after his death. In recent years efforts have been made to gather and publish them. All are not yet recovered; doubtless many have been destroyed.

Other minds of the Renaissance indulged in much inquiry, against the express warning of ecclesiastical authority. It seems odd that Christians should get into trouble with their Church when refuting the ideas of the great pagan Aristotle. But so it appears to have been. The trial of Galileo Galilei (1564–1642) of Florence before the seven cardinals in 1633 would be absurd today were it not tragic revelation of what can happen at any time as a result of conflict between what one has learned from others and what one observes for oneself, and also of the ability of powerful, intolerant ignorance to suppress facts. Galileo showed the fallacies in Aristotle's statements that the velocity of a falling body depended upon its weight and that there were two kinds of motion, "natural" and "violent." From his observations Galileo properly held that there was one kind of motion only; the forces which caused the motion might be different but not the motion itself. Nevertheless, the often repeated story of his experiments in dropping weights in 1590 from the top of the Leaning Tower of Pisa is now believed to be without foundation in fact.

Scientific study of motion brought Galileo to the conclusion that Aristotle had been wrong also with regard to the movement of the earth and that the Polish German Koppernigk, or Copernicus, was correct. Copernicus had written in 1543 that the earth revolved about the sun. But in this discussion Galileo was taking exception to Holy Scriptures concerning the physical relation between Heaven and Earth. The cardinals, "inquisitors-general throughout the whole Christian Republic," declared that the proposition of Copernicus was "absurd, philosophically false, and, theologically considered at least erroneous in faith." [1] This is not the place to examine the torture of the great scientist's mind as Galileo stood condemned for undermining the tenets of the religious faith he professed. Galileo publicly retracted his statement concerning the motion of the earth, protesting *sotto voce*, it is said, as it were to himself, "It moves for all that." We had best leave the issue with Leonardo da Vinci's judgment that whoever appeals to authority applies not his

[1] William T. Sedgwick and Harry W. Tyler, *A Short History of Science,* The Macmillan Co., New York, 1917, p. 415.

intellect but his memory and quote Leonardo: "The man who blames the supreme certainty of mathematics feeds on confusion, and can never silence the contradictions of sophistical sciences which lead to an eternal quackery." [2] Certain it is that, notwithstanding persecution and bloodshed, ideas were expressed in the sixteenth and seventeenth centuries, and devices made to apply them, which materially changed the foundations of society.

Like the birth of Greek science, the rise of modern science in the seventeenth century was to have a profound effect on engineering. The great scientific advances which began in the seventeenth century were based on Greek science. However, there is a new ingredient in modern science, experimentation, which distinguishes it from Greek science in which there were very few experiments except in biology. Experimental science began particularly with the work of Robert Grosseteste in the thirteenth century, but in the sixteenth and seventeenth centuries scientists began to adopt the empirical experimentation that engineers and artisans were already using.

Many scientists of the period visited with and learned much from engineers, and Galileo opened his *Dialogues Concerning Two New Sciences* by saying, "The constant activity which you Venetians display in your famous arsenal suggests to the studious mind a large field for investigation, especially that part of the work which involves mechanics; for in this department all types of instruments and machines are constantly being constructed by many artisans, among whom there must be some who, partly by their own observations, have become highly expert and clever in explanation." [3] Not only did the scientists adopt the experimental attitude in large part from engineers and other technologists, but they also began to use the products of engineering, such as pumps and balances, as well as telescopes and microscopes. Engineers started to use some of the knowledge of the new science in the seventeenth century, but it was only in the nineteenth century that engineers could consistently apply science to achieve advances in some fields. Engineering helped to stimulate the rise of modern science in the seventeenth century and was in turn changed in character by the birth of applied science in the nineteenth.

[2] Leonardo da Vinci, *Literary Works*, 2 vols., Oxford University Press, London, 1939, vol. 2, p. 241.
[3] Galileo Galilei, *Dialogues Concerning Two New Sciences*, Northwestern University Press, Evanston, Ill., 1939, p. 1.

Tools of Engineering

Many of the concepts so essential to modern engineering had been germinating a long while before the time of Leonardo da Vinci. The thirteenth-century author, possibly Jordanus Nemorarius, of *De ratione ponderis* had thought of investigating the weight of an object resting upon an inclined plane. This idea the Netherlander Simon Stevin (1548–1620) developed in 1586 into the "parallelogram or triangle of forces," demonstrating that three forces in equilibrium at a point can be represented both in magnitude and in direction by the sides of a triangle. Stevin, an old man in Holland when Pilgrims of the *Mayflower* were still living at Leyden, described his triangle as "a wonder, yet no wonder." With it he established some of the principles in the science of statics, or solids in equilibrium, which engineers must apply if they would know beforehand that their structures are economically designed and will stay upright and in balance.

In 1586 Stevin published the results of his experiment of dropping two lead balls, one ten times as heavy as the other, from a height of 30 feet onto a board. The two balls hit the board as one, contrary to the opinion of Aristotle who had held that the heavier ball would strike the board first. Using inclined planes, Galileo further investigated the laws of falling bodies and reported in his *Dialogues* that irrespective of weight they accelerate uniformly and that their velocity is proportional to the square of the time—facts of great importance to engineers. Huygens later computed the value of the acceleration constant. Galileo investigated also the strength of materials, and his were the first crude testing machines made for that purpose. He hung weights on copper rods and on a wooden beam or cantilever jutting from a wall. He measured the strength of his "copper," almost certainly an alloy, with apparently reasonable accuracy. His test of the cantilever was not so successful because he failed to take into account the fact that the fibers on the underside of the projecting beam were in compression while those on the upper side were in tension, as Edmé Mariotte (ca. 1620–1684) proved in 1680. Meanwhile Robert Hooke had shown that any substance under stress is deformed in some degree.

Inspired by Galileo, his secretary Evangelista Torricelli (1608–1647) took a great step forward. In studying hydrostatics Stevin had found that the pressure of water was proportional to its depth. Torricelli now linked hydrostatics and dynamics to show that under hydrostatic pressure a fluid like water passes through an orifice practically as fast as if it were falling from a height equal to the depth of the fluid over the orifice. The next step was easily taken about 1650 by Torricelli's contemporary, the French humanist Blaise Pascal (1623–1662). He reduced the propositions to the law which bears his name and is, or should be, familiar to every student of elementary physics: the pressure put upon an enclosed fluid is transmitted equally in all directions without loss, and it acts with equal intensity upon equal surfaces. Applying this principle, the hydraulic presses of today deliver overwhelming force or the lightest touch, as the need of industry requires.

Torricelli and, later, Pascal were responsible for developing the barometer, which balances the weight of the air by means of a column of liquid under a vacuum in a tube closed at the top. As they did this, they exposed the fallacy in Aristotle's belief that nature abhorred a vacuum. They proved that nature abhors only such vacant space as it has the power to fill. There remained at the top of Torricelli's tube of 1643 a space into which the weight of the atmosphere was not able to force the mercury. And in 1648 Pascal demonstrated that at sea level the mercury rose higher in the tube than it did on a mountain top, where the balancing column of atmosphere was lighter. For the same reason, it is not possible to raise a column of water more than 33 or 34 feet at sea level merely by exhausting the air above it with a suction pump. Such a column of water is the equivalent in weight to the weight of the atmosphere above it. While also experimenting with air pressure and its absence, the German Otto von Guericke (1602–1686) in 1650 developed into an air pump the water pump which had been known for centuries. With it he was able to exhaust so much of the air within his famous Magdeburg sphere that the pressure of the atmosphere held its two sections together against sixteen horses, eight pulling on either side. He could of course have exerted the same force with eight horses pulling against an anchor. The way was rapidly clearing for the discoveries of Savery, Newcomen, and all others who worked on engineering problems which dealt with subatmospheric pressures.

Two French mathematicians, Pierre Fermat (1601–1665) and René Descartes (1596–1650), independently discovered analytic geometry, which has had a notable influence on engineering. Since Descartes' exposition of the new geometry was more systematic and influential than that of Fermat, it is Descartes who usually receives the principal credit for the discovery, which is sometimes called Cartesian geometry in his honor. Descartes' great achievement was in joining algebra to geometry and thus introducing into geometry an analytical method to represent curves by algebraic equations or vice versa. The great value of analytic geometry to the engineer is that it facilitates an analysis of the relation between such variables as temperature and pressure, speed and power, and countless other groups of variable quantities.

Four men born in the second quarter of the seventeenth century made exceptional scientific contributions which proved to be of great value to engineering. Two of them lived on into the eighteenth, but the work of all four belongs to the century of Stevin, Galileo, Torricelli, Pascal, and Descartes. Three were British; one was a Netherlander. Their names were Robert Boyle, Robert Hooke, Christian Huygens, and Isaac Newton. Robert Boyle (1627–1691) studied the elasticity of air and formulated in 1661 or 1662 the law which bears his name. It is credited also to the Frenchman Edmé Mariotte, who discovered it independently. At a constant temperature, a given quantity of a perfect gas varies in volume inversely as the pressure exerted upon it; that is, the volume decreases as the pressure increases, while the product of the two remains constant. Thus Boyle and Mariotte established a fundamental principle for engineers— though engineers never work with "perfect" gases. And with Hooke as his assistant in the laboratory at Oxford, Boyle made the first air pump of modern design.

Robert Hooke (1635–1703) experimented on his own account at Gresham College in London with the elasticity of watch hairsprings, in part his invention, and formulated the physical law which carries his name. Material under tension lengthens in proportion to the pull exerted upon it; under compression, it shortens in the same manner. This relation holds true only up to the elastic limit, which varies with different materials. Some materials, like glass and certain kinds of steel, have well-defined elastic limits; others, like putty or even copper, have not. In any case, a limit of elasticity of any given material has to be found in actual testing. Hooke analyzed the several forces acting in an arch, and he developed also the "universal joint" which had been suggested by Leonardo da Vinci. Through it, the power in a shaft can be transmitted to another shaft, connected to it but at an angle with it.

Meanwhile, Christian Huygens (1629–1695), a more serene fellow than Robert Hooke and more versatile, was developing the watch spring into a spiral, inventing the pendulum clock, and using the pendulum to measure the acceleration of gravity. The pendulum of a given length swings, as Galileo had observed, with a constant period regardless of the weight or of the range of its swing. Huygens brought mathematical genius to the study of mechanics. It was he who derived the formula for determining the radial or centripetal force necessary to cause a body to move in a curved path, an essential contribution to engineering as well as science. Colbert, great Minister of Finance for France and patron of the arts, summoned the Dutch scientist to the French Academy to be honored at its founding in 1666.

The last and greatest of these four scientists was Isaac Newton (1642–1727). Few engineers leave their work long enough to read the important sections of his *Principia*, but the profession would be much poorer without its fundamental thinking. His German contemporary Gottfried Wilhelm Leibniz (1646–1716) also deserves great credit for his part in inventing the differential calculus, or what Newton called fluxions—the computation of rates or proportions at which variables increase or decrease. Nevertheless, it was this modest Englishman, born to be a farmer, who worked through the whole maze of concepts, speculations, and surmises, to state the three basic laws of motion. These, engineers can never ignore: "[1] Every body continues in its state of rest, or of uniform motion in a right line, unless it be compelled to change that state by forces impressed upon it. [2] The change of motion is proportional to the motive force impressed; and is made in the direction of the right line in which that force is impressed. [3] To every action there is always opposed an equal reaction: or, the mutual actions of two bodies upon each other are always equal, and directed to contrary parts." [4]

Seventeenth-century instruments for surveying were much the same as those of the Romans. The *dioptra* was employed with little change, except that the astronomer's graduated arc with a movable sighting device had been adapted to it by 1616. Wooden poles were still used for measuring short distances. Dimensions of a field or the length of a road were often obtained by a perambulator or wheel of known circumference. The method, except in flat country, was of course inaccurate for map making —a map is a surface projection. A surveyor's wire chain of nine links, appearing in England about 1600, gave way after 1620 to Gunter's chain of 100 links, 66 feet long. The compass, used as early as the twelfth century, perhaps earlier in China, continued to aid both navigators and sur-

[4] Isaac Newton, *Mathematical Principles of Natural Philosophy*, University of California Press, Berkeley, Calif., 1947, p. 13.

veyors. A young Dutch mathematician, Willebrord Snell van Roijen (1581–1626), was the first to employ a system of triangulation in measuring the earth's surface as he laid out a series of triangles for a distance of some 80 miles in 1615; he was the first, therefore, to practice geodetic, or spherical, surveying as distinguished from plane surveying. Pierre Vernier (ca. 1580–1637) invented in 1630 a device so useful to engineers that his name has entered the language. The vernier consists of two different scales of measurement side by side, making it possible by sliding one upon the other to determine smaller divisions precisely. The telescope, though invented in 1608 and used in his 1669 triangulation of the region between Paris and Amiens by Jean Picard (1620–1682) who added the cross hairs of spider web, did not become part of ordinary surveying equipment until after 1800.

There were decided improvements in calculation. Excepting the Moslem and Hindu invention of zero to express the difference between equal quantities and the introduction of Arabic numerals, methods of computation had changed very little since Egyptian times until Stevin wrote a treatise on decimal fractions in 1586. With it he helped engineers along the way to speed as well as accuracy in their reckoning and laid the groundwork for the metric system of two centuries later. John Napier (1550–1617), strict Protestant and Scottish laird, invented logarithms about 1594 and published his table in 1614. Henry Briggs (1561–1630), friend and disciple of Napier and first professor of mathematics at Gresham College, set them upon a base of 10. The logarithm of a given number is the number of times the base must be multiplied by itself to raise it to the given number. Thus 3 is the logarithm of 1,000 (10 × 10 × 10) in the common or Briggs system. William Oughtred (1574–1660), clergyman in the English Church, a "pittyful preacher" but very much of a mathematician, applied logarithms about 1622 to the invention of a mechanically crude slide rule, making it possible by mechanical addition and subtraction to multiply and divide. Oughtred also taught Christopher Wren, architect of St. Paul's, and wrote books which Isaac Newton studied. Both Pascal and Leibniz constructed calculating machines.

Engineers today plan swiftly and effectively because they know how to measure and calculate, thanks in large part to the scientists of this period. Distance and weight had been measured from at least 3000 B.C. Dependably accurate measurement of time began with Huygens and his pendulum clock and with Hooke and his watch spring. The science of the strength of materials dates from Galileo and Mariotte. Stevin showed how to add forces. The measurement of water and air pressures goes back to Stevin, Torricelli, and Boyle. Among very many others, Napier, Briggs, Oughtred, Descartes, Pascal, Newton, Leibniz, all contributed to the intricate system of calculations upon which modern engineering depends.

These residents of Europe, citizens of the Western world, lived in frequent and increasing contact with one another. They intermingled and corresponded freely. Their ideas interchanged rapidly, cross-fertilizing one another, often rousing bitterness, but germinating still newer ideas in controversy. They formed learned societies after the fashion of Plato's Academy, the Lyceum of Aristotle, and the Museum at Alexandria. Leonardo is said to have founded an academy. Giovanni Battista della Porta started one at Naples in 1560, named the Academy of the Secrets of Nature, but he closed it under clerical suspicion. The Academy of the Lynx, so named for the animal presumed to see much that others miss, was established in 1603, and although once closed, still exists. Its membership included Della Porta and Galileo. Its plan was to place "non-monastic monasteries in the four quarters of the globe." The Royal Society of London, growing out of informal and secret meetings, obtained its charter in 1662. Among its members with Boyle, Hooke, and Newton, were Huygens, the microscopists Malpighi and Leeuwenhoek, Papin (a French experimenter with steam), and other foreign celebrities. The French Academy began in 1666; the Berlin Academy was founded by Leibniz in 1700.

There was no lag in printing. Books on almost every subject multiplied after Gutenberg's Bible had appeared about 1454. Prominently revealing the interest of the age, despite ecclesiastical frowns, were the publications of the scientists and their beneficiaries, the engineers. The notes on construction which the Florentine architect Léon Battista Alberti (1404–1472) had accumulated thirty years before were published in 1485. During the following year the works of Vitruvius were put into print. Relying upon Vitruvius, John Shute (fl. 1550–1570) produced in 1563 the first book in the English language on architecture. Vannoccio Biringuccio's (1480–ca. 1539) *Pirotechnia* on metallurgy was published in 1540; sixteen years later in 1556 appeared the more widely known *De re metallica* on mining and metallurgy by the German Georg Bauer or Georgius Agricola (1494–1555). Both of these publications were posthumous.

Stevin's work on statics came in 1585, Gilbert's on magnetism in 1600, Napier's and Snell's writings in 1614 and 1617. Descartes published his discourse on reasoning at Leyden in 1637, and Galileo's *Dialogues* were also printed there in 1638. The works of Boyle, Pascal, Huygens, Leibniz, and Newton followed in succession to 1687. Only the notes and drawings of Leonardo da Vinci had failed to get into print near his own time. Meanwhile, there were coming off the presses illustrated books on machinery, some fanciful, others practical. Among them were Münster's *Cosmographiae universalis*, 1550; Besson's *Theatrum instrumentorum et machinarum*, 1578; Ramelli's *Le diverse et artificiose machine*, 1588; Veranzio's *Machinae novae*, 1595; Zonca's *Nova teatro di machine*, 1607; and Branca's *Le Macchine*, 1629. Few of these books have been reproduced in modern times, but many in their original editions may be seen in the world's great libraries.

Buildings and Bridges

Changes in thought are slow to penetrate daily habits. Architects and engineers throughout the sixteenth and seventeenth centuries continued to use familiar ideas and practices. Though contemporary with Stevin and Galileo, Domenico Fontana (1543–1607), eminent at the age of forty-two in a noted family of Italian engineers, skilfully applied methods from past experience rather than attempting innovations. The story of Fontana's celebrated removal of Caligula's obelisk from the Circus of Nero, where Christian martyrs had died, to the plaza before St. Peter's (Figure 6.1), some 800 feet distant, gives insight into methods of the Italian engineers and information in detail regarding procedures during the Renaissance. As in the past, the engineering of the time was involved with politics and religion.

Fontana proposed to raise the shaft vertically off its base, lay it flat on a cradle, draw it to the new site, lift it again to a vertical position, and lower it into place on its new pedestal. He received authority in October, 1585, to commandeer workmen and equipment, to make use of timber, to gather food, to decide what houses should be torn down in clearing a path from the Circus to the Piazza. All of these requisitions were subject, of course, to compensation, but no one could refuse to sell to Fontana or could molest him and his workmen without incurring fines and the displeasure of the Pope. Any one who should "speak, spit, or make any loud noise" during the actual moving would be liable to punishment by death. Nobody appears, however, to have suffered the extreme penalty.

The account written by Fontana in 1590, which has been preserved in

Figure 6.1 Fontana moving the Caligula obelisk in 1586 (Courtesy Biblioteca Vaticana)

full, explains that he had designed a very heavy rope especially for the task and estimated carefully beforehand the stress which it would have to withstand; he had distributed the load among his 40 capstans and their tackle in proportion to their tensile strength, which also he had determined in advance. He had provided 50-foot levers capable of lifting and holding the entire weight of 375 tons so that the hoisting tackle would not have to remain under stress, and he had wedges ready to be driven under the obelisk as it rose. He had not taken adequately into account the shearing stress which came upon the metal bands about the shaft nor anticipated correctly the tension upon the bolts and eyes of the lifting rods. Sooner or later all of these had to be reinforced by rope lashings.

For his derrick, Fontana built a tower around the shaft. Each of the four corner posts, 40 inches square, was constructed of bolted timbers so that the tower could be dismantled and taken to the new site. The posts were stepped into a heavy timber platform and raised to clear the top of the shaft. They were braced and trussed to prevent buckling and further stayed with eight shrouds from the top of the tower. To support the full length of the shaft as it was laid flat and then up-ended again, Fontana provided at one side of the derrick a movable strut capable of sustaining the whole weight and under control by tackle at its feet. From the location in the Circus to the top of the new pedestal in the Piazza, there was a decline of only 4 feet, so that between the sites Fontana could have a practically level roadway or viaduct which he made of earth in a wooden crib.

By April 30, 1586, two hours before dawn, all was ready for the first move. The 907 men and 75 horses were at the 40 capstans, and the 5 levers and the wedges were held in reserve; food had been prepared, the crowd kept back, confessions made, blessings given, two masses said, and God invoked as Fontana and all his workmen knelt in prayer. Then from his platform overlooking all, Fontana gave the word, and the trumpeteer sounded the blast. Men and horses moved, ropes strained, the tower creaked, and according to Fontana the earth trembled as the obelisk came up nearly to the true vertical position (it was leaning a bit). The bell rang. All stopped, Fontana inspected the work, found the top iron band broken and made it secure with rope lashings. By four in the afternoon, after a dozen lifts, the shaft had been raised the 2 feet above the pedestal necessary for placing the cradle. A cannon announced this to the city, and its batteries responded with a salute. There were eight days of further preparation, another of lowering, and the shaft lay horizontal upon the cradle, ready to be conveyed along the viaduct to the Piazza.

Months passed while the new foundations were being prepared, the pedestal removed to the new site, and the apparatus assembled again. The subsoil of the Piazza had proved unsatisfactory. So Fontana excavated an area 45 feet square to the depth of 24 feet and drove 9-inch-diameter oaken piles down 18 feet farther. These he closely spaced and capped with a chestnut floor to resist moisture. Upon this base he laid a concrete bed of broken stone, brick, and pozzuolana cement.

By September 10, 1586, all arrangements had been made in the Piazza. The crowd was as great, the auspices as well taken. This time Fontana used 800 men and 140 horses at the 40 capstans. In thirteen hours with 52 moves, they brought upright and intact Caligula's pagan trophy of Roman conquest where it now stands before the Christian Church of St. Peter. Thirteen centuries preceding the birth of Christ, it had been the monument of the Egyptian Meneptah to himself or to his deities in gratitude for the recovery of his sight. A stranger mixture of religious symbol and human foible will be hard to find in history. The contemporary accounts of Fontana's moving the obelisk indicate clearly and in much detail the methods used in heavy construction in sixteenth-century Italy. It must be remembered, however, that in A.D. 41, more than fifteen centuries earlier, this same obelisk had been moved by Roman engineers from its original site in Egypt, where it had stood for ten centuries, to the Nile, loaded on a barge, conveyed through the Mediterranean Sea to the coast of Italy, unloaded, transported overland to Rome, and set up in the Circus of Nero where Fontana found it—all this without damage. There are practically no records to indicate how this was accomplished.

The architects and engineers of the Renaissance were willing to adorn and embellish but were satisfied to leave the fundamentals of construction much as they had been. There had been no developments in the technique of building that might be called significant changes in engineering, except for the renewed use of the full dome upon pendentives, as had been done in the Church of Santa Sophia at Constantinople, and the example set in 1436 by Filippo Brunelleschi's two concentric domes with cross-bracing which, with the same amount of material as in one shell, had given greater stiffness to the dome of Santa Maria del Fiore in Florence. Fontana with Giacomo della Porta (1541–1604) had completed the dome of St. Peter's, which Bramante had designed and Michelangelo had begun magnificently but left unfinished at his death in 1564. There was much variation in form. Growing tired of the Gothic, architects returned to Rome for inspiration. Brunelleschi (1377–1446) and Donatello (1386–1466) had taken the Pantheon as their model for Santa Maria del Fiore. Others followed the example, and the architecture of the Renaissance differed accordingly from both Romanesque and Gothic. The curve of variation rose from the simple and direct toward the elaborate and involved as designing in the Renaissance became excessively ornate. Then it turned back to the simple and restrained. Thus came the baroque, or "grotesque," period in the latter part of the sixteenth century, followed by a general revival of classic architecture.

Engineers sometimes have more than they realize in common with artists. This was certainly true of those who built the Taj Mahal in 1630 for Shah Jahan at Agra, India. Its pointed dome rises 200 feet, some 20 feet higher than Santa Sophia. It does not rest, however, upon pendentives; it is a true arch set on a square base. The outward curve from its base was obtained by corbeling. Other structural details are as simple and as ancient in origin, but the most distinctive feature is the consummate skill with which simplicity of engineering detail was coordinated with intricate and ornate design. Like the Gothic cathedral, the Taj Mahal revealed in material form the inner thoughts and hopes of men. It was the ultimate expression of Shah Jahan's love for his favorite wife, Mumtaz Mahal. It was also the artistic response of its builders to the expectations which they were to satisfy.

Much the same aesthetic demands were made of the bridgebuilders during the two hundred years of the Renaissance. Many structures varying in design were built upon the established principles of arched masonry. Among these were the Ponte della Trinità in Florence, the Ponte di Rialto in Venice, the Pont Neuf of Paris, and the Rote Sachsenhaüser Brücke at Frankfurt. Quite a number are still in use. But there were new ideas about bridges. There was experimenting. Andrea Palladio (1518–1580), architect famous for reviving the classic style and creating the Palladian window, was apparently the first to introduce the truss into modern bridgebuilding (Figure 6.2). He said that he had heard of a truss bridge in Germany; whether it actually existed is not known. Engineers had long understood that a triangle cannot be deformed without changing the length of at least one of its sides and had used its stability to support roofs. Palladio's work, originally published in 1570 at Venice, illustrated four wooden truss bridges which, regardless of his modesty, he himself most likely designed. Of the one across the Cismone River near the border of Italy and Germany, he wrote that it was constructed with-

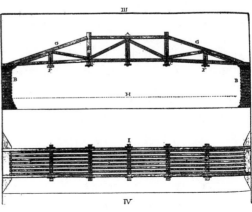

Figure 6.2
Palladio bridge truss (From
*Four Books of Andrea
Palladio's Architecture,*
1736)

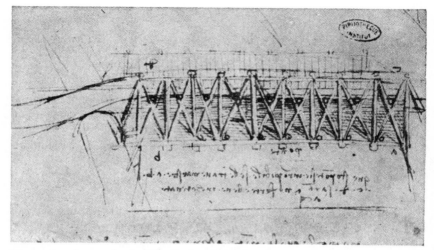

Figure 6.3 Leonardo da Vinci's drawing of a truss (From his *Manoscritti e i disegni,* 1941; courtesy La Libreria dello Stato, Rome)

out posts in the water to avoid the violence of the current and the shock of the stones and trees brought down from the mountains. Bridges so constructed, he said, were strong "because all their parts mutually support each other." [5]

Palladio's bridges were a distinct innovation. They were built of short pieces framed together to make long spans. They did not need the heavy abutments which were necessary to withstand the thrust of arches. It has been declared that Palladio erred in leaving the central panels with only one diagonal brace; the stress there might have become too great as loads moved across the structure. On the other hand, it is to be said that the second diagonals, or counterbraces, are not necessary where single (uncrossed) diagonals in a panel can carry both compression and tension. Although Palladio may not have mastered the theory, he deserves nonetheless full credit for inventing and constructing the truss bridge; his bridges stood up in service reasonably well.

Many years before Palladio, Leonardo da Vinci had apparently interested himself in applying to bridges the wooden-truss principle which had long been common in building construction. A sketch in one of his notebooks (Figure 6.3) clearly indicates a wooden truss bridge of relatively modern appearance with the inscription "this bridge is unbreakable

[5] Andrea Palladio, *The Four Books of Andrea Palladio's Architecture,* 4 vols., Isaac Ware, London, 1738, Book 3, p. 65.

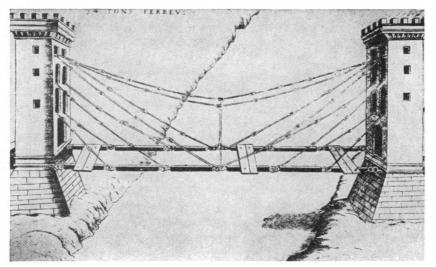

Figure 6.4 F. Veranzio's drawing of a Chinese suspension bridge (From his *Machinae novae*, 1595)

if the master beams *ab*, *cd* are strong and well bonded." [6] In 1595 and again in 1617 Fausto Veranzio (1551–1617), Dalmatian bishop, illustrated in print his conception of a truss bridge made with metal rods and eyebars. It was never constructed, and truss bridges of wood or iron, excepting Palladio's, were apparently not built for many years to come. Veranzio also planned a suspension bridge with eyebar chains and level floor (Figure 6.4).

François Blondel (1618–1686), French mathematician and engineer, improved upon the Roman method of founding bridges in 1665. His construction for protecting bridge foundations from being undermined was unique; it was a kind of false river bottom. The Roman structure at Saintes was in ruins because the Charente River had undermined the piles. Blondel made trial borings with socketed iron rods and had to go down, he said, 66 feet before reaching good clay. So he used cofferdams to hold out the water and placed a grillage of oak beams in squares from bank to bank 7 feet below the bed of the river. He filled these squares with stone laid in mortar. Then he bolted platforms of oak to the grillage and built piers of masonry 5 feet high upon them. The current did not wash away this substructure, and the bridge stood until it was taken down

[6] Leonardo da Vinci, *I Manoscritti e i disegni*, 5 vols., La Libreria dello Stato, Rome, 1949, vol. 5, p. 43.

in 1845. Blondel's method was too cumbersome and costly to become general practice. François Romain (1646–1735), a Dominican priest, native of Ghent, improved greatly upon it when he laid foundations for the Pont Royal at Paris in 1685. Romain employed the open caisson still favored for some construction under water. He built a chest of heavy timbers and laid the masonry of his pier inside it. As the pier rose, its increasing weight forced the caisson down into the soil to hard bottom. Romain appears before this to have used dredging for pier foundations at Maastricht in Holland.

Canals and Harbors

Italian, Flemish, Dutch, and French engineers at this time were digging canals with an earnestness that is best ascribed to patriotism. Their sites were strategic, their economic purposes national in scope. Netherlanders struggled to keep open the way to the sea for Bruges, once the port of all northern Europe, as clay filled the Zwyn River. The Sluys Canal became useless; they tried in 1622 to reach deep water at Ostend, then in 1640 by way of Dunkerque. But the decline of Bruges could not be stopped. The port for the Lowlands developed on the Scheldt River after the Willebroeck Canal gave Brussels access to the sea by way of Antwerp. Begun in 1531 and finished in 1561, 17 miles long, this canal was made large enough for small coasting vessels. To overcome a rise of 40 or 50 feet, it had five locks. Mitered gates, held tight by the pressure of the water, were in full use by this time, moved either by manpower or by animals.

Adam de Crapponne (ca. 1525–1576), who won fame in engineering for France, was born shortly after Leonardo da Vinci came to the end of his career in the service of François I. Crapponne enlarged the irrigation system in the southern part of the nation. Tapping the Durance River north of Marseille in 1554, he drew its water to the west and south 40 miles into what had been the desolate Plaine de la Crau. He planned many other canals for Henri II and was the first, it is thought, to propose the Canal du Centre joining the Loire with the Saône and forming today a major link in the national system of inland navigation. Crapponne may even have directed the early operations. The canal was not completed, however, until 1792, after Émiland-Marie Gauthey (1732–1806) had put nine years' work upon it with a large detachment of the French Army.

Crapponne died in 1576, poisoned—it has been both asserted and denied—by contractors whose work he had condemned.

Another in the French system of navigable canals joins the Loire with the Seine just south of Paris. This, known as the Briare Canal, was begun in 1605 under the eye of Sully, zealous minister for Henri IV, and finished in 1642 at the direction of Cardinal Richelieu—both of them national statesmen whom the French people will esteem long after they have forgotten their kings. The Briare was the first canal in France to have locks, introduced by François Andréossi (1633–1688), who had spent some years in Italy, and it was the first in the Western world to cross a watershed. Marco Polo had seen the Grand Canal of China in 1290, "a wide and deep channel dug between stream and stream, between lake and lake, forming as it were a great river on which large vessels can ply." [7] Without doubt, it crossed more than one divide between Hangchow on the Yangtze River and Tientsin on the Hai.

Colbert, another farsighted statesman for France, persuaded Louis XIV to construct the third arterial waterway of the nation. The Canal du Midi, or Languedoc Canal, from the Garonne River at Toulouse to the Mediterranean, crossed the heights near the old fortified town of Carcassonne. Louis was to pay half and the province of Languedoc the other half. Estimates of the original cost have varied from 2½ to 6 million dollars. The engineer and promoter, Pierre-Paul de Riquet (1604–1680), who owned lands along the way, was to have title to the canal. But he died in 1680, the year before it was finished, and his family received no dividends until 1724. Andréossi also had much to do with planning and building this canal, but he seems to have had no share in the ownership. The inhabitants of ancient Carcassonne resented the intrusion of this public utility, privately owned, perhaps because it was to be so controlled, perhaps also because of the taxes which they had to pay for its construction, and so the original line of the canal was diverted to avoid the town. Today it runs through the northern outskirts.

As first constructed, this Canal du Midi rose 200 feet in 24 miles to its summit near Carcassonne and then dropped gradually 620 feet to the Mediterranean at Sète. There were 100 locks in the 148 miles of Riquet's day. Several aqueducts carried the canal over intervening streams. The tunnel at Malpas, 500 feet long, was the first in which gunpowder was

[7] Marco Polo, *The Book of Ser Marco Polo*, 2 vols., Charles Scribner's Sons, New York, 1926, vol. 2, p. 174.

used extensively for blasting. The reservoir at St. Feriol excited attention in England as "among the Wonders of the World, both for the Contrivance of its admirable Structure, and for the prodigious quantity of Water, it is to contain." It was 7,200 feet long, 3,600 wide, and 132 deep, supplying for the canal "every minute of an hour, for six moneths of the Year, more than eight Cubic feet of water." The English also learned that it required an "incredible quantity of Earth, which 2000 Woemen daily carry" to build the great wall of the reservoir! [8] In later years the Languedoc Canal was extended down the Garonne to Bordeaux and lengthened near the Mediterranean. It thus became an inland waterway of more than 300 miles across southern France, and French vessels up to 100 tons could avoid the long sea voyage around the Spanish peninsula, its peril of fogs, Barbary corsairs, and English men-of-war.

In the midst of the Thirty Years' War, Gustavus Adolphus of Sweden took time for thought of something besides crushing Wallenstein and enlarging his own kingdom. He was personally interested in building the Arboga Canal west of Stockholm to give central Sweden a waterway to the Baltic Sea. It was designed to have 12 locks. The King's letter from camp at Wittenberg in 1631 directed his Minister Palatin to see to it that the peasants worked diligently and, if the engineer did not understand locks, Gustavus would have "one or two from Holland to look at it." He would "pay their travelling expenses, and something for their trouble, so that the work may become lasting, and posterity may reap the benefit and use of it." [9] Other canals in Sweden, built under the supervision of Thomas Telford, born in Scotland, were later to connect the North Sea at Göteborg with the Baltic at Stockholm, and thus to provide a national system of waterways free from Danish tolls and other foreign inconveniences.

The youthful Peter the Great, too, had the national fervor of the 1600s. After his stay in Holland and visit to England in 1698, he summoned the young British engineer John Perry (1670–1733). Perry's job in Russia was no less than to build a grand system of inland navigation from Peter's new capital, St. Petersburg, by way of Lake Ladoga and its streams, to the upper Volga and so to the Caspian Sea. And then he was

---

[8] "Of the Conjunction of the Two Seas, the Ocean and the Mediterranean, by a Channel, Cut Out through Languedoc in France," *Philosophical Transactions of the Royal Society*, vol. 4, p. 1127, Feb. 17, 1669–70.

[9] "Navigation Inland," *The Edinburgh Encyclopaedia*, 18 vols., Joseph and Edward Parker, Philadelphia, 1832, vol. 14, p. 274.

to link the ancient capital, Moscow, with the Don River and the Black Sea. John Perry expected his dams to stand "so long as the World endures," but he was in trouble from the start. Local authority balked at Peter's innovations just as the monks of Salzburg had muttered over Charlemagne's canal. "God had made the Rivers to go one way . . . it was Presumption in Man to think to turn them another." [10] John Perry had to flee from Peter's Russia with no pay for his services. Peter had the canal finished on the route from Ladoga to the Volga regardless of local feeling, but the watershed between Moscow and the Black Sea was not crossed by canal until recently. Work on the Volga-Don Canal began in 1939.

Britain, like Holland, was sending abroad engineers trained in canal building. Unlike the Netherlands, they were constructing few canals at home; there was little need at the time. The Romans in their day had dug a few ditches in Britain, and during the sixteenth century a small canal with sluices had restored Exeter's connection with tidewater. The famous Bridgewater Canal was not to come until after 1700. The development of deepwater harbors was more important to Englishmen, who were taking command of the sea. Bristol had set an example in 1239. The need for greater harbors increased with the tonnage and draught of ocean-going vessels as the Atlantic nations of Portugal, Spain, France, and England outstripped the Mediterranean cities of Venice, Milan, Florence, and Genoa and fought among themselves for new continents. The Elizabethan ports of Plymouth, Southampton, Portsmouth, London, and Bristol —whence sailed Hawkins and Drake, Sir Walter Raleigh, Captain John Smith, Winthrop and Bradford—all kept pace with the needs of shipping on the high seas.

The effort of France to compete with England, however, has drawn particular attention to the harbor engineering of this period. François I, rival of Elizabeth's father, Henry VIII, wanted a seaport at the mouth of the Seine. Rouen was too far inland. Honfleur and Harfleur were small; besides, Harfleur was filling with sand from the tides and the Lézarde River. A site 5 miles west on the north bank of the estuary and at the mouth of the little Grâce River lay on the open sea, but it might be enclosed and protected from storms. Work at Le Havre began in 1516. François' engineer, Guyon Le Roy, died in 1533, but when his original plan, extended somewhat, was completed at the end of the century, the

10 John Perry, *The State of Russia under the Present Czar*, Benjamin Tooke, London, 1716, pp. 7, 10.

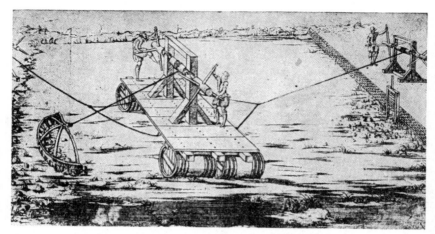

Figure 6.5 Dredging in the late sixteenth century (From Jacques Besson, *Théâtre des instruments,* 1596)

Lézarde River had been brought 4 miles from Harfleur in a 60-foot-wide canal to a new outlet at Le Havre; there was a coordinated system of jetties, shore protection, lock gates, and basin that overcame a tidal range of 25 feet. The shoaling which had occurred was effectively checked by gates on the river entering the harbor; they could be closed at high tide and reopened at low to let the rush of water scour the bottom of the harbor. Jetties of stone 25 feet high, laid in cement and held with iron clamps, reached into the sea over 100 feet and spread apart 200 at the entrance. When London installed its wet dock in 1660, Le Havre responded in 1667 with another.

Cofferdams, kept fairly dry by pumping, were used for the work at Le Havre, even for excavating, as the tides were too much for the dredges of those days. Leonardo da Vinci, prolific conjurer of visions, had turned his thoughts from the smile of Mona Lisa to dredging machines of many forms. It took an Englishman, however, to make one actually work; John Gilbert was in 1618 apparently the first to obtain a patent on a dredger—his "water plough" for deepening the Thames River. But dredges of several kinds, scoops or spoons operated by hand, had long been in use throughout Europe. The clamshell type, designed to close, as its name implies, and so common today, was illustrated in a book on machines published in 1596 (Figure 6.5). With modern power and grappling devices, it is better for deep water than the scoop. In the seventeenth century, when operated by hand, the clamshell dredge was slow and clumsy, and it was practically useless in water more than 6 feet deep as at Le Havre. Probably the machine which the traveler of 1663, Balthasar de Monconys, saw at work on the lower Rhine was more effective. This was an iron chain of buckets rolling over wheels down to the river bottom between two boats and operated by manpower. Before the end of the seventeenth century, there were many other types of dredges powered by men, horses, and even by the current of the stream.

Municipal Engineering

Cities and towns continued as in medieval times to have their own traits, their distinction, and their aloofness from one another. But their maintenance and improvement were becoming matters of national concern. On June 16, 1510, Louis XII, though aiming at Paris, ordered everyone in the cities of France to lay a pavement before his house and keep it in good repair. Specifications even for the sizes of the blocks and the method of setting them were beginning to appear in royal ordinances. François I issued precise regulations in 1540 for their enforcement and threatened to punish severely those who scamped the work.

François and his engineer Le Roy fully appreciated that jetties and wharves, tidal basins, lock gates, and control of silting alone could not make a national port of Le Havre. There had to be people with businesses and homes; conditions would have to be attractive, or they would not come to the new town. The King proclaimed on October 8, 1517, that for ten years he would exempt from royal taxes all who lived within the district; he would grant also free salt for their fishing and for their personal use. When they did not come as he wished, he extended these privileges three years later "irrevocably and in perpetuity." The King and Le Roy appreciated also that there had to be a good water supply. For the sum of 3,000 livres, Le Roy undertook under separate contract to bring water in clay pipes from a spring 3 miles distant. Porters would then carry it from the fountain to the dwellings. As Le Havre grew, François I had an Italian engineer lay out a new residential district in 1541 and still another within two years.

There were in those days few attempts to keep streets clean and lighted. The effort met with little cooperation from householders. Much greater interest was taken in municipal water supplies. People usually can be expected to think of their own wants before relieving or accommodating their neighbors. The water system of Augsburg in Germany stirred

Figure 6.6 Toledo waterworks pumps (From Ramelli, *Le Diverse et arti-ficiose machine*, 1588)

the curiosity and envy of visitors as early as 1548. Le Roy's pipeline at Le Havre delivered water by gravity like the Roman aqueducts, but the citizens of Augsburg drew their supply up from the river. Their water towers rose to 130 feet above the stream. It is not clear how the water was raised to them in 1548. One conjecture has been that there were undershot wheels in the stream to turn a series of seven Archimedean screws, or spiral pumps, and these lifted the water to the towers. In 1705, the traveler De Blainville saw mills in the river working pumps night and day to force water through lead pipes up to the towers. They held so large a supply that all public fountains were kept flowing, and a thousand houses besides, he said, received 120 "pretty large measures" [11] every hour. What constituted such a measure can only be surmised.

Twenty years after the Augsburg system was installed, Toledo in Spain had a method of raising water that has mystified engineers, not so much to learn how it worked as to discover if it worked at all (Figure 6.6). The ancient aqueduct went to ruin following the expulsion of the Moors in 1502, and the inhabitants had to haul water by donkey from the

[11] De Blainville, *Travels through Holland, Germany, Switzerland, but Especially Italy*, 3 vols., John Noon and Joseph Noon, London, 1757, vol. 1, p. 250.

Tagus River, half a mile or so from the city. The climb of 300 feet to the Alcazar Palace, however, was too much, and so Juanelo (Giovanni) Turriano (ca. 1501–1575), Italian clockmaker, devised a wondrous mechanism. An undershot wheel in the stream of the Tagus was supposed to raise the water to the loading trough and to operate the cogwheels, rods, and rocker arms of the apparatus. The idea was to lift on its rocker arm each T-shaped scoop in succession and transfer the water into the next. So the water was to go on up in the series of scoops and rocker arms to Alcazar. How much water was lost en route will never be known. Juanelo Turriano's ingenious device did not work for long, if ever. The poor folk of Toledo were soon again pushing and hauling their water donkeys.

There is no such doubt about either the effectiveness or the mechanical details of the municipal water system which Peter Morice built for London in 1582 when Elizabeth was Queen. Whether he was originally German or Netherlander does not matter; Morice was skilled in pumping water to heights from low-lying sources. He knew how to make use of the power of the tides that were sweeping through the narrow arches of London Bridge (Figure 6.7). His undershot wheels generated more than 100 horsepower and ran piston pumps that raised some 4 million gallons

Figure 6.7
Part of the London Bridge waterworks (From R. Thompson, *Chronicles of London Bridge*, 1839)

daily through a 12-inch main to the height of 128 feet. As in Augsburg, the force of gravity then took the water from reservoirs through lead pipes to the dwellings. Morice's "most artificiall Forcier" gave Londoners a thrill by raising water above the steeple of St. Magnus Church to "the highest ground of all the Citie." [12] But his water supply was not sufficient to stop the great fire of 1666, just as modern supplies have proved inadequate in Chicago, Boston, San Francisco, and elsewhere. For his services to the city of London, Morice and his heirs were granted for five hundred years, at an annual rent of but 10 shillings, the pumping right, the use of five arches of the bridge, and the necessary land. His successors enjoyed the privileges and obligations until an Act of Parliament removed the waterworks in 1822. How many fortunes came from this famous lease, or were lost in fulfilling it, would be interesting historical data for either side of the argument regarding the use of private property in public services.

Hugh Myddelton (ca. 1560–1631), wealthy goldsmith and friend of Sir Walter Raleigh, spent his fortune from 1608 to 1613 on an aqueduct to bring a purer supply of 13 million gallons daily to London. For a half interest James I advanced more than half the total cost of nearly $2,500,000. The water came from the springs near Ware in Hertfordshire, 21 miles to the north, under 60 culverts and 160 bridges in a canal 18 feet wide and 5 deep, wandering through the countryside some 40 miles on the way with an average fall of but 3 inches to the mile. A contemporary who "diligently observed" and admired the construction recorded that "the depth of the Trench (in some places) descended full thirty foot, if not more; whereas (in other places) it required as sprightful Art againe, to mount it over a valley in a Trough, between a couple of hils, and the Trough all the while borne up by Woodden Arches, some of them fixed in the ground very deepe, and rising in heighth above 23. foot." [13]

The reservoir of Myddelton's New River supply was placed at Islington, and from there 58 pipes of elm or lead were laid into the city. Houses renting for 15 or 20 pounds a year had smaller lead pipes brought directly into them. Tenements in courts and alleys were provided generally with a common pump. For the moment, probably, the chronicler Stow was accurate in exclaiming that there was "never a city in the world so well served with water." But many Londoners had still to rely upon local wells often contaminated by seepage from outhouses and refuse

[12] John Stow, *The Survey of London*, Nicholas Bourne, London, 1633, pp. 12, 206.
[13] *Ibid.*, p. 13.

Figure 6.8 Samaritaine pumps of Lintlaer's on the Pont Neuf, Paris, 1608

in the streets. Greatly enlarged and improved, the New River supply is part of London's water system today. At one time it included 400 miles of wooden pipes. John Smeaton provided a steam engine for it in 1767, and a Watt engine was installed in 1787 to give higher pressure. Soon after 1810 the wooden pipes were entirely replaced with cast iron. Although Hugh Myddelton had spent his fortune, the value of the stock in the New River Water Company increased to fifteen times the original investment.

Paris had depended upon the old aqueducts from distant springs to supply some public fountains in nearby squares and furnish its abbeys and priories with drinking water. The Seine was notorious for its "Griping Quality." Nonetheless Henri IV, following the example of London, imported a Flemish engineer in 1608 to make greater use of the river. Jean Lintlaer placed 16-foot wheels under the Pont Neuf (Figure 6.8) to supply the King's palaces of the Louvre and the Tuileries. These pumps, known as the Samaritaine, were used until 1813 when Napoleon had them dismantled because steam pumps had been installed elsewhere on the river. In 1624 Marie de' Medici, mother of Louis XIII, thinking of pleasanter if not safer drinking water, had the Aqueduc d'Arcueil built to bring it from the springs of Rungis, 8 miles to the south. This supplied her palace and gardens of the Luxembourg with some to spare for 14 public fountains.

The showpiece of hydraulic engineering in France was the water-

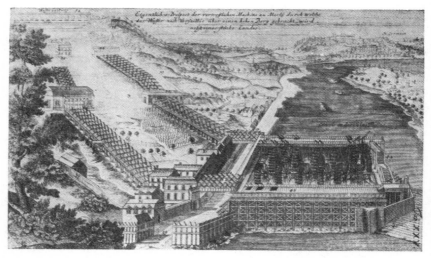

Figure 6.9 Sualem's Marly machine, supplying water to Versailles, 1682 (From J. Leupold, *Theatri machinarum hydraulicarum*, 1724–1725)

works of Marly (Figure 6.9) built for Louis XIV in 1682 by the Flemish engineer Rennequin Sualem (1645–1708). It was to supply water for the gardens at Versailles. There above 80 million French livres, over 4 million pounds sterling, were spent upon machines that could raise no more water than one of the larger English "fire engines" of 1744 whose cost did not exceed £10,000. There were 225 pumps at Marly set in three steps up a ¾-mile incline from the river. Twenty-five others were operated at the Seine. Fourteen undershot wheels provided the power. They were connected with the pumps on the hillside by chains or jointed rods which clanged and clattered like "waggons loaded with bars of iron, running down a hill with axles never greased." [14] Altogether there were 500 valves in the first ⅜ mile to contribute their share of leaks. The cost of maintenance and repair must have been high; an English estimate in 1749 placed the annual upkeep at £25,000. The efficiency was low; 95 per cent of the power, it is said, was expended in moving the rods and chains. But the machine did raise water 533 feet above the Seine into the reservoir at Versailles. It remained in operation until 1804.

The works at Marly have significance in engineering beyond their complex mechanism and their inefficiency. Sualem used cast-iron pipe—so far as is known for the first time. Cast iron had been employed almost

[14] "Machine de Marly," *Knight's Penny Magazine*, vol. 4, p. 240, June 20, 1835.

exclusively for cannon. The process of making iron tubing was fairly well established, but expensive. Louis XIV, however, avoided no expense on the beauty of Versailles, whether or not there was water for public fountains; he seems to have had no doubt that France could afford cast-iron pipe for the royal gardens. Marie Antoinette, of later date, is better remembered for this attitude toward the more common people of France: if meantime they had no bread, they could eat cake. Cast-iron pipe did not appear in American construction until 1817. By then it had become so economical that it was everywhere rapidly replacing wood.

Water and Fire

Men had many ideas about the use of water as they worked with it, ideas that were impractical but nevertheless ingenious and prophetic. Before 1578 Jacques Besson had thought of extinguishing fire with a jet of water thrown by compressed air. In 1617 Jacob de Strada planned a horizontal water wheel rather like a modern turbine. Some time after 1620 the Dutch scientist Cornelis Jacobszoon Drebbel (1572–1634) is said to have made a boat which "swam under the water" of the Thames from Westminster to Greenwich. By 1629 Giovanni Branca (d. 1629) had conceived of an impulse turbine driven by steam and geared to apparatus for pounding substances (Figure 6.10). It was a fanciful device, but it was significant. Active experimenting with steam had begun. Few of these machines were ever built. For one reason, necessary tools had to be developed. For another and more important reason, nobody felt as yet any pressing need for such devices. One comes upon firmer ground when approaching the question from the standpoint of simple wants of men in their daily tasks that forced open the way of discovery of the power in fire and its greater usefulness when joined with water. The pressure of daily needs, in fact, may best indicate the ultimate cause for all departures in engineering.

Presenting accumulated knowledge in his *Pirotechnia* of 1540, Vannoccio Biringuccio stressed the great value of water to metallurgy: ". . . of all the inconveniences, shortage of water is the most to be avoided," he said, "for it is a material of the utmost importance in such work, because wheels and other ingenious machines are driven by its power and weight. It can easily raise up large and powerful bellows that give fresh force and vigor to the fires; and it causes the heaviest hammers to strike, mills to turn, and other similar things whose forces are an aid to men (as you can see), for it would be almost impossible to arrive in any

Figure 6.10
Branca's conception of a
steam turbine (From his
Le Macchine, 1629)

other way at the same desired ends because the lifting power of a wheel
is much stronger and more certain than that of a hundred men." [15]

From the experiences of metalworkers and miners, Biringuccio commented, too, upon the efficiency of labor in terms of the working day.
There should be shifts every six or eight hours of "new and rested men."
This was a suggestion far beyond Roman treatment of slaves or medieval
procedure with free labor; it is one that even yet is not being followed
everywhere. Certainly no contemporary paid any attention to him. And
he remarked that mining "rather than warfare with all its annoyances"
was the business for men who wished to have wealth. Mining was superior
even to commerce, for commerce went about "outwitting the world and
perhaps doing other tiresome things which may be illicit for honest
men." Biringuccio's descriptions in detail show that the processes of smelting and steelmaking were well developed. The medieval smith was a
skilled craftsman with his forge and hammer, his bellows and fires, bring-

[15] Vannoccio Biringuccio, *Pirotechnia of Vannoccio Biringuccio*, American Institute
of Mining and Metallurgical Engineers, New York, 1942, p. 22.

ing his metal "to whatever end he wishes to make of the work." But he was too engrossed for much else. "As you can understand," said Biringuccio, "the unhappy workmen are never able to enjoy any quiet except in the evening when they are exhausted by the laborious and long day that began for them with the first crowing of the cock. Sometimes they even fall asleep without bothering about supper." [16]

Improvements had been made in the mines of Europe since Roman times. Greater care was taken in sinking shafts, bracing and shoring galleries. There were better tools. The first railway trucks, four-wheeled cars, even on rails (Figure 9.16), had been introduced underground by 1520, increasing output and incidentally relieving human backs. Hoisting gear was often operated now by horse whims, or drums, about which the rope was wound. There were elevators, too, for the men. Mines had ventilating shafts and fans of a sort turned by men or animals. And at last in 1627, although its power had been well known since 1403 when used to blast the walls of Pisa, gunpowder was at work in the mines beside the ancient method of cracking rock with fire and water. But pumps, even if better than the Roman tympanum or cochlea, were not equal to the underground streams which poured into their galleries and drifts and often drowned the work.

Water pumps had been developed long before 1600. Vitruvius had described the pump of Ctesibius with great care, and Vitruvius's writings, printed in 1486, had been among the earliest works to appear in print. In his De re metallica of 1556, Georg Agricola described a number of pumps, suction, ball and chain, and bucket; one elaborate installation at Chemnitz lifted water 660 feet in three stages with pumps of the ball-and-chain design. It was operated by 96 horses, working four hours and resting twelve, in teams of eight to a pump. Another design illustrated in the 1589 edition of Hero's Spiritalia, had four suction pumps moved by cranks set at right angles to each other and operated by one horse. These early pictures reveal cranks and rods changing reciprocating to rotary motion, upon which patents were allowed in England two hundred years or so later. But the pumps were slow and feeble and did not meet the needs of the time. Their users could not have calculated the ratio between the amount of energy expended and the work obtained. They did know that even with teams of big Flemish horses working night and day (Figure 6.11), their pumps were not doing what they needed done in their mines.

[16] Ibid., p. 370.

Figure 6.11
A "4-horsepower" pump
for draining a mine (From
Salomon de Caus, *Les
Raisons des forces
mouvantes*, 1615)

Ctesibius and Hero, two men of Hellenistic times, will always be acclaimed for their investigations into the properties of steam. It is said that Ctesibius invented the piston and cylinder before 200 B.C., and Vitruvius attributed to him the pump in which they were used. Some of Hero's writings survived; they were translated from Greek in the sixteenth century, first into Latin by Federico Commandino in 1575 and then into Italian by Bernadino Baldi in 1589. Hero described various devices, among them machines which are now called hot-air, or steam, engines. One in particular has been honored as an early-type reaction steam turbine. There is no telling how much influence the publication of Hero's ideas had upon practical thinking. Still, by 1601 Giovanni Battista della Porta (ca. 1535–1615) had noted that a vacuum developed as steam condensed and that it could be used to draw up water. Here was the fundamental idea of the first practical steam engine of a hundred years later. In 1615,

Salomon de Caus (1576–1635) described an arrangement for heating water in a hollow sphere and making the steam rise into a tube. He wrote in some detail about a device for raising water by condensing steam, but he never made it. David Ramsay, Scotsman (d. ca. 1653), received a patent in 1630 on a device "To Raise Water from Lowe Pitts by Fire," but there is no information concerning what he had in mind or what he did with it. Bishop John Wilkins (1614–1672) published in 1648 his *Mathematicall Magick*, the first book on mechanics printed "in the vulgar tongue" of English, and in it he spoke of "concave vessels" containing water from which the air issued "with a strong and lasting violence" when they were heated. They were frequently used, he said, "for the moving of sails in a chimney corner, the motion of which sails may be applied to the turning of a spit, or the like." [17]

In 1659 there appeared from the print shop of Thomas Leach, in London, an obscure little book on *The Elements of Water Drawing* by R. D'acres. It was published, its title page declared, "for the improving the service of the Mineral World, for supplying our most necessary wants of firing, for raising of water for Cities and Towns, and for watering and draining of Grounds." It was the earliest work exclusively on the subject by an English writer, and its author was the first Englishman to describe a heat engine in detail. Possibly he was Robert Thornton (1618–1679), who lived near the Warwickshire coal mines which are known to have had trouble with water at that time. Robert D'acres, or Thornton, points out that most "Distillatory vessels" are perfect patterns of "Coelistial performings" for evaporation and that "two foot and a half and some odd inches of a tube of quicksilver, equiponderates with 32 feet of water . . . betwixt the Earth and Atmosphere."

Let him say it in his own words: "The best *heating*, is by the incensed *Air* of a close *furnace;* The speediest *Cooling* is by water. . . . For the speedier Intercourse of these two *contraries*, the one may be applyed *within side:* the other *without side* the *Cilinder*, or *Region* of the *Air;* The Cooling water may not enter, for then it necessarily frustrates the ascent of the water; the heated *Air* out of the Furnace may enter (by the turning of a Cock) into the *Boul*, and so the heat is acted in an instant; Then the *materials* and *Globe* being all overhead in a *Pond* or *Cistern* of water, they (after the heat by the returning of the Cock is diverted) do as speedily cool, and so the rarified Air condensing, the water ascends, and having a *brazen Sucker* or *Clack* in the bottom, it can not go out

[17] John Wilkins, *Mathematicall Magick*, Ric. Baldwin, London, 1691, p. 149.

again, but then by turning the Cock, the uppermost *water* issues forth, by the sucker in the spout, which now the descending water thrusts open, and in the same *Act*, the enflamed *Air* follows after; return the Cock, and the water ascends as before." [18]

More famous than D'acres, or Thornton, though surely no more entitled to credit for developing the steam engine, was his contemporary Edward Somerset (1601–1667), Marquis of Worcester. In his well-known account of his achievements, *A Century of the Names and Scantlings of Such Inventions As at Present I Can Call to Mind to Have Tried and Perfected*, published in 1663, Worcester discussed one hundred heterogeneous ideas. And he secured from Parliament in 1663 a monopoly that was to last ninety-nine years upon a "water-commanding engine." Worcester praised his own invention as "the most stupendious Work in the whole world." Samuel Sorbière, historian to the French king, appears to have supported Worcester's claim for he wrote, "One Man, by the Help of this Machine, raised Four large Buckets full of Water in an Instant, Forty Foot high, and that through a Pipe of about Eight Inches long." [19] But this performance was not above the strength of one man, and Sorbière made no specific mention of steam. The remainder of the evidence is too circumstantial and hypothetical to prove that Worcester should be hailed as the one who invented the steam engine. On a wall of Raglan Castle in Monmouthshire, residence of the Marquis for some years, there are marks that have often been supposed to be the remains of a steam pump. They are puzzling, but in view of the fact that they are apparently a structural part of the fourteenth-century wall, it is difficult to think of them as the marks of Worcester's steam engine. Moreover, the castle was not reoccupied after its burning in 1646 during the civil war.

Samuel Morland (1625–1695), "Master of Mechanicks to Charles II," later experimented in hydraulics and constructed the first steam table to show the relation between pressure and temperature. Denis Papin (1647–ca. 1712), French scientist, invented a "digestor," or what is now called a pressure cooker. He was wise enough to put a safety valve on it, the first in history. It is said that Papin also came close to the idea of the centrifugal steam pump and suggested the pneumatic caisson. He was the first to propose that a vacuum be produced under a piston by condensing

[18] R. D'acres, *The Art of Water-Drawing*, published for the Newcomen Society by W. Heffer & Sons, Ltd., Cambridge, 1930, pp. 6–7.
[19] Samuel Sorbière, *A Voyage to England*, J. Woodward, London, 1709, p. 29.

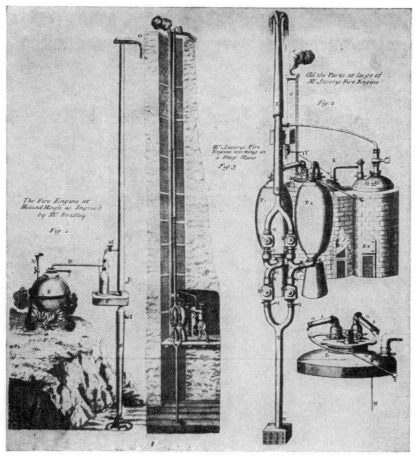

Figure 6.12 Savery's steam pump (From his *Miner's Friend* of 1702)

steam. None of these men from Della Porta to Papin, however, deserve as much credit as Thomas Savery (1650–1715) for inventing the steam engine because they did not put their ideas into effective use. Savery not only obtained in 1698 a patent upon a device for draining mines (Figure 6.12), serving towns, and working mills, but he also set up the world's first steam-engine factory in Salisbury Court, London, during 1702, and started his advertising with a book entitled *The Miner's Friend*.

Savery's invention had no moving parts other than cocks and check valves. These had to be turned rhythmically by hand to let steam from a boiler into a vessel, which then was chilled by water from the outside somewhat as D'acres, or Thornton, had suggested. For each cycle Savery poured on cold water which condensed the steam, created a not too perfect vacuum, and set in motion the suction which Della Porta had described. Thus water was drawn up the suction pipe into the vessel and driven on from there by the further introduction of steam. In spite of great losses of heat in condensing the steam, the cycle of operation could be repeated about five times in a minute. The modern pulsometer with no moving parts, patented in 1872, is still used by contractors; it is essentially an improved Savery pump.

Expensive of time and fuel though it was, and strictly limited in capacity and speed, Savery's steam engine, or pump, of 1702 had completed the pioneering. The effectiveness of the energy in the weight and mobility of water which had been used for centuries was about to be surpassed many times over. Men were now to set fire upon water and generate steam.

Bibliography

Agricola, Georg: *De re metallica*, The Mining Magazine, London, 1912.

Belidor, Bernard F. de: *Architecture hydraulique*, 4 vols., F. Didot, Paris, 1810.

Biringuccio, Vannoccio: *The Pirotechnia of Vannoccio Biringuccio*, American Institute of Mining and Metallurgical Engineers, New York, 1942.

Branca, Giovanni: *Le Macchine*, Iacomo Manuci per Iacomo Mascardi, Roma, 1629.

Butterfield, Herbert: *The Origins of Modern Science*, The Macmillan Company, New York, 1951.

Dartein, Fernand de: *Études sur les ponts en pierre*, 4 vols., C. Béranger, Paris, 1907–1912.

Duhem, Pierre M. M.: *Les Origines de la statique*, 2 vols., A. Hermann, Paris, 1905–1906.

Leonardo da Vinci: *The Notebooks of Leonardo da Vinci*, Reynal & Hitchcock, Inc., New York, 1939.

Palladio, Andrea: *The Architecture of A. Palladio*, 2 vols., A. Ward, London, 1742.

Parsons, William B.: *Engineers and Engineering in the Renaissance*, The Williams and Wilkins Company, Baltimore, 1939.

Savery, Thomas: *The Miner's Friend*, W. Clowes, London, 1827.

Timoshenko, Stephen P.: *History of Strength of Materials*, McGraw-Hill Book Company, Inc., New York, 1953.

Wolf, Abraham: *A History of Science, Technology, and Philosophy in the 16th and 17th Centuries*, new ed., George Allen & Unwin, Ltd., London, 1950.

Worcester, Edward S.: *The Century of Inventions of the Marquis of Worcester*, J. Murray, London, 1825.

SEVEN

The Industrial Revolution

Thomas Savery's engine marked in 1702 the opening of a historic period that is separate and distinct though it is variously named and its limits are shifted with differing views. Historians who stress political phase rather than economic change prefer to concentrate upon the French Revolution at the end of the century. It is the revolution governing all subsequent events; not even the American revolt from British rule takes precedence in the minds of these historians although it antedated the upheaval against the *ancien régime* in France. For those to whom literary and philosophical trends are more important, the period has quality and distinction as the eighteenth century, the century of rational thinking and scientific enlightenment. Others do not believe that the turning of the calendar from one year to another makes any great difference. It is convenient but not especially significant to characterize history by centuries; thinking and investigation do advance but not necessarily in centenary rhythm. Nor will such historians concede that philosophical trends determine rather than accompany events. To these historians, also, the practical applications of scientific discoveries and the economic consequences for society are far more important. Accordingly, they prefer to carry the period well into the nineteenth century and to call it the Industrial Revolution, if not more precisely the Age of Steam.

However named or limited, this period of history opened with so great a change in the relation of men to the resources of nature that it should be given place with the Middle Ages, the Roman Empire, the Age of Bronze, among the epochs in the course of civilization. Sharing their theories and discoveries, Britons and Frenchmen worked together to bring about this tremendous development for the material benefit of mankind. True to the irony in history, they fought one another as they did so. It was a second hundred years of war that for magnitude and spread would have amazed the contemporaries of Jeanne D'Arc. The strife this time was not merely over land and prestige in Western Europe; it was for empire the world around, in America, India, and the Far East. It did not end until the monarchy of France lay ruined and the blood of the French Republic had been shed for Napoleon from Africa and Spain to the plains before Moscow.

We do not presume that the invention of the steam engine and industrial machinery alone made it possible for the small insular kingdom to defeat again and again a continental nation of more than twice its population. It has been estimated that the numbers in Great Britain rose from 7 millions in 1660 to some 20 millions in 1820, while the population of France, approximately 19 millions in 1660, increased to 30 millions in 1820. But, together with their teeming population and their strength on the sea, the advance of the British people in industrial organization and resource had much to do with creating their empire. When they came to grips at last with Napoleon, their factories were a very great help in warding off his attempt to crush them as he endeavored to make himself supreme.

This is the period, too, in which the mercantile theory of Western nations had to give way. Though perhaps rightly accredited to the French statesman Colbert, it was advocated and applied by Englishmen, Hollanders, Spaniards, and all who would have their countries self-sufficient at home but in economic hostility toward their neighbors while they held dominions over lands and peoples beyond the sea. This monopolizing of colonial resources—while selling everything possible to neighboring countries and buying from them as little as possible, and then only what could not be produced at home—was supposed to keep a country independent, safe, and strong. In Britain, at least, the mercantile theory was obliged to make room for the doctrine of Adam Smith, published in 1776, stating that the real wealth of a nation lay not in exploiting colonial areas in a fixed trade but in a free and expanding commerce with all parts of the world.

This change of view in Britain may have come from the overexpansion of its domestic industry and the need of foreign markets to take off the surplus goods. But the American subjects of the British king had demonstrated before 1776 that the enterprises of a growing people overseas cannot readily be kept at a primitive level, certainly not by the simple decision in the governing circles of the mother country that they ought to be restrained. It was futile to insist that the colonies should continue to supply only raw materials from extractive industries and leave to the mother country the business of manufacturing goods for consumers. There were other causes of the American Revolution and the creation of the United States, not the least of which was interference with the expansion of the colonies from the seaboard into the interior of the continent.

Food had to be obtained from remote sources of supply as working people crowded into urban communities. Spinners and weavers in the cottages, and rural ironworkers too, had been able to supply themselves in part and in poor times to fall back upon subsistence farming. There was nothing now to take care of urban unemployed. There was no social legislation; there were no public agencies to provide relief. As the system of wages developed, Elizabethan poor laws succumbed to the idea of *laissez faire*, and it was complete *laissez faire*. Those who now received wages in ready cash were entirely free to handle their money as they saw fit, to look out for themselves and to suffer if they did not. But there was a catch in such freedom. There was no telling how long, if the factory were to shut down, the wage earner might have to look out for himself. It would matter little how provident he might have been in setting aside funds for ordinary slack times. Any day, without notice, because of economic depression, his wages might cease for weeks and months on end.

Statistics of population and industrial growth, of investments, payrolls, and production, of increases in consumers' goods in relation to the purchasing powers of communities, even of hardships for masses of men at any given time may perhaps be compiled. It is quite another task, however, to determine the variations between different periods of history. The lot of peasants and laborers had been improving with the introduction of mechanical aids to human strength and the use of new kinds of power. Nevertheless, humble folk were still a long way from the fabled land of milk and honey, ease, and plenty, even though the life in cottages, where besides raising and preparing their food they made candles and soap and spun and wove their own clothes, sounds picturesque. All things considered, it would be an interesting speculation as to whether those who worked in the dust and murk of factories and mines after the Industrial Revolution, though they had more consumers' goods, were better off than the peasants and day laborers in the fields and quarries of the Middle Ages. If evolution were to be made solely in terms of labor-saving machinery, the conclusion would be that the life of the working men was growing steadily better with the progress of engineering. This does not mean, however, that the margin widened between what they received for their efforts and what they paid to keep alive; it does not indicate that their real wages increased with their release from back-breaking labor. The manufactured products were more plentiful and cheaper. Shoes, stockings, clothing, and pots and pans were more readily obtained by those in humble circumstances. However, it cannot be asserted that the workers who made them were able to enjoy a larger share of the produce from their efforts than they had before.

The steam engine, as first developed by Newcomen, freed thousands of men and horses from the hard physical labor of keeping mines clear of water, but these men were not released to hours of leisure. The margin of profit for workmen in the time and place gave no such assurance; they were released from one dull job to others perhaps no more pleasurable. For the moment they may have been out of work altogether, even though the steam engine made possible reopening many mines which could not be kept clear of water without it and soon increased the demand for labor in the pits. Their descendants, and society as a whole, gained a greater profusion of goods and services. Over a period of time, despite the increase of population which accompanied the mechanization of industry and the development of medical science, particularly preventive medicine, there was an even greater demand for labor and a marked rise in the general standard of living. For this material advance, much credit is due the engineers who devised the improved instruments of production. In general, the immediate profits from the Industrial Revolution (or the Age of Steam) accrued, however, not to the workers but to the relatively few who, by accumulating and risking capital, owned the tools which the workers used.

Steam Engines

Thomas Newcomen (1663–1729), ironmonger and preacher on occasion, completed his engine in 1712 after years of experimenting. His first engine was installed at Dudley Castle in Staffordshire to pump water from a mine. Newcomen's partner, John Calley or Cawley, a plumber and glazier of Dartmouth, provided the manual skill. Newcomen's contribution was an ingenious mixture of familiar devices with ideas of his own (Figure 7.1). He took the copper kettles and furnaces of brewers to supply the steam for the same kind of cylinder and piston which had been used for lifting water as far back as Roman times and condensed the steam below the

Figure 7.1 A Newcomen engine (From J. T. Desaguliers, *A Course in Experimental Philosophy*, vol. 2, 1744)

piston within the cylinder to create there a partial vacuum which would allow the weight of the atmosphere above to force the piston down. Instead of using Savery's method of creating the vacuum by dashing cold water on the outside and thus condensing the steam, Newcomen sprayed water directly into his vertical cylinder. This condensed the steam, created the vacuum, and allowed the atmospheric pressure to force the piston down. To put this downward pull of the piston to work, he connected its rod by a chain to a horizontal and centrally pivoted working beam, or walking beam as it was later called, and fastened the other end of the beam by a chain and rod to the piston or plunger in the pump cylinder. As new steam was admitted below the first piston and the vacuum broken, this piston was raised again by the weight at the other end of the working beam, and the piston of the pump was lowered to its original position. Thus the reciprocating motion, up and down, was complete.

At the ends of the beam were circular arcs over which the chains rolled as the beam rocked and let the rods themselves rise and fall in vertical lines. To open and close his valves at the proper moments, Newcomen attached a device which controlled them from the motion of the engine. Thus he eliminated the operation of the valves by hand and increased the frequency of his power cycles, it is said, from 6 or 8 to 15 or 16 to the minute. As the story goes, Humphrey Potter, Newcomen's boy who was supposed to open and close the steam valve, attached what he called a "scoggan," so that the beam would trip the valve each time and do his work while he went fishing. It's a diverting story, but one which has no foundation. The Newcomen engine was a great improvement over the Savery engine. Newcomen's working beam was for some time an essential feature in stationary reciprocating engines. But the rods worked only under tension. The engine operated at or below atmospheric pressure, making no use of the driving force in steam. Its thermal efficiency, or the ratio between fuel consumed and power produced, was something to be more thoroughly investigated and improved by his successors.

The scope of Savery's patent was so broad that Newcomen did not try to get a patent for himself. Instead, a company or partnership took over Savery's rights about the time of his death in 1715 and controlled the manufacture of steam engines in Britain for the next eighteen years. Though details remained to be improved with experience, the popularity of the Newcomen engine for clearing mines was instant. It could be operated from the surface to lift water from far greater depths than Savery's suction pump. The diameter of the steam cylinder was necessarily considerably larger than that of the cylinder of the pump at the other end of the working beam. The atmospheric pressure applied to the larger area thus would lift a proportionately taller column of water.

Word got abroad that the British had something very good. A Newcomen engine was installed in 1722 for a mine in Hungary, and another set up at Passy in 1726 raised water from the Seine for the city of Paris. By the time of Newcomen's death in 1729 his engine was in general demand both at home and on the Continent. It had practically revolutionized the mining industry. The Swedish observer Mårten Triewald (1691–1747) wrote in 1734 that the machine in the mines at Dannemora was doing as much work "as 66 horse-whims, each drawn by 4 pairs of horses, or altogether 528 horses in 24 hours." [1]

By 1739 the colliery at Fresnes in France had a Newcomen engine with a 30-inch-diameter cylinder and a 9-foot piston stroke that lifted water at the rate of 15 strokes per minute from 90 feet below the surface. It had to work only forty-eight hours a week with very little attendance where previously 50 men and 20 horses, in shifts throughout the week, had been necessary twenty-four hours a day to keep the mine clear. The French engineer Bernard Forest de Belidor (ca. 1697–1761) was ecstatic. "It must be avowed that this is the most marvellous of all machines," he said, "and that there is not a single other of which the mechanism has so much resemblance to that of animals. Heat is the cause of its motion, a circulation takes place in its different tubes like that of blood in the veins; it has valves that open and close at the proper moment; it feeds itself, it rejects what it has used at regular intervals, it draws from its own work everything that it required for its support." [2] This was just praise at the time, but James Watt was soon to introduce changes that would make the Newcomen engine obsolete.

As it spread from mines to ore smelters, iron works, textile mills, Newcomen's engine revealed many of its own shortcomings. He himself had placed a leather disk on the piston and put a water seal above that to check leakage. But his early cylinders were not bored; they were roughcast, and although they were polished by hand, there was much friction and loss of power. Variations in diameter often caused the piston to bind or the vacuum to fail. Though cylinders of brass could be made thin, they were increasingly expensive as they grew larger with the demand for more and more power. By 1765 there were engines with cylinders 72 inches in diameter, more than 10 feet long, and weighing nearly 7 tons. Cast-iron cylinders cost much less, but they had to be made thick, too thick for the rapid changes in temperature which were necessary. For in the Newcomen process, the cylinder walls had first to be cooled to condense the contained steam and so create a vacuum. Immediately the admission of fresh steam heated them again, the complete cycle occurring within a few seconds. A very great deal of heat was wasted.

A Newcomen engine had no rotary motion; its moving parts were all

[1] Mårten Triewald, *Short Description of the Atmospheric Engine Published at Stockholm, 1734,* Courier Press, London, 1928, p. 33.
[2] Bernard F. de Belidor, *Architecture hydraulique,* 4 vols., C. A. Jombert, Paris, 1737–1770, vol. 2, pp. 324–325.

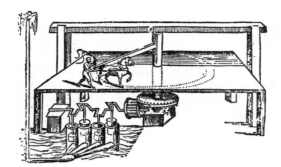

Figure 7.2
Four-cylinder pump show-
ing rotary to reciprocating
motion (From *Transac-
tions, Newcomen Society*,
vol. 16; original in Hero,
Spiritalia, 1589; courtesy
Newcomen Society)

reciprocating. There was need for rotating shafts to do millwork, grinding, turning of lathes, raising of coal and ore by whim or windlass. Cranks
had long been used to change rotary into reciprocating motions in pumps
driven by water power and other machines operated by hand (Figure 7.2),
foot, or animals. Reciprocating motion, produced by hand or foot power,
had also been changed by cranks into rotary motion for making pottery,
milling flour (Figure 7.4), kneading dough, etc., but neither Savery,
Newcomen, nor any others adapted the crank and flywheel to the steam
engine for the purpose of changing its reciprocating into rotary motion.
To get rotary motion they were in some instances actually using the
reciprocating engine to pump water into overhead tanks which then
poured their contents down upon overshot water wheels!

In spite of its advance over its predecessors, the Newcomen engine
was slow and inefficient. As has been said, it operated at atmospheric
pressure or less. It used steam only to create by condensation a vacuum
which allowed the atmosphere to press down upon the piston. Its working beam was balanced to rock again as the vacuum broke and steam re-
entered the cylinder. The expansive force in the steam was not used to
push or pull anything. The engine consumed a huge amount of coal in
comparison with the work accomplished; critics declared that it took an
iron mine to build a Newcomen engine and a coal mine to keep one
going. The wastage of fuel was over 99 per cent, and yet Thomas Newcomen had opened vistas into mining, manufacturing, and transportation
for the engineers who came after him.

John Smeaton (1724–1792), five years old when Newcomen died in
1729, made a working model of the fire engine as a boy and became so
absorbed in mechanical things that he abandoned his father's profession
of the law. Smeaton set out to be an instrument maker, but at the age
of twenty-nine had established himself with wide and varied interests in

engineering. In fact, he was the first to call himself a civil, as distinguished from a military, engineer. Though more famous in his time for his canals, the Eddystone Lighthouse, his investigations of cement, and his experiments with water power, Smeaton built a number of Newcomen engines at the Carron ironworks in Scotland and greatly increased their mechanical efficiency by more precise cylinder boring, a better proportioning of parts, and general improvement in shopwork. He attained, it is said, the highest performance of which this type of engine was capable in practice. One of his largest engines went to Russia in 1775 to pump out Catherine II's dry docks at Kronstadt. It replaced two enormous windmills, 100 feet high, which Dutch engineers had installed in 1719. It was said that these mills had required a year for the task; the Newcomen engine built by Smeaton did the job in about two weeks.

Within forty years of Newcomen's death, James Watt (1736–1819) made changes so fundamental and important that he, together with Savery and Newcomen, is also given credit as the originator of the steam engine. Like John Smeaton, Watt started as an instrument maker and went on to be a practicing engineer. His big idea came to him while he was assisting in the laboratory of Glasgow University where the Scottish scientist Joseph Black (1728–1799) was lecturing on heat.[3] Watt noted the waste of heat between strokes in the model of the Newcomen engine, which he was repairing, because the walls of the cylinder had to be cooled and reheated with each cycle. He continued to use steam at atmospheric pressure and a partial vacuum, but he provided a separate chamber for condensing the steam, connected with, but apart from, the cylinder around which he put a steam jacket to keep its walls hot. Thus he saved three-fourths of the fuel which Newcomen had required.

Between 1765, when he invented the separate condenser, and 1769, when he obtained his first patent, Watt made two other important improvements in his engine. He added an air pump to maintain a vacuum in his condenser by pumping out its contents of water, condensed steam, and air. He also closed the open upper end of the cylinder, built around the piston rod what today is called a stuffing box, and introduced steam rather than air to push the piston down. In his 1769 patent he said, "I intend in many cases to employ the expansive force of steam to press on the pistons, or whatever may be used instead of them, in the same manner in which the pressure of the atmosphere is now employed in

[3] Henry W. Dickinson, *A Short History of the Steam Engine*, The University Press, Cambridge, 1939, pp. 68–69.

Figure 7.3 An 8-horsepower double-acting engine of the Watt type (From M. A. Alderson, *An Essay on the Nature and Application of Steam,* 1834)

common fire-engines," or Newcomen engines. The four novel ideas in this patent are the separate condenser, the heat jacket for the cylinder, the air pump, and his use of the expansive force of steam.

Like Newcomen's engine, this Watt engine was single-acting and useful only for pumping. In 1782, however, Watt patented a double-acting engine (Figure 7.3) in which steam and vacuum were applied to opposite sides of the piston alternately. This new engine subjected its piston rod to compression as well as tension. The Newcomen assembly of a chain which joined the piston rod to the working beam could therefore no longer be used. Instead Watt constructed an assembly of parallel rods to keep the piston rod in virtually straight alignment as its motion was transmitted to the beam. He was prouder of this invention, which he patented in 1784, than of any other he ever made, he said, but it by no means exhausted the fertility of his imagination. He made improvements of much greater merit.

James Watt produced also a mercury steam gauge, a glass water gauge, and a poppet valve to admit and release steam. He developed a stroke counter and a device to show graphically on an indicator card the pressures in the cylinder at all points throughout the stroke. He devised a method of recessing the edge of the piston and packing the recess with hemp in tallow to reduce both friction and leakage. For controlling the steam throttle and thus the speed of his engine, he adapted the centrifugal, or flyball, governor from the flour mills where it was used to regulate the "distance of the top millstone from the bed stone." He even designed and patented in 1782 a device for cutting off the steam supply early in a stroke so that the steam's expansion might be utilized. This mechanism is used today in most reciprocating steam engines, although Watt himself did not apply it effectively in his low-pressure engines. It is economical only at higher pressures than he would allow.

The elimination of wood was to be desired wherever possible as the size of the engines increased. A contemporary description of a huge working beam, constructed of 20 pieces of fir in sets 10 deep, gave its dimensions as 2 feet wide, 5 feet thick at the ends, and 10 feet at the center. By 1787 Watt had replaced such clumsy parts with cast iron; henceforth he made his engines entirely of metal. The iron working beams could be of trussed- or diamond-shaped pattern so that all of their material was usefully employed. Unnecessary metal could thus be left out.

Matthew Boulton realized that the single-acting, reciprocating Watt engine had a relatively limited future since it could be used only for pumping. Boulton also knew that "the people of London, Manchester and Birmingham are *steam mill mad*," as he put it in a letter to Watt in 1781. He therefore badgered Watt into developing a rotative engine. The most obvious way of converting reciprocating to rotary motion was by the use of a crank and a flywheel to carry the crank through the dead center. This arrangement had long been known, as Figure 7.4 shows; indeed, the earliest known crank of 850 (Figure 5.1) was attached to a grindstone which also served as a flywheel. No one, including Watt, apparently realized that a flywheel would keep the engine rate practically uniform throughout one revolution. This lack of understanding is not surprising, for a flywheel acts as a reservoir of energy, releasing energy as the crank passes through dead center and storing energy during the rest of the stroke, but it was not until the middle of the nineteenth century that scientists evolved the concept of energy. It was unknown to Watt.

In any event, by the time Watt began seriously to produce a rotative engine, James Pickard of Birmingham had already patented the crank as a method of applying steam engines to "the turning of wheels." Watt was quite right about the crank in asserting that "applying it to the engine

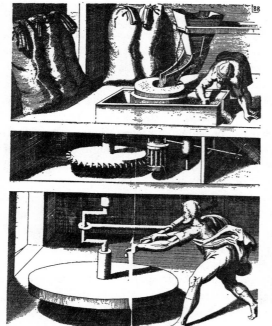

Figure 7.4
A suggested hand-operated grain mill using a crank (From Strada, *Kunstliche Abriss*, 1617–1618)

was like taking a knife to cut cheese which had been made to cut bread." Since the crank had been used for centuries, Watt might well have paid no attention to the patent claims.

It is possible that Watt feared for some of his own patents if he contested Pickard's claims. Watt's claims were exceedingly broad and gave him a monopoly in the field of the steam engine. In any case Watt chose to avoid controversy. Until Pickard's patent should expire in 1794, Watt tried to use the so-called "sun and planet" system of intermeshed gears, one of which was rigidly fastened to the rod connected with his working beam, the other to the rotating shaft. He seems never to have come to terms with Pickard. The planetary system had one advantage; with wheels of equal diameter the shaft made two revolutions for each cycle of the engine. But the device consumed much power in friction and was very noisy. There is evidence, however, that the crank was used on a number of engines at this time, whether or not Pickard received royalties; Boulton and Watt themselves appear to have applied it to a limited extent before the patent actually expired.

James Watt lacked the funds necessary to get his engine into production, but he had found in Dr. John Roebuck (1718–1794) of the Carron ironworks another inventor who was able and glad to provide the capital in return for a two-thirds interest. When Roebuck met hard times and withdrew, Watt was fortunate again to find in Matthew Boulton (1728–1809) of Birmingham a man of means who was eager to take over the financing. The firm of Boulton and Watt, established in 1775, prospered for years and continued successfully under the management of their sons well into the nineteenth century, even though Watt's basic patent and the monopoly had expired in 1800.

The first Watt engine, set up as an experiment at the firm's factory in Soho near Birmingham and never put to work, was single-acting with a cylinder only 18 inches in diameter. The commercial engines which the firm built in 1776 for the Brosely ironworks and the Bloomfield colliery had cylinder diameters of 38 and 50 inches, respectively. As the iron-master John Wilkinson had developed in 1774 a machine which greatly improved the process of boring, Watt was soon able to guarantee a cylinder of 72 inches that would be "not farther distant from absolute truth than the thickness of a thin sixpence at the worst part." Watt's 6-foot-diameter cylinder may have seemed then the ultimate in the production of power. Pistons and cylinders were in fact not to increase much in size; an 8-foot-diameter cylinder was considered large even after 1900. It was the steam pressure and the speed at which reciprocating steam engines operated that increased their capacities to degrees hardly conceivable in the days of James Watt. Of course, Watt had no idea how far turbines and internal-combustion engines would supplant his reciprocating double-acting engines.

The use of steam had become so common by the turn of the century that definite standards had to be set for determining the capacity of engines. Familiarity with horsepower made it the natural unit of measurement. Savery had remarked that a steam engine was capable of doing the work not only of the horses which it replaced at a given moment but also of those which would have to be maintained for continuous operation twenty-four hours a day. Smeaton had estimated more precisely that the mechanical effect which a horse could produce amounted to 22,916 pounds raised one foot high in one minute against the force of gravity. Desaguliers increased the amount to 27,500 foot-pounds. And Watt found by experiment in 1782 that a "brewery horse" was able to produce 32,400 foot-pounds per minute. The next year Boulton and Watt standardized the figure at 33,000 foot-pounds per minute in order to classify their engines for sale. By 1809 this was generally accepted as equivalent to 1 horsepower, and so it is today.

There were some 500 Boulton and Watt engines in service by 1800 when Watt's basic patent expired. Thirty-eight per cent were engaged in pumping water, sixty-two per cent in supplying rotary power for textile mills, iron furnaces, and rolling mills, flour mills, and other industries. So alert an industrialist as the famous pottery maker Josiah Wedgwood (1730–1795), though a personal friend of James Watt and Matthew Boulton, had been slow to use the new source of power; it was not until 1790 that Wedgwood installed a steam engine at his Etruria works, but others like the ironmakers John Wilkinson and Henry Cort had been more eager. That Watt's engines were having tremendous influence was evident from the throng seeking to enter into the competition of producing and making improvements in the steam engine.

The limitations of the Watt engine were apparent. It was, after all, slow; it was crude and cumbersome. Its parts were large and difficult to keep in alignment; many of them were merely rough castings which had to be chipped and filed by hand before they would fit. Forgings were made as close as possible to desired size and were not cut in machines; gears were cast and not machined at all. The consequent loss of power through friction was great. The engine worked at low pressure and, like the Newcomen engine, did not take advantage of the expansive force in the steam which it consumed. James Watt was aware of the advantages in higher pressures but refused to use them for fear of explosion. An American was among the first to design and put to work an engine which used steam at higher than atmospheric pressure. As an apprentice to a wagon maker about 1772, Oliver Evans (1755–1819) of Philadelphia thought of propelling carriages without animal power, and according to his own account in 1812, the neighboring blacksmith's boys had given him the idea of the "elastic power of steam." They had filled a gun barrel with water, rammed it tight with a wad, and put its breech in the fire. "The crack," said Evans, "was as if it had been loaded with powder." The Marquis of Worcester had noticed the same characteristics in steam many years earlier.

Evans petitioned the legislature of Pennsylvania in 1786 for exclusive use of his improvements in "flour mills" and "steam waggons" in that state. He was granted, in March, 1787, the rights to a truly remarkable mechanism for continuously handling and milling grain, but the legislators took no notice of his "steam waggons." His representations, he said,

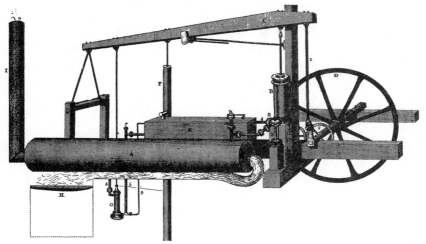

Figure 7.5 Evans's Columbian steam engine, grasshopper type (From *Niles'*
Weekly Register, vol. 3, Addenda, 1813)

made them think him "insane." By 1801, however, he had returned to his
ambition to propel carriages and boats by steam and had enlarged his
aim to include developing an engine that would apply the "elastic power"
of steam to industry. A tablet at Market and Ninth Streets in Philadelphia marks the site of the workshop where he constructed a small,
noncondensing, high-pressure steam engine at a cost of $3,700, all the
funds that he could "command."

He gave the engine (Figure 7.5) a 6-inch cylinder, an 18-inch piston
stroke, and a wooden flywheel 7½ feet in diameter, but it was not the
flywheel that made Evans's engine significant. The distinctive feature
was that the engine made some 30 revolutions to the minute under 50
pounds of pressure to the square inch, exhausting the steam to the air,
thus eliminating the condenser. To obtain this pressure, Evans encased his
copper boiler in wood bound with iron hoops, and he ran a horizontal
flue lengthwise through the center of the boiler so that the heat of his
fire reached the water more effectively. He set the engine upright so that
the piston rod drove directly to the beam above. As this beam had its
fulcrum, or point of support, not in the center but at one end and as the
support of the fulcrum was on a rocking column, there was no need for so
complicated an arrangement as Watt's parallel rods. Evans's working beam
aligned itself with the piston rod. This design became known as the
grasshopper type and continued long in use.

Benjamin Henry Latrobe (1764–1820), versatile engineer from England who built for Philadelphia the first steam-pump waterworks in this country and who later designed parts of the Capitol at Washington, reported in 1803 to the American Philosophical Society on the state of steam engineering in America. Josiah Hornblower had imported the first Newcomen engine in 1753 for a copper mine at Belleville, New Jersey. Engines "of the old construction," said Latrobe, presumably meaning Newcomen engines, had been introduced some forty years before 1803 from England, but then, "during the general lassitude of mechanical exertion which succeeded the American Revolution, the utility of steam engines appears to have been forgotten." There had followed, he said, "a sort of mania" for impelling boats by steam engines, but it had largely subsided; there were now in 1803 only five steam engines "of any considerable power" at work in America. One of these, he said, belonging to the Manhattan Water Company, was a Boulton and Watt engine. Another in New York was sawing timber for Mr. Nicholas J. Roosevelt. Two, which had been built by Mr. Roosevelt, were pumping water for the city of Philadelphia. The fifth, Latrobe had heard generally, was engaged in some kind of manufacture at Boston. He would notice, he said, in his second report, "the improvements made by the very ingenious Dr. Kinsey" who had erected an engine in New York upon a "new principle" should it be found "on experiment" to fulfill the "intended purpose." [4]

Unmistakably, the audience was to gather that this country was far behind Britain. As thirty years had not yet passed since the United States had won independence from the mother country and had broken from the restrictive legislation of Parliament and since England prohibited emigration of technicians, this was hardly surprising. But Latrobe was constrained to remark further, "Nor ought I to omit the mention of a small engine, erected by Mr. Oliver Evans, as an experiment, with which he grinds Plaister of Paris; nor of the steam-wheel of Mr. Briggs." Indeed he ought not. And Evans later took him to task for it in the nation's news magazine of those days, *Niles' Weekly Register*.

By 1807 Oliver Evans had established the Mars ironworks. By 1812 he was reporting that he had "succeeded perfectly," that he had ground 300 bushels or 12 tons of plaster in twenty-four hours and had cut

---

[4] Benjamin H. Latrobe, "First Report . . . in Answer to the Enquiry of the Society of Rotterdam, 'Whether Any and What Improvements Have Been Made in the Construction of Steam-Engines in America?'" *Transactions of the American Philosophical Society*, vol. 6, pt. 1, pp. 91–92, 1804.

marble at the rate of 100 feet in twelve hours. It was not "speculative theory," he said; there were two of his engines in Philadelphia and three on the Mississippi, two of them driving sawmills; three or four at Pittsburgh were milling grain and rolling iron; there was one at Marietta, Ohio, and another in Lexington, Kentucky; still another in Middletown, Connecticut, was running a cloth factory. In 1815 Evans installed an engine with a 20-inch-diameter cylinder, 5-foot stroke, and four boilers designed to deliver steam under 200 pounds pressure for the new Fairmount waterworks of Philadelphia. This plant was overambitious; its operation proved so costly that in 1822 it was replaced by water power under the direction of Frederick Graff, Latrobe's assistant and successor at Philadelphia. But Oliver Evans had 50 of his steam engines in operation along the Atlantic Coast by 1819 when he died, the year of the financial panic and depression following the War of 1812 with Britain. Doubtless this rapid expansion was only part of the growth which came with the introduction of machinery from England and the development of American manufacturing, particularly the textiles, behind the blockades and embargoes of the Napoleonic Wars in Europe and then this second war with Britain. Even so, Oliver Evans was swiftly catching up with the British engine makers.

Others came as close as Evans did to Watt in mechanical skill and engineering imagination. The impetuous Cornishman, wrestler, and strong man, Richard Trevithick (1771–1833), ignored Watt's alarm over high pressures just as he sought a way to evade the Watt patent. He had constructed a puffer, or noncondensing engine, about 1798, before Evans built his in America, and by the time news of Evans's work had reached England, Trevithick had developed his design into the first direct-acting engine in history. Trevithick had first placed his cylinder upright inside the boiler; thus the boiler served also as the steam jacket. Then he did away with the working beam. This he first accomplished by guiding the piston rod upward to a transverse horizontal beam, or crosshead, which then acted downward through connecting rods at either side upon the crankshaft beneath the boiler. The whole arrangement was self-contained, compact, and—more significantly—portable as a unit. His direct-acting engine came when he laid the cylinder horizontal and connected its piston by means of a connecting rod to the crank. Trevithick's next step was to be the construction of a locomotive.

An explosion when the operator tied down the safety valve of his boiler led Trevithick to construct two safety valves. One of these was beyond the control of the operator. The safety valve worked upon the principle that a known weight set upon a disk covering an opening of specific dimensions will rise and uncover the opening and thus allow the steam to escape whenever its pressure reaches a certain point. The safety valve was then the only means of determining the degree of pressure above a few pounds. Gauges for higher pressures were not introduced until about 1848. Trevithick also used the fusible plug of lead or similar material which, inserted into the boiler plate, would melt if and when the water in the boiler got too low for safety. This device acts on the principle that the temperature of the water in a boiler does not rise above the boiling point at any given pressure so long as it remains water; when it flashes into steam, however, this protection is removed. The parts of a boiler directly exposed to the fire, therefore, have always to be in contact with water or there is immediate danger. So long as it is covered by water the fusible plug remains intact, but if it is uncovered, it melts

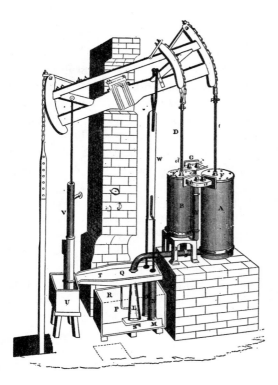

Figure 7.6
Compound steam engine,
Hornblower type (From
Edinburgh Encyclopaedia,
vol. 18)

and allows steam to escape through the opening thus made, giving audible and visual warning of low water.

The startling results obtained by Newcomen, Watt, Evans, Trevithick, and less famous men led many others into designing and constructing steam engines. The next significant advance in steam engineering came with higher pressures by the addition of another cylinder to use the steam again as it expanded. This new type was known as the compound engine. Jonathan Carter Hornblower (1753–1815), familiar with the Newcomen engine, had created the first of the compound engines in 1781 by adding another cylinder larger than the first (Figure 7.6). This second cylinder received the exhaust from the first or high-pressure cylinder and, like it, was connected to the crankshaft. Thus the steam performed additional work in the second cylinder as its volume increased with its decreasing pressure. For Hornblower's own profit, the operation was too nearly like that in Watt's single cylinder and separate condensing chamber; Watt claimed that his patent had been infringed. It would seem that Watt's condensing chamber did no more than cause the vacuum in the cylinder and performed not a bit of additional work, but Watt declared that Hornblower's arrangement was simply his own double-cylinder engine operating on his principle of expansion. He convinced the court, and Boulton and Watt won their suit in 1799. Hornblower died in proverbial disappointment and poverty.

Like Watt's device for cutting off steam early in a stroke, Hornblower's idea of using the expansive force of steam in a second cylinder was destined to be of great value. In his own day, it gave little advantage because the pressures then prevailing were too low, and proper proportioning of cylinder diameters was not well understood. In 1803 Arthur Woolf (1766–1837), a millwright who had worked with Hornblower, constructed a compound engine (Figure 7.7) that in time was successful, notwithstanding the fact that the dimensions of its high- and low-pressure cylinders were not at first properly proportioned to the expansion of the steam as it declined in pressure. With Humphrey Edwards, Woolf perfected an engine which became generally known in France. Despite its high cost and complicated operation, it was popular there because it saved expensive fuel.

Although the early development of the engine itself has been emphasized in the preceding paragraphs, there had to be many accompanying improvements in materials, manufacturing methods, and supplementary devices to permit progress both in engine and in steam supply. New-

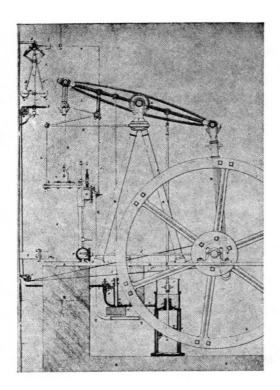

Figure 7.7
Woolf's compound engine
erected in Paris, 1815
(From *Transactions, New-*
comen Society, vol. 13;
original in *Bulletin de la*
Société d'Encouragement,
vol. 17; courtesy New-
comen Society)

comen's boilers were adapted from brewers' kettles with closed tops;
their brick foundations were so constructed that the flames and gases of
the fires swept around them into the chimneys. The metal first used for
boilers was copper. After 1800, Evans's 15-foot boiler with inner flue
was made of copper with cast-iron heads held together by wrought-iron
rods or stays running lengthwise through the boiler. Wrought-iron plates,
riveted together, had come into use by 1725, but boilers encased by
wooden staves like a barrel and with cast-iron fireboxes continued to be
used nevertheless into the nineteenth century.

The first boilers were very inefficient. It took a long while to heat
the water sufficiently to create steam; a great deal of heat went up the
chimney. Fuel may have been plentiful, but transportation was expensive.
The desire to save fuel led first to lengthening the boiler in order to
extend the heating surface, and so there developed the "waggon" with flat
surfaces as distinguished from the beehive or haystack boilers. Pressures
of only a few pounds above atmosphere were likely to distort, if not
burst, the flat plates of the "waggon" boilers because their structure was

weak. Cylindrical forms were more stable, and the demand for higher pressures and more steam was growing. The objective thus became one of increasing the heating surface of cylindrical boilers, which was accomplished by the adoption of the inner flue which Evans had used. Trevithick designed a cast-iron cylinder in 1812, 30 feet long and 6 in diameter, that had an inner flue 3½ feet in diameter. He made the ends of flat wrought iron, bolted to the cylinder. Though strengthened by the inner flue, the ends were relatively weak because they were flat, but they were able to withstand the pressures which Trevithick imposed upon them at the time.

The next step was to increase the number of flues or fire tubes. George Stephenson (1781–1848) built a double flue, anticipating the Lancashire boiler with fires at both ends, which William Fairbairn (1789–1874) developed in 1845 and which remained in use throughout Great Britain as late as 1939. The American John Stevens (1749–1838) of Hoboken had constructed a boiler with multiple tubes in 1825. Working with two locomotives which Stephenson supplied to him, the French railroad builder and industrialist Marc Seguin (1786–1875) made practical experiments which established the use of multiple tubes in 1829. Aided by Henry Booth, George Stephenson and his son Robert were simultaneously applying the principle in their famous locomotive the *Rocket*. Peter Cooper (1791–1883) used gun barrels in 1830 for the fire tubes of his tiny locomotive *Tom Thumb*.

The idea had come much earlier that the tubes should carry not the gases from the fire, but the water which was to be turned into steam. Water had been so heated in Roman times. Several engineers tried water-tube boilers, notably John Stevens, who built what happened to become the first sea-going steamboat, and Goldsworthy Gurney, who made steam carriages for the English highways. The scheme was not generally employed, however, until late in the nineteenth century. For one reason, water-tube boilers were too expensive until the manufacturing process had been greatly improved. For another, the tubes were not easily kept clean and steamtight. Robert Stephenson found that they "became furred with deposit, and burned out." [5] It was difficult to secure proper circulation of water in the boiler and to prevent the formation of dead spots which got hot and caused burnt tubes.

Newcomen's engine of 1712, when compared with other sources of

[5] Samuel Smiles, *The Life of George Stephenson*, Ticknor and Fields, Boston, 1858, p. 262.

power in its time, was entitled to the praise it received. But as has been said, the Newcomen engine wasted more than 99 per cent of the heat in its fuel. Good coal has been found to have the value per pound of approximately 14,000 British thermal units. Each such unit represents the amount of heat required to raise the temperature of one pound of water at about 39°F (when it is densest) one degree Fahrenheit; it has the equivalent mechanical value of about 778 foot-pounds. If the Newcomen engine had been 100 per cent efficient, it could have raised 10,892,000 pounds of water 1 foot for every pound of coal consumed. It is recorded, however, that one engine raised only 43,000 pounds of water 1 foot per pound of coal consumed. Its efficiency was therefore less than 0.5 per cent. Improvements had been made by 1774, so that Smeaton found a Newcomen engine delivering 105,000 foot-pounds of work per pound of coal. His own changes in the mechanism raised the figure to 120,000. Some of Watt's engines are said to have delivered per pound of coal as much as 320,000 foot-pounds of work and thus to have attained a thermal efficiency between 2 and 3 per cent.

Other things being equal, though they seldom are, efficiency would rise as pressure and temperature increased. James Watt was fully aware of this fact. He chose, however, to develop his machines at low pressures rather than to take the risk of putting the materials then available under high pressures. The Evans and Trevithick engines, operating near 50 pounds pressure, might be expected to have efficiencies considerably higher. However, as they were not condensing engines and exhausted their steam at relatively high pressures and temperature, they lost a great deal of their potential efficiency. The Hornblower compound engine gave little practical advantage at the start because it operated at such low pressures. Many factors enter into the determination of economic value, and it is not fair to say that these early high-pressure and compound engines were more valuable to industry in their time than Watt's single-cylinder, low-pressure engine.

During the next fifty years the stationary reciprocating steam engine with a number of improvements gradually increased in size, speed, and efficiency. For some decades the wide variety of types served mainly to furnish power for industrial machinery and in pumping. In 1849 George Henry Corliss (1817–1888) of Providence, Rhode Island, developed a novel quick-acting valve to replace the slide valve which had been generally used on stationary engines. The Corliss engine became popular in the United States and was quite widely copied by European engineers. The

Figure 7.8 The Corliss engine at the Centennial Exhibition (From *Harper's Weekly*, May 27, 1876)

famous engine installed at the 1876 Centennial Exhibition in Philadelphia was a noteworthy example of Corliss design (Figure 7.8). It was a beam-type engine. With two high-pressure cylinders 40 inches in diameter and a 10-foot piston stroke at 360 revolutions per minute, it produced under normal conditions 1,400 horsepower. Erected centrally in Machinery Hall, it became a focal point for sightseers, with two shafts 108 feet long at right angles. Power taken from these shafts by leather belts drove the building's machinery.

The Centennial Exhibition was the first world's fair at which the practicability of generating electricity by steam was demonstrated, an art that shortly developed with great rapidity. At first the electric generators, or dynamos, being essentially high-speed devices, were driven through belts. Before long it became possible, by increasing the number of poles, to connect dynamos directly to the engines. As engines and dynamos grew in size the steam pressures were gradually stepped up to 200 pounds, and triple-expansion engines were common. A popular arrangement made use of vertical high-pressure and horizontal low-pressure cylinders to drive the shaft. Such a combination design was common until the recipro-cating engine was superseded by the steam turbine early in the twentieth century.

Water Power

Steam penetrated rapidly into the established industries of milling flour, sawing wood and stone, pumping water, hoisting coal, mining and crush-ing ores, and smelting and forging iron, especially where the best loca-tions for the enterprises were remote from waterfalls. It made its way, too, among newer activities such as rolling and slitting metal plates, draw-ing wire, spinning and weaving textiles by machinery. But the engineers of the eighteenth century did not abandon water power simply because they were amazed and delighted with the marvels of steam. Water wheels, it was clear, notwithstanding the progress with steam, could generate power just as satisfactorily for many industries and more cheaply for some where the dams did not cost too much and local conditions were favorable. This was particularly true if several factories were clustered about a site where the flow of water was ample and reasonably reliable—an important consideration, because so many rivers run low or well-nigh dry at times. Moreover, human nature is resistant to innovation; many people cling to the old and familiar because it has served them and their fathers.

There was no such inertia in John Smeaton (1724–1792). As he made improvements in Newcomen's engine, he was thinking of further possibilities of improving water wheels. Smeaton's curiosity was great. He was an "incessant experimenter" in several fields and upon his death in 1792 was entitled to acclaim among the foremost engineers of his time. His widely read studies have remained classic to this day in the engineering profession, especially his paper before the Royal Society in 1759 on "the natural powers of water and wind." [6]

As a result of Smeaton's demonstrations and work as a consulting engineer, overshot wheels of increased dimensions and with improved regulation came into use in Britain, Europe, and America. Centrally located, their power could be transmitted by shafts and belting to many buildings near the power site. One of these wheels, 50 feet in diameter and 6 feet wide, erected in 1800 near Merthyr Tydvil, Wales, generated some 50 horsepower for blast furnaces. Another Smeaton wheel made in London and sent to Italy about 1852 was 76½ feet in diameter and was rated at 30 horsepower. The 72½-foot Laxey water wheel on the Isle of Man (Figure 7.9), installed in 1854 for mine pumping, has about 200 horsepower. Acknowledging his own indebtedness, Oliver Evans quoted extensively in his writings from Smeaton's paper before the Royal Society; Smeaton, he said, was the only author whom he had read who "joined practice, and experience with theory." [7] During Evans's lifetime and after, many improved overshot wheels were built in America, one of the largest being in the Cumberland Gap of eastern Tennessee.

Another water wheel, known as the pitchback, was common in the United States. The water poured from above upon the near side, as millwrights used to say, turning the wheel toward the flume, rotating the lower part of the wheel with the current of the stream below the fall or at worst in the dead water of a pool. But the pitchback, like the overshot wheel, could never use more than a fraction of the head of the falling water between the flume and the tailrace. The loss of power was obvious. Moreover, all overshot or pitchback wheels were large and cumbersome even when made of iron. They might be lightened if built

[6] John Smeaton, "An Experimental Inquiry Concerning the Natural Powers of Water and Wind to Turn Mills, and Other Machines, Depending on a Circular Motion," *Philosophical Transactions of the Royal Society*, vol. 51, pp. 100–174, 1759–1760.
[7] Oliver Evans, *The Young Mill-wright and Miller's Guide*, 5th ed., H. C. Carey and I. Lea, Philadelphia, 1826, p. iii.

Figure 7.9
The great Laxey water
wheel on the Isle of Man,
1854 (From *Scientific
American*, Jan. 11, 1873)

smaller; there would be gain in efficiency if no part of the wheel had to
turn in dead water. But equally apparent, any reduction in the diameter
of the wheel would sacrifice part of the fall, or available head of water,
with consequent loss of power.

Jean Victor Poncelet (1788–1867), Alsatian mathematician and engineer who had served under Napoleon, hit upon a device about 1820 to
compensate somewhat for the loss (Figure 7.10). He constructed a sort
of penstock which confined the water to a narrow stream and delivered
its force upon curved buckets in the lower part of the wheel. This type
of wheel became popular in France, especially for low falls. One erected
in 1849 near Montserrat in Catalonia, Spain, was nearly 17 feet in diameter
and 30 feet wide. Driven by a waterfall of only 6½ feet, it delivered 180
horsepower. An ordinary breast wheel in the same location would have
had to be three times as wide to produce the same amount of power,
but it was still evident that the full potentialities in water power had not

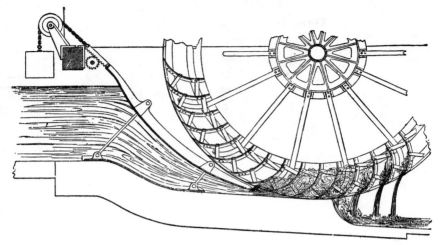

Figure 7.10 A Poncelet wheel (From J. Glynn, *Power of Water*, 1852)

yet been attained. Poncelet himself continued to experiment and make further contributions. A younger French engineer, Benoît Fourneyron (1802–1867), was to take the next great step.

The idea of a water wheel rotated horizontally on a vertical shaft is very old, but its development into the turbine is relatively new. A primitive water mill of India had its stream directed horizontally against the blades as if they were set in a natural current. However, the principle of reaction or recoil, as distinguished from yielding to the weight and velocity of water or some other substance such as steam, was introduced with the reaction turbine credited to Hero. Jacob de Strada's illustration of 1618 shows what might be considered an impulse turbine (Figure 7.11), though it appears to be hardly more than a horizontal wheel.[8] The turbine seems not to have been put into practical use before the middle of the eighteenth century, when about 1740 this ancestor of jet propulsion appeared in a form very much like a rotary garden sprinkler; it was called Barker's mill (Figure 7.12). Doctor Barker's device stirred the Swiss mathematician Léonhard Euler (1707–1783) and his son Albert to investigation; Albert exclaimed optimistically that he had learned how to make hydraulic reaction machines that would produce "the whole effect"

[8] Jacob de Strada, *Kunstliche Abriss allerhand Wasser-, Wind-, Ross-, und Handt Muehlen beneben schoenen und nuetzlichen Pompen,* 2 vols., getruckt durch P. Iacobi, in Verlegung Octauii de Strada, 1617–1618, vol. 1, plate 6.

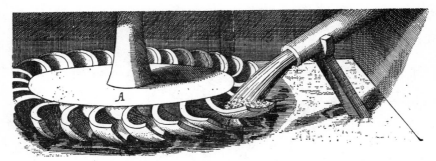

Figure 7.11 Impulse wheel, vertical axis (From Strada, *Kunstliche Abriss,* 1617–1618)

of which water was capable, an effect "equal to the moment of impulsion." [9] He erred in concluding that the width of the wheel necessarily should be one-half the height of the column of water, but the fact was soon established that the water should enter the wheel without any turbulence and leave spent of all velocity save enough to get it clear of the wheel. Professor Claude Burdin (1790–1873), who coined the word "turbine" from the Latin *turbo* meaning "I spin," and many other investigators experimented with these essentials, but no one could find a practical way of applying them to perform work. It was Burdin's pupil Fourneyron who succeeded.

Fourneyron was secretive about his experimenting, especially in regard to the curves of his blades and the way he lubricated the bearings of his machine. He had so perfected the turbine that soon after 1837 he could use a head of water exceeding 350 feet, delivered to a wheel through a cast-iron penstock or pipe 16½ inches in diameter, at a pressure near 160 pounds to the square inch. The wheel was little more than 12 inches in diameter, weighed less than 40 pounds, and was made of bronze. It developed 60 horsepower, with an efficiency of more than 80 per cent of the potential in the waterfall, and it made an estimated 2,300 revolutions per minute. No such speed had ever before been approached in any mechanism. The flow was outward from the center of the rotor, and the wheel was propelled by the reaction of the water escaping from its rim. A contemporary German engineer stood amazed before Fourneyron's installation at Saint Blaisien in the Black Forest. As he looked up to the height from which the water descended upon the little wheel, he marveled that

[9] Marcel Crozet-Fourneyron, *Invention de la turbine,* Libraire Polytechnique Ch. Béranger, Paris, 1924, p. 15.

it did not burst in the spiral masses of water which rushed from it and threatened to destroy the surrounding walls. These spirals of water, incidentally, were wasting a lot of energy that might have been giving more power. Fourneyron nevertheless had made his name "forever historical" in the technical and scientific world, exclaimed Moritz-Rühlmann, who also noted that the turbine was driving the 8,000 spindles and all accessory machines in a cotton factory. Fourneyron designed two other turbines that worked under a head of only 16½ feet and yet developed 220 horsepower each, for a textile mill of 30,000 spindles and 800 looms in Augsburg. Thereafter the turbine, though varying often in detail from installation to installation and from designer to designer, was thoroughly established as a means of industrial power.

The reaction turbine was superior at that time to any type of water wheel rotating upon a horizontal axle. A vertical turbine could run en-

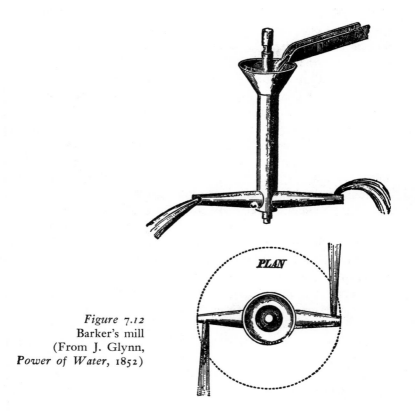

Figure 7.12
Barker's mill
(From J. Glynn,
Power of Water, 1852)

tirely submerged under any head of water from a few inches to hundreds of feet. Its speed could be varied considerably with little change in the effectiveness of its work. Most important of all, its capacity was astounding. The older types of water wheels developed horsepower only in the hundreds; improved reaction turbines have produced more than 50,000 horsepower. Fourneyron's invention gained wide use in America during the next decade, and from then on, as improved by James Bicheno Francis (1815–1892) and others, had a marked effect upon the country's industry. The water turbine, however, did not make so much headway in Britain, perhaps because the steam engine was well established there, rivers often ran low, and coal was relatively cheap. The impulse turbine, adapted for use with high heads, came later in the century.

Formulas

Mathematicians and scientists of the period were carrying the study of stresses into elaborate theoretical concepts of elasticity, and at the same time they were clarifying knowledge of the strength of materials for engineers in their daily tasks. Though some deplore dependence upon formulas and cite instances where it has resulted in disaster, most engineers today are grateful for the work of that remarkable genius, Thomas Young (1773–1829). His modulus of elasticity, first defined in 1807, is a measure of the resistance of materials to deformation under stress. The modulus of elasticity is the pull in pounds which would be required to stretch a bar one inch square to double its original length if it remained uniformly elastic during the stretching. As almost all materials reach their elastic limit long before they have been stretched to double their original length, the modulus of elasticity of a material is in practice determined from the percentage of elongation of a test piece under a stress which had not been sufficiently great to bring it to its elastic limit. For structural steel it has been found that the modulus of elasticity is about 30 million pounds per square inch.

In 1729, a century after Galileo, Pieter van Musschenbroek (1692–1761), professor at the University of Leyden, constructed machines to test small iron rods, stone blocks, and wood under tension (Figure 7.13), compression, and bending. It was the French engineer Jean Rodolphe Peronnet (1708–1794), most eminent of his time, who made the first modern testing machine of practical capacity, some 18 tons. He did this in 1768 while building his famous bridge at Neuilly, France. Tradition

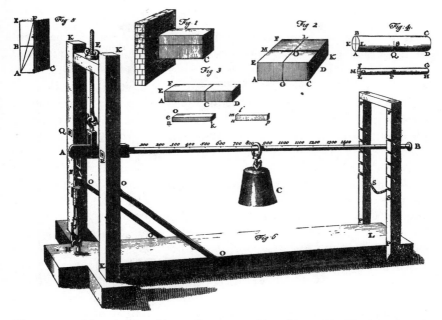

Figure 7.13 Musschenbroek's tensile-test machine (From his *Physicae experimentalis et geometricae*, 1729)

has it that he did the testing so that he might know in advance whether his unusually flat arches would stand. His associate Jean Baptiste Rondelet (1743–1829) went further in 1787 and eliminated a great part of the inaccuracy resulting from friction in Peronnet's machine. Rondelet set the lever on knife edges like grocers' scales and applied the loads by turning a screw. This was the first testing machine to use all the essentials of lever, counterpoise, and screw power, but it was capable of testing compression only, until improved. A machine with a capacity of 100 tons was soon built for Le Havre and used to determine for the first time the strength of materials of large cross sections.

Practice in the field and scientific theory were coming abreast of each other as industry expanded and the construction of railroads and bridges increased in the nineteenth century. Experimental scientists, engineers, and manufacturers cooperated in the effort to determine allowable stresses upon chains for ships' anchors and wire or chains for suspension bridges, the stability of cast-iron columns, the strength of railway rails and of riveted joints in boilers. There was no longer so much need for the Roman method of supersafe construction; margins of safety could be calculated with assurance. The materials for Telford's great bridge across the Menai Straits, Wales, in the 1820s were tested in advance and a large amount of material saved during its construction. Following a series of boiler explosions in the United States during the 1830s, a committee of the Franklin Institute in Philadelphia made extensive tests of wrought iron, using a machine that is still on exhibit in the museum of the Institute. But risks were still to be taken and disasters suffered.

The men of theory were proceeding also with studies in hydraulics that were to have inestimable value for engineers. Daniel Bernoulli (1700–1782), one of eight mathematicians and physicists in a Swiss family, laid the most thorough foundations in 1738 for all subsequent work on the flow of water, light oils, gasolines, and other liquids of low viscosity. His theorem with regard to pressures and velocities conforms with the law of the conservation of energy; in a word, the total head, or energy, of a liquid flowing at any point in a pipe is equal to the total at any other point, if there is no friction. In designing pipes such as the Big Inch or Little Inch in the United States or aqueducts like those which supply New York and Los Angeles, engineers today rely on the Swiss mathematician's reasoning and plan their conduits accordingly.

A French contemporary of Bernoulli was making independent studies in hydraulics along lines that had been laid down by Torricelli a century earlier. Henri Pitot (1695–1771), twenty years old before he took any interest in science, rose rapidly and became a member of the Académie des Sciences. It was as a hydraulic engineer, however, that Pitot gained his right to fame. Using Torricelli's theory of velocities at orifices, he made in 1732 an instrument to measure at any given depth in an open or closed channel the speed of a flowing liquid. It consisted of an open tube, with its lower end bent upstream at right angles, that could be held vertically in the stream. The height of the liquid in the tube is an indication of the velocity of flow. The Pitot tube is invaluable to engineers today, saving time and accurately measuring the velocities of water, steam, gases in pipes, or the flow of air in wind tunnels for the aircraft industry. A Pitot tube also operates the air-speed indicator in the modern airplane.

Iron

Advances in metallurgy came with the steam engine. The Darbys, Wilkinson, and Cort, who developed the iron industry, had as much to do with changing the course of history as the engine makers Savery, Newcomen, and Watt. The manufacture of iron, and even of fine steel, had been an art for centuries. The blades of Damascus and of Toledo in Spain and the knives of Sheffield in England made from Swedish ore were famous in medieval times. The Romans had used iron for domestic as well as military purposes, but now its manufacture in both Europe and America was to become a business of vast extent and economic consequence.

Early craftsmen had found that wood burned in contact with certain ores changed them into metallic substances of great usefulness. Workers in iron knew how to add carbon and to heat, hammer, and reheat the mixture until they had removed the impurities and had made it harder or softer, stronger, more flexible or more brittle, as they wished. They understood the difference between wrought and cast iron, and they knew steel. They had found that certain ores were better than others, although they may not have known why it was that some were more easily worked and why those containing sulfur and phosphorus made poor steel. They had discovered that for producing iron there was no fuel as good as charcoal. The consequences were twofold. The best craftsmen of earlier times fairly often obtained reliable products in small quantities. Wherever the industry took hold, charcoal burners ruined the forests of the land.

It was this calamity rather than the success of the early ironworkers which led to the next great advance in metallurgical engineering. The smelting of iron by the use of coal has been credited with having as much historical importance as the Norman conquest of England, King John's signing of Magna Charta, the discovery of America by Columbus, or the defeat of Napoleon Bonaparte. Americans may wish to pause for brief rebuttal on behalf of Columbus or to add a few more events to the record, but the day in 1709 when the first Abraham Darby (1677–1717) actually smelted iron ore with coal on a commercial scale was a day of revolution, not simply in metallurgy. It determined the subsequent history of Britain, and, in fact, of the world.

The ruin of Britain's forests had been foreseen and deplored. Admirals and statesmen were alarmed for the future of the Royal Navy and British supremacy upon the seas if their oak timber were destroyed. Raleigh's expedition of 1585 had the production of iron from the ores and immense forests of the new continent as one of the purposes of the colony to be established in America for Elizabeth. When Virginia at last had been founded in 1607, one furnace and two small forges were erected at Falling Creek, 66 miles above Jamestown. It was the lighting of these that provoked the massacre of March 22, 1622. The Indian method of preserving his forests was to wreck the furnaces and slaughter the workmen.

Meanwhile, to save timber at home, the British government had issued to Simon Sturtevant, a German miner, a patent for reducing iron ore with coal. The government later exiled Sturtevant, and John Rovenzon took over the project. Rovenzon seems to have done no more than write about what he was going to do and coin the name "sow" or "pig" iron to describe the metal from its appearance as it cooled after running from the furnace into the first trough or mold and from there into short branches. Then came Dud Dudley (1599–1684), fourth of Edward Lord Dudley's 11 illegitimate children by the same mother, the daughter of a collier. It would seem that this collier's daughter deserved something more than she has received from the historians, but this is not the place to venture upon discussing such human problems. Dud Dudley was "fetched from" Oxford in 1619 at the age of twenty to look after his father's iron-works in the forests of Worcestershire. As he too became anxious over Britain's timber, he began trying to smelt iron with coal.

Years later, in 1665, Dudley claimed that before he had gone off to war as military engineer for Charles I against Cromwell, he had succeeded in producing "good merchantable iron" with coal at a price from 30 to 40 per cent below the price of iron made with charcoal. And this he had done, he declared, in spite of the hostility of "riotous persons" who on one occasion had cut his bellows. But how he had changed the coal to coke, if he had actually done so, and how he had "reduced" or deoxidized ore with his fuel to produce iron, Dudley did not set down for the readers of his famous pamphlet on the metal of Mars (*Mettalum Martis*).[10] The consensus among experts is that Dudley was a wishful thinker, contentious, given to boasting, and that he never did do what he had set out in youth to do, even though he returned to the task after the civil war and worked at it doggedly in spite of illness and poverty until his death in 1684. Thirty years elapsed between his death and the first Abraham Darby's "charcking cole."

By 1713 Darby and his son-in-law Richard Ford were producing from 5 to 10 tons of iron a week at Coalbrookdale on the Severn in western England, using coke to supplement charcoal, and they could make their

[10] Dud Dudley, *Mettalum Martis; or Iron Made with Pit-coale, Sea-coale, &c.*, printed by T. M. for the author, London, 1665.

cast iron hard and brittle or soft and tough as they pleased. The production on a commercial scale of cast iron from ore smelted with coal had begun the very next year after Newcomen's engine had appeared on the market. By 1721 the Darby works were turning out heavy castings, and within a few years plans were being made to expand them when the Savery-Newcomen patent would expire in 1733, opening the manufacture of engines to competition. Steam power was rapidly increasing the demand for iron, and industry was increasing the demand for engines.

Stronger air blasts brought higher temperatures in the furnaces, releasing impurities in both fuel and ore and making effective the use of coke to reduce the iron without the help of charcoal. By 1745 the second Abraham Darby (1711–1763) and his son-in-law Richard Reynolds were using coke regularly to obtain a superior cast iron. By 1775, the third Abraham Darby (1750–1791) with the ironmaster John Wilkinson (1728–1808) had so improved the coke furnace that they were able to cast the arched ribs for the world's first iron bridge. Wilkinson, noted among the first of Britain's millionaires, made other contributions to both industry and engineering. His cylinder-boring mill, patented in 1774, introduced the guide principle into machine tools. It was much better than Smeaton's and indispensable to Boulton and Watt in manufacturing their steam engines. Wilkinson ordered one of the first Watt engines in 1775 and pioneered in applying steam power to the rolling of iron. He launched the first iron vessel in 1787, a barge 70 feet long built of riveted iron plates.

British industry was not capable of meeting the country's own demands when the first Abraham Darby made his contribution to the metallurgy of iron. Supplies came from Sweden, Norway, and Russia. Then there was a great change. As the mercantilists of the 1740s advocated strenuously that pig iron should be obtained from the American colonies to free the mother country from dependence upon foreign supplies and to save Britain's forests, the second Abraham Darby successfully reduced iron ore with coke. Britain thus advanced swiftly toward independence of both American and Baltic supplies. Parliament, nevertheless, passed the Iron Act of 1750 to encourage the importation of pig iron from the colonies and to restrict colonial manufacture in favor of Britain's industry. The act was as unnecessary for Britain as it was futile in America; Britain's forests were safe, and British pig iron was now adequate. The third Abraham Darby developed further the process of smelting iron ore with coke in 1775 as the American rebels began their war upon King and Parliament.

By 1788, British production exceeded importation. Two-thirds of the domestic supply was being smelted with coke. Henry Cort (1740–1800), with his puddling process, had freed the British forests entirely from the curse of iron and the industry, too, from the chief obstacles to its expansion. Great Britain was on the way to the industrial supremacy over the world which it maintained until rivaled by Germany and surpassed by the United States in the twentieth century.

The achievement was appreciated at once. Following the peace of 1783 which had meant the end of British domination in North America, Lord Sheffield proclaimed in his *Observations on the Commerce of the American States*, "Mr. Cort's very ingenious and meritorious improvements in the art of making and working iron," together with the Boulton and Watt engine and Lord Dundonald's new process of making coke at lower costs would be, if they all succeeded, "more advantageous to Britain than Thirteen Colonies." They could give "the complete command of iron trade to this country, with its vast advantages to navigation." It may seem that Sheffield was making the best of an unhappy event. He was, in fact, shrewdly prophesying the future of the British Empire. When comparing the loss of the North American colonies with the industrial development which Lord Sheffield foresaw, one must not think of the United States of today. There were fewer than 4 million people in the 13 American states of 1784, and their settlements reached but 100 miles or so up the river valleys from the Atlantic seaboard. A very few pioneers had gone over the Appalachians into the great mid-continent. The prospect of the Colonies' economic strength and political future was magnificent and their purpose determined, but their immediate value to the mother country was about as Sheffield asserted. What he did not foresee was that the development to come with the more and more effective linking of coal, iron, and steam would generate much besides British power. It would create also the huge productive machinery of the United States in which Britons were to invest profitably many hundred millions of pounds sterling.

Henry Cort had bought an ironworks near Plymouth harbor in 1775 with funds he had saved while ten years an agent of the Royal Navy. He tried to get a Boulton and Watt engine in 1779, but it was still so new that it had not been adapted to this kind of heavy machinery. Watt had not yet perfected his rotary motion; Cort continued using water power. In 1783 he obtained his patent for grooved rolls. Now he could squeeze, as well as hammer, impurities from his heated metal, and he could roll it into plates, bars, rods, and other commercial shapes. Then in 1784 Cort patented an improvement on an invention of 1766 by the Cranege brothers (Thomas and George, of Coalbrookdale) which is known as the puddling process. The flame and gases in his furnace swirled over the molten pig iron which rested on a bed of sand so that the metal did not come in contact with the fuel and absorb so much of its impurities. Together, this reverberatory furnace and Cort's rolls made it possible for him to remove silicon, phosphorus, and other impurities as slag from the iron as he rolled the spongy mass, cut, piled, hammered, and rerolled it to suit his purposes. Wrought iron could now be obtained directly from pig iron with coke and without any charcoal whatever.

Joseph Hall improved upon Cort's process in 1838 by lining the furnace with iron oxide. As the oxygen combined with the carbon to make carbon monoxide, thus lowering the carbon content of the iron, the melting point of the iron rose so that it was changed from a liquid to a pasty substance and the slag and cinder were more readily separated from the metal in the rolling machine. In this so-called boiling process, Hall was able by means of the oxygen in the lining to reduce the phosphorus and silicon content and thus obtain a much superior metal. The oxides of silicon and phosphorus produced by this process pass into the slag.

Like Dudley before him, Henry Cort had been stubbornly opposed by ironmasters of the old school. He was deceived by his partner Adam Jellicoe, who as his share in the business had put up money he had stolen from the Navy while he was a deputy paymaster. Under the investigation following Jellicoe's sudden death in 1789, Cort was deprived of his patents and forced into bankruptcy, losing property worth by that time £25,000. He tried again and again in the next ten years to get reemployment with the government but received only a pension of £160 annually and died in 1800 disgraced, although the value of his contributions to Britain was beyond question. He had made great innovations in metallurgical engineering. He had established the iron industry upon its modern foundations. The next significant advance was to come with the commercial manufacture of steel, later in the nineteenth century.

Cement

The use of pozzuolana in cement ended temporarily with the Roman Empire, and there was probably no construction under water in which it was used again until recent times. However, its fame did not die with the Empire and engineers wished they might come upon something as good as "Roman cement." John Smeaton (1724–1792) experimented with various materials as he built the Eddystone Lighthouse in 1756. He found that the fitness of lime for underwater work was not determined by the hardness of the original limestone or the depth of the stratum or the bed in which it lay but by the presence within the stone itself of clay, and so he was able to make a lime that hardened very satisfactorily under water. His lighthouse stood for 123 years until 1879; its foundations then were undermined, not broken by the sea. Smeaton's conclusions about the value of clay overturned many ancient theories, but they were not published until after his death in 1792. Meantime he had become more interested in steam engines, harbor improvements, the Forth and Clyde Canal which was under construction from 1768 to 1790, and many other engineering projects and plans—so versatile and so restless were his talents. Others finished the work on cement.

Inspired by Smeaton, it is said, James Parker observed in 1796 that small hard lumps of clay which he found in the estuary of the Thames, when burned in a kiln and ground to powder, produced a material which hardened rapidly either in air or under water and which he called Roman cement at first. The name was soon changed to natural cement, as other kinds, artificially prepared, came into use and men learned more about their composition. The engineer Louis Joseph Vicat (1786–1861) began systematic study of the thousands of limestones in France soon after 1809, proceeding much as Smeaton had done in his search for a material that would harden under water. Vicat's conclusions were similar to Smeaton's with regard to the function of clay. By the time Vicat published his comprehensive results in 1839, however, chemistry had progressed so that he could distinguish compounds containing oxygen from other elements, calcium, silicon, aluminum, and he could establish definite chemical formulas. He attained wide recognition as the founder of the French cement industry, and his writings were quoted for years in Britain and the United States. German authorities give place beside Vicat to a less widely known contemporary, Johann Friedrich John (1782–1847) of Berlin. Cement rock, a clayey limestone containing variable proportions of silica, alumina, and iron oxide, was discovered in the United States early in the nineteenth century and quarried in many localities, especially along the lines of early canals like the Erie and the Delaware and Hudson. It was calcined, much as lime is, in kilns and then ground to a fine powder. This cement was named for its locality; Rosendale and Louisville were the better known brands. Practically no other type of cement was used in the United States until the 1870s; in fact this natural hydraulic cement continued in general use until the end of the century. Concrete made from it was used in break-water construction, dams, and heavy foundations like those of the Brooklyn Bridge.

Engineers were still a long way from comprehending all of the uses for cement as a building material, but the final step in developing its manufacture to the present time had come within a few years after Vicat's investigations. Joseph Aspdin (1779–1855), a bricklayer of Leeds, England, made the first artificial cement about 1824. His product was a carefully proportioned mixture of limestone and clay, containing there-fore silica and alumina, which he seems to have calcined and ground. He named the product portland cement because when hardened it re-sembled a popular building stone (oölite) quarried on the peninsula, or Isle of Portland, on the coast of the English Channel. Aspdin's cement was used about 1828 as a kind of stucco in Brunel's Thames Tunnel; it was tested in 1848 as a mortar for brickwork and thereafter its success was assured. At present portland cement is made from a variety of carefully selected calcareous and argillaceous materials which are intimately mixed in definite proportions, ground, calcined to a very high temperature to incipient fusion in long, nearly horizontal, rotary kilns, after which the resultant clinker is ground to an impalpable powder. Portland cement began to be imported into the United States from England, Germany, and Belgium in the 1870s. For many years it was regarded as an ex-pensive luxury to be used only sparingly. The American portland-cement industry dates from 1875 when manufacture began in the Lehigh dis-trict of eastern Pennsylvania. During the following generation its pro-duction increased enormously, and early in the twentieth century it had practically replaced the cheaper but less dependable natural cement in most localities and for most purposes.

Bibliography

Bathe, Greville, and Dorothy Bathe: *Oliver Evans*, The Historical Society of Pennsylvania, Philadelphia, 1935.

Belidor, Bernard F. de: *Architecture hydraulique* (see Chap. 6).

Dickinson, Henry W., and Rhys Jenkins: *James Watt and the Steam Engine*, Clarendon Press, Oxford, 1927.

Dickinson, Henry W., and Arthur Titley: *Richard Trevithick*, The

University Press, Cambridge, 1934.

Girvin, Harvey F.: *A Historical Appraisal of Mechanics*, International Textbook Company, Scranton, Pa., 1948.

Hammond, John L. Le B., and Barbara Hammond: *The Rise of Modern Industry*, 5th ed., Harcourt, Brace and Company, Inc., New York, 1937.

Jenkins, Rhys: *Links in the History of Engineering and Technology from Tudor Times*, University Press, Cambridge, 1936.

Pasley, Charles W.: *Observations on Limes, Calcareous Cements, Mortars, Stuccos and Concrete, and on Puzzolanas, Natural and Artificial*, J. Weale, London, 1838.

Wolf, Abraham: *A History of Science, Technology and Philosophy in the Eighteenth Century*, The Macmillan Company, New York, 1939.

EIGHT

Roads, Canals, Bridges

The Industrial Revolution in England stimulated important improvements in inland transport, first by roads and canals and later by railroads. In the early part of the eighteenth century, most British roads were incredibly bad and transport canals did not exist. Since there had been no effective program for building or repairing roads after the departure of the Romans from England, it is not surprising that the roads were poor. Writing about the Northern Road from London through Dunstable, Leicester, and Nottingham in 1724, Daniel Defoe reported: "it is perfectly frightful to Travellers, and it has been the Wonder of Foreigners, how, considering the great Numbers of Carriages which are continually passing with heavy Loads, those Ways have been made practicable; indeed the great Number of Horses every Year kill'd by the Excess of Labour in those heavy Ways, has been such a charge to the Country, that the new Building of Causeways, as the *Romans* did of old, seems to me to be a much easier Expence." He emphasized the need on most roads for new bridges, "which not only serve to carry the Water off, where it otherwise often spreads, and lies as it were, damn'd up upon the Road, and spoils the Way; but where it rises sometimes by sudden Rains to a dangerous Height; for it is to be observ'd, that there is more Hazard and more Lives lost, in passing, or attempting to pass little Brooks and Streams, which are swell'd by sudden Showers of Rain, and where Passingers expect no Stoppage, than in passing great Rivers, where the Danger is known, and therefore more carefully avoided." [1]

Obviously, such roads were not adequate to transport the increasing

[1] Daniel Defoe, *A Tour thro' the Whole Island of Great Britain*, 2 vols., printed for Peter Davis, London, 1927, vol. 2, pp. 518, 530.

number of goods produced toward the end of the eighteenth century, although, as Defoe pointed out, not all English roads were as bad as the ones he described. Nevertheless, the roads in general were most unsatisfactory even for the traveling merchants of the earlier decades who took their goods from town to town during the summer on the backs of pack animals. By the middle of the century, some of the "turnpike trusts" had managed to improve old roads and construct new ones. Indeed the turnpikes had accomplished some improvements by the time Defoe wrote, but, as he put it, "the Inland Trade of *England* has been greatly obstructed by the exceeding Badness of the Roads."

It was not until the last half of the century that important engineering advances in road construction were made, and they were based largely on previous French experience. The construction of English canals, which also began in the last half of the century, was patterned after the work of French engineers. With the new roads and canals came also more attention to bridges.

Road builders did not use cement for their road surfaces until the latter part of the nineteenth century. Nor were they led to think of asphalt, although Sir Walter Raleigh had observed it on a visit to Trinidad in 1595 and had praised it as a material for caulking ships. It did not come into use for roads until two and a half centuries later. The engineers of the Industrial Revolution followed practices which had varied little from Roman through medieval times. They relied still upon compression under the wheels of passing vehicles for consolidation of their roads. Only recently, as self-propelled vehicles with rubber tires replaced horses, have road builders turned to reinforced concrete. Since horses require elastic footing, our modern hard-surfaced roads would have been quite impracticable in the days of horse-drawn traffic.

The great overland European routes had been maintained during the Middle Ages, but hardly more than maintained. As men broke from the feudal regime, ambitions became nationwide in purpose. Swifter communications were necessary for the defense of the larger areas brought again under central control. Growing cities had to have better transport from more distant sources of supply and better access to other distributing points. National interest in highway engineering spread slowly, but at length, as the need became unmistakable, France established its Corps des Ponts et Chaussées in 1716. A Frenchman fixed the principles of modern road making which endured until highway engineers came at last to the use of cohesive materials. Pierre Marie Jérôme Trésaguet (1716-

1796) began to formulate his technique in 1764; in that year the famous Scottish builders John Loudon McAdam (1756–1836) and Thomas Telford (1757–1834) were boys of eight and seven.

There were some roads of broken stone before the time of Trésaguet, Telford, and McAdam. The writings of the French Hubert Gautier, first published in 1693, indicate that small pieces were laid on edge and tamped with finer material. At this time treatises on road making appeared in England and the German States. There was construction of a similar nature in Sweden and in Russia. But it was Trésaguet who systematically improved upon the foundations of the Romans and finished his roadway with broken stone carefully graded in size. Lessening the total depth to some 10 inches, Trésaguet laid the first course on edge so that the individual pieces could be packed with a hammer more closely than if they were laid flat (Figure 8.1). The second layer was similarly placed by hand and wedged with hammers so that no crevices remained. Both of these courses were crowned or curved below the surface to parallel the finished crown of the road. The third and last course, 3 inches thick, was made of hard stone broken to the size of walnuts. Since he relied upon this layer to consolidate the road, Trésaguet sought the hardest stone available. He insisted also upon keeping the surface under systematic maintenance and

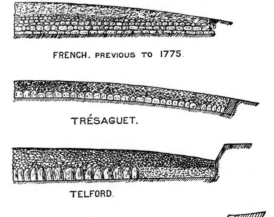

FRENCH. PREVIOUS TO 1775.

TRÉSAGUET.

Figure 8.1
Various highway cross sections (From Byrne, *Treatise on Highway Construction*, 1892)

TELFORD.

MACADAM

repair. Although he had given his country the best roads in the world, Pierre Trésaguet enjoyed national respect only as Inspector-General of Bridges and Highways for Louis XVI. Thanks to the Revolution, it was his fate to die in poverty and be all but forgotten by the time of Napoleon.

Thomas Telford's roads were built with the same intention of having the lower courses so stable that they would carry the weight of the traffic even though the soils beneath them were spongy and yielding. The Scot laid his lowest course flat, possibly because he had begun as a mason; he obtained his crown by making the upper courses thicker at the center than on the sides (Figure 8.1). The middle layer was 7 inches of broken stones small enough to pass through a 2½-inch ring. For the top course he used an inch or so of gravel. Two of Telford's best roads were the highway between Glasgow and Carlisle, nearly 100 miles long, and the thoroughfare from London to Holyhead in Wales, some 300 miles. The latter was built from 1815 to 1828 as part of the mail route to Ireland at a cost of 3½ million dollars. Where it crossed the Menai Straits, Telford erected his remarkable suspension bridge.

The roads of McAdam made him the most renowned of these three engineers; so well known that his name has become part of the English language. His construction was thinner than Trésaguet's or Telford's, much less expensive, much easier to build. McAdam's unique contribution was his insistence that the subsoil must be thoroughly drained. It had to be dry and kept dry; if it were dry, it would be firm, and there would be no need for deep foundations. The thickness of the road, he said, "should only be regulated by the quantity of material necessary to form such impervious covering, and never by any reference to its *own* power of carrying weight." [2] Neither Trésaguet nor Telford had stressed this observation.

A macadamized road was hardly ever more than 8 inches thick, but its broken stones, each no larger than could be held in one's mouth and no heavier than 6 ounces, had to be sharp-edged and so tightly packed that rain could not penetrate. The surface had to be so well crowned that puddles would not stand on it (Figure 8.1). In a climate such as England's, where frosts do not strike deeply nor thaws heave the soil, McAdam's roads were superior except for the heaviest use. Until the stone covering itself was worn through, traffic only smoothed and bound the

[2] John L. McAdam, *Remarks on the Present System of Road Making*, 9th ed., Longman, Rees, Orme, Brown, and Green, London, 1827. p. 47.

surface more firmly. English travelers yearned for the macadamized roads of the homeland and loathed the ruts and mud in America, much as they recoiled from American table manners.

None of the roads in the American Colonies prior to their revolt from the mother country was the work of engineers. Only two of those built in the thirty years following the war deserve attention. This lack was not because people were satisfied with the intercolonial routes which they had; contemporary accounts were full of the hardships and dangers of such overland journeys as they felt obliged to make. It was rather because those who lived close to the seaboard could travel by coastal waters, sounds, bays, and rivers if they cared to visit or trade with one another. There was always local demand for better roads into the back country. Colonial statutes give full evidence of it. Writing in 1652 about Salisbury in Essex County, Massachusetts, Edward Johnson stated that "the constant penetrating farther into this Wilderness, hath caused the wild and uncouth woods to be filled with frequented wayes, and the large rivers to be over-laid with Bridges passeable, both for horse and foot." [3] By 1725 stagecoaches were running on schedule between the larger settlements, in particular across New Jersey from New York to Philadelphia. John Mercereau advertised his vehicles in 1771 as "Flying Machines" which made the trip in 1½ days. He neglected to state that the traveler from New York had to leave home the night before, sail to Perth Amboy, and take his seat at three o'clock in the morning in the "Imitation of a Coach" which had no springs.

Early in the nineteenth century Timothy Dwight referred in the record of his travels to "the great road from Georgia to Maine," but no one should think of it as a national highway, conceived, constructed, and maintained in 1794 according to a general plan. It was, like the present U.S. Route 1, no more than the linking up of a series of local roads made by chance according to immediate terrain and convenience. Local interest grew into interstate rivalry and national purpose as the frontiersmen moved up the valleys beyond tidewater and crossed the mountains into the forests and prairies of the mid-continent.

The Lancaster Turnpike, completed in 1794, reached 62 miles through the rich farming country of Pennsylvania from Philadelphia to Lancaster, largest of the country's inland towns. It was essentially a local street serving a thickly settled community. Like so many of its British pro-

[3] Edward Johnson, *Wonder-working Providence, 1628–1651*, Charles Scribner's Sons, New York, 1910, p. 234.

genitors, it was built by a private corporation hoping to profit from the tolls, and until the railroad superseded it, the investors received from 6 to 15 per cent annually. For the first time in America, but not the last, the state exercised its power of eminent domain to aid private capital. As land was condemned for the right of way, there was a torrent of complaint. There was a flood also of subscriptions for stock in the corporation. Of the 2,275 applicants, only 600, chosen by means of a lottery wheel, were permitted to subscribe. The cost proved to be about $7,000 a mile. One contractor at least resented the implication that this was too much. Harassed by the "Mallice" of onlookers "who prostitute truth at the Shrine of Malevolence," Matthias Slough protested: "I can lay my hand on my heart, and declare that I, in no instance, wantonly sported with one shilling of the Company's money." [4]

The scientist David Rittenhouse had charge of the original surveys. Hasty construction with stone and boulders of every shape and size led to disaster as the rains opened gullies and made sinkholes in which horses broke their legs. However, the advice of more experienced builders like William Weston, recently arrived from England to construct neighboring canals, helped to produce in the end a solid roadway, which the British traveler Francis Baily in 1796 called "a masterpiece." The Lancaster Turnpike was 20 feet wide and 17 inches deep, paved the whole distance of 62 miles with stone overlaid with gravel, so that it was never obstructed, said Baily, "during the most severe season."

The only other early thoroughfare in the United States to deserve attention was the Cumberland or National Road, from its conception designed to be publicly owned and operated. Albert Gallatin, Secretary of the Treasury for President Jefferson, fostered a national policy of local improvements at general expense in 1803 as part of the plan for the admission of the state of Ohio into the Union. The Cumberland, as the first of these roads, was to run from the Potomac River in Maryland over the mountains into the Ohio Valley. For some distance it followed the route of General Braddock's military road in 1755 through the wilderness toward the French position, then Fort Duquesne, now Pittsburgh. However, the rising industrial city where the Allegheny joins the Monongahela to form the Ohio River was not intended to be the terminus of the National Road, which turned away to cross the river into the forest lands of the Ohio country at Wheeling, now West Virginia. The eventual destination was St. Louis beyond the prairies of Illinois on the farther bank of the Mississippi River.

The original requirements as reflected in the authorizing Act of Congress, March 29, 1806, called for a roadway 66 feet wide and paved in the 20-foot carriage path with stone, earth, or gravel and sand. The right of way was later increased in Ohio to 80 feet. There were to be drainage ditches on both sides. The grade in no instance was to be inclined more than 5 degrees with the horizon, which meant a maximum gradient of 8.75 per cent, moderate for those times although at least twice as steep as the acceptable maximum today on high-speed superhighways. The route was to be laid out by "three discreet and disinterested citizens of the United States." To be both disinterested and discreet in the midst of local jealousies is indeed the mark of distinction for a public servant. Albert Gallatin, near whose lands it was bound to pass, was discreet and wise; he made sure that President Jefferson himself should undertake the selection and direction of the commissioners. Jefferson chose Thomas More and Eli Williams of Maryland with Joseph Kerr of Ohio, and they selected Josias Thompson, "a surveyor of professional merit." Together these men located the road so that it crossed not only Virginia and Maryland but also a corner of Pennsylvania. Doubtless, terrain had most to do with fixing the route as the road climbed 2,300 feet through the mountains; it was good politics as well to include the three states. The work was assigned in sections to local contractors, although this procedure did not necessarily make for uniformity in construction or quality.

According to their report in 1806 the commissioners expected the cost from Cumberland in Maryland to the Ohio River at Wheeling, about 112 miles, to be $6,000 a mile, without including the bridges along the way. They still believed in 1808 that the road could be "properly shaped, made and finished in the style of a stone-covered turnpike" for that amount. When completed to the Ohio in 1818, however, the National Road had cost practically $13,000 a mile. Some of this increase, of course, was chargeable to unexpected delays and to extra expenditures with which engineers are all too familiar. How much resulted from carelessness, incompetence, scamping, or downright theft is not known in detail, but when the Army came to repairing the road in 1834, the engineer in charge reported no less than eight different ways in which local contractors were defrauding the government. The more obvious were altering the gradient, failing to break the stone as specified, diminishing the amount of materials delivered, and charging twice for those which were supplied.

Stages, carriages, trains of Conestoga wagons thronged the National Road for a decade before the Erie Canal opened in 1825 to compete with it altogether successfully. Some five thousand wagons were unloaded at Wheeling in 1822 with about $400,000 paid in costs of transportation. Every tenth wagon, it is said, passed through to Ohio. Returning loads brought such Western produce to Eastern markets as flour, whisky, hemp, tobacco, bacon, and wool. The population of Wheeling doubled. From four to six weeks had been required to travel from Baltimore to the Ohio before the road was opened; by 1832 both time and cost had been cut in half. These figures are merely indicative. There is no measuring with statistics the full effect of the National Road upon the life of the American people as they moved westward with their household goods in relative ease along its stretches toward the Ohio country and beyond. The traffic was so heavy that the first roadbed soon needed major repairs.

In July, 1832, General Charles Gratiot, Chief of Engineers, detailed Lieutenant J. K. F. Mansfield (1803–1862) to supervise this rebuilding east of the Ohio River and gave explicit orders to have it done according to the principles in McAdam's *Remarks on the Present State of Road Making*, first published in 1820. Mansfield's successor Captain Richard Delafield (1798–1873) carried out the instructions with even more deference to McAdam. These officers had learned their engineering at West Point. Thanks to the influence of Gallatin and Lafayette, Claude Crozet (1790–1864), veteran of Napoleon's march to Moscow and the "hundred days" to Waterloo, had been West Point's professor of engineering from 1816 to 1823, and the textbook used was that of Joseph-Mathieu Sganzin (1750–1837), once Inspector-General of Bridges and Highways in France and a younger colleague of Pierre Trésaguet. The influence of French road building upon American practice therefore was certainly direct. Its fusion in America with the principles of McAdam becomes more interesting as one speculates upon the amount of Trésaguet's technique which Mansfield and Delafield had absorbed in their studies at West Point.

The War of 1812 accentuated the need of strategic lines of communication for purposes of defense and brought the policy of local construction at general expense to the fore as the major issue of "internal improvements." The United States had gained not only economic independence from Britain but also uncontested control over the Mississippi Valley to the Rocky Mountains. It was natural that statesmen like Henry Clay and John C. Calhoun, young with the nation, should advocate a policy in keeping with the opportunity which lay before it. But the proposal that the government of the United States should survey, construct, own, and operate a national system of roads and highways hardly survived the fervor of the war with Britain which had made the suggestion possible.

Opponents of Federal expenditures upon local roads were numerous among taxpayers in sections of the country where no immediate benefits from those improvements would be available. Devotees of state sovereignty construed the Constitution strictly to find in it no authorization for such appropriations from the United States Treasury, and of course private companies and states themselves wished to keep the business of providing the means of transportation. Turnpikes and toll bridges were lucrative. The far view of Albert Gallatin in 1803, and of others who favored a national road into the West to be open and free for everybody, was blocked by the nearer objectives of private corporations like the Lancaster Pike and by state enterprises such as the Erie Canal in New York. These were improvements no less national in ultimate purpose, but they were meant to produce immediate and local profit.

Yielding to constitutional scruples and political expediency, Presidents Madison, Monroe, and Jackson vetoed successive measures of Congress for local improvements at Federal expense. The Federal government did not respond to the overture from the State of New York for a grant-in-aid; New York dug its own canal. Pennsylvania and other states pushed their own systems of roads and canals toward the Western country. Private promoters were everywhere feverishly competing to supply the ways and to collect tolls from the pioneers as they passed in the greatest overland migration the world had seen. By the 1830s the concept had changed from government ownership and operation of a national system to Federal grants of land from the public domain to private or state enterprises. Even the National Road was never completed to its destination at St. Louis. By 1856 its sections had all been turned over to the respective states to maintain as they saw fit. This they did after a fashion, collecting tolls, until privately owned railroads had absorbed the public interest and had made the National Road a relic for the time being.

Canals

Better transportation by water accompanied road building. In the United States, canals with long-range objectives promised to be the means of communication with the Mississippi Valley even more than the National

Road—until the railroad came to surpass both. As William Weston, James Renwick, and others from Britain planned or directed successive projects, the influence upon American construction, including the Erie Canal, was preponderantly British. And this influence stemmed from the remarkable work of James Brindley (1716–1772) in the Midlands of England.

Francis Egerton, third Duke of Bridgewater, traveling as a young man in Europe, marveled at the Languedoc Canal of France. Its locks, aqueducts, and tunnel made transportation by water possible through mountainous country. He had coal to get down from his mines 10 miles back of Manchester, across the Irwell Valley, and up again to the industries in the city. To do the job, he called upon a local mechanic virtually without education, some twenty years older than himself. All that James Brindley knew about canals when he started was what he had heard from the Duke; before he finished, Brindley was a celebrated engineer. The Duke and his heirs gained considerable profit. As surveyor, engineer, contractor, and foreman, Brindley acquired 3 shillings 6 pence a day, and fame.

More impressed by the Duke's report on the aqueducts of the French canals than by their locks, Brindley made what he described as an "ochilor servey or a ricconitoring" and prevailed upon his employer to let him keep to one level, crossing the Irwell River on a stone bridge 40 feet above the stream. The result was the Barton Aqueduct of 1761 (Figure 8.2) which has influenced builders of water conduits from Brindley's younger contemporary Thomas Telford to engineers of today. It was an exacting task to carry the embankment over an extensive bog without settling and cracking. This Brindley accomplished by first digging drainage ditches to increase the bearing power of the soil; then he spread the load and distributed the weight of the embankment over a large enough area for proper bearing. It was even more difficult to make the channel watertight. But he learned, if he did not already know, how to puddle clay with sand in successive layers so that the mixture after it dried was impervious to water. When Telford came later to build his aqueducts on the Ellesmere Canal, he used on the first a cast-iron bottom with sides of brick masonry laid in Parker's cement, which had just come on the market. For his second aqueduct, Telford built a complete iron trough.

Brindley's canal for the Duke of Bridgewater was only 18 feet wide, 4½ deep, and 10 miles long, but it halved the price of coal at Manchester when the first little boatload arrived in 1761 and created good business for the Duke. Its cost he never revealed. Raising his sights, he took aim at the freight rates between Manchester and the port of Liverpool; they were 40 shillings a ton over the road, 12 by the river and tidewater route. Brindley extended the Duke's canal on a winding route along the valley, still keeping it level, and then let it down in a series of 10 locks 82 feet into the tidewater of the Mersey at Runcorn, 15 miles southeast of Liverpool. The total cost of the Bridgewater Canal has been estimated at £220,000. Its completion reduced by 50 per cent the freight charges between Liverpool and Manchester. It survived for some time even the competition of the Liverpool and Manchester Railway which was built

Figure 8.2 Barton Aqueduct (From Smiles, *Lives of the Engineers*, vol. 1, 1862)

in 1830. The canal was more economical at first than the railway. There were gentlemen's agreements, too, with respect to rates and division of tonnage.

Today the Manchester Ship Canal, completed in 1894, follows a shorter route in the valley. Brindley's aqueduct has been replaced by a swing span where the Bridgewater Canal crosses the Irwell River and the Ship Canal. The towpath has been carried over in a gallery 9 feet above the water level in the swing span. To conserve water, the span was constructed as an iron tank 6 feet deep and 19 wide. Four movable yet watertight bulkheads had to be provided, two on the movable span and one on each shore end. The joints had to be so closely fitted that they were watertight whether the bridge was open or closed. Coal barges were often swung with the tank as the span opened to allow ocean-going vessels to pass.

With the success of the Duke's canal, Brindley was in great demand to build others. The first of these, known as the Grand Trunk, connected the Mersey with the Trent and Severn valleys, linking Manchester and Liverpool by water with Hull and the North Sea to the east and with Bristol and the Atlantic Ocean to the southwest. It was designed to be the main stem of a network which should serve the industries of central England. Leaving the Bridgewater Canal on the north it passed southeastward through the Cheshire salt district and close to the potteries of Staffordshire. Josiah Wedgwood actively promoted this canal, for he had been obliged to bring much of his flint and clay in and carry his products out by pack horses. The saltworks of Cheshire had obtained their coal from Staffordshire in the same expensive way. The economic advantages in having the canal were obvious. John Wesley observed that it also improved the people. Samuel Smiles, biographer of Brindley, declared that the population in the area of the potteries increased from 7,000 partially employed in 1760 to 21,000 "abundantly employed, prosperous, and comfortable" [5] twenty-five years later.

James Brindley had to bore a tunnel at the summit level of the Trent and Mersey Canal. He had previously bored a tunnel into the Duke's coal mine at Worsley, but this one at Harecastle was the first real canal tunnel in Britain (Figure 8.3). It took eleven years to complete as the rock was difficult and the workmen were at first inexperienced. When finished in 1777, it was 1⅔ miles long, but it was only 12 feet wide, and there was no towpath. The men and women who pushed the tiny barges through

[5] Samuel Smiles, *Lives of the Engineers*, John Murray, London, 1862, vol. 1, p. 448.

the tunnel did so by "legging," or pressing their feet against its roof. Sixty years later, Thomas Telford completed a parallel tunnel of larger bore in three years. There was a towpath, adequate, however, only for human "tractors."

During the last decade of the eighteenth century, there came with steam power in manufacturing a veritable mania for building canals in Britain. This new means of transportation reduced costs sharply and speeded the growth of such factory centers as Manchester, Birmingham, and Leeds, which soon had cross-country communications by water with Liverpool, Hull, Bristol, and London. The development on the Continent was slower, as a result, doubtless, of the derangements which had been caused by the revolution in France and the succeeding wars of Napoleon. By 1808, however, Thomas Telford (1757–1834), outstanding among British engineers, was designing the Gotha Canal for Sweden, a project which had long been germinating. With its generous dimensions, big locks, and length of 120 miles—considerably greater than any of the canals

Figure 8.3 North entrances to Harecastle tunnels: left, Telford's with tow-path; right, Brindley's (From Smiles, *Lives of the Engineers*, vol. 1, 1862)

in Britain—the Gotha Canal deserves its rank among the achievements of transportation engineering prior to the railroads.

In the Netherlands, supplementing a system of canals whose origins in earlier centuries have been described in Chapter 6, the North Holland Canal, 50 miles long, was completed in 1825 to give Amsterdam a much better connection with the North Sea. In the same year, after visiting England and Scotland, Charles Dupin tried to arouse his wealthy French compatriots to action. But it was not until 1835 that French engineers again took up the work at which they had been expert and started the canal to link the Marne with the Rhine. As completed some years later by way of Nancy and Strasbourg, 200 miles distant, it crossed two water-sheds with 177 locks and about 5 miles of tunnels.

The American canal first to be planned and actually completed was the Santee, built to give Charleston by way of the Santee River a water connection with Columbia, the new inland capital of South Carolina. It was chartered in 1786 and built by a private corporation at a cost of $750,000. It was 35 feet wide at the surface, 4 feet deep, and 22 miles long; there were 13 locks of brick and stone with wooden gates. Most of the work was done by slaves, of whom fully a third were women. When opened to traffic in 1800, it had taken the best part of seven years to finish. Lack of water often left 5 miles of the upper canal dry, but the Santee served until replaced by the railroad in 1850. George Washington's "Potomack Company" loomed in his clear vision of "our rising empire" beyond the Appalachians. The canal was chartered in 1785 and incorporated later into the Chesapeake and Ohio Canal. In a determined effort to fulfill Washington's great hope, the engineers built an aqueduct of nine arches across the Potomac at Georgetown, and they drove a tunnel ⅗ mile long through the ridge beyond Cumberland, Maryland. At last, after an expenditure of 15 million dollars, the Chesapeake and Ohio Canal was open in 1850 as far as Cumberland, 184 miles; but it never reached the Ohio. It got no farther into the West than the beginning of the National Road. The traffic was wholly inadequate for so great an investment in a canal. The Baltimore & Ohio Railroad had already doomed a venture which never had enjoyed the fair prospect of canals farther to the North.

The promoters of the Middlesex Canal in eastern Massachusetts planned to reach the trade of Canada. Their route was to be, from tidewater at Boston to Chelmsford on the Merrimac River and then northward by way of Lake Sunapee in New Hampshire and the upper waters of the Connecticut. The immediate objective, however, was the firewood, timber, and other produce of New Hampshire for the Boston market. As the first determination of the relative elevations of the Mystic and the Merrimac Rivers proved inaccurate, the directors called upon the English engineer William Weston (1753–1833?), then in Philadelphia. Work began in 1794 and ended in 1803 after an expenditure of nearly $600,000; common labor had cost from $8 to $10 a month and keep. The width of the channel at the water line was 30 feet and the depth 4. There were 7 wooden aqueducts over intervening streams and 20 locks of solid masonry, 90 feet long and 12 feet wide. Three of these in the last 6 miles at the northern end let the canal down 28 feet to the Merrimac. Some excavations along the way were 20 feet deep. The boats were drawn by two horses at 3 miles an hour, passing the whole length of the canal in twelve hours. In his report on the roads and canals of the United States, Secretary Gallatin called attention in 1808 to one raft of timber, a mile long and weighing 800 tons, which had been drawn through stretches between the locks of the Middlesex Canal by a pair of oxen at 1 mile an hour.

Repairs and alterations were constantly necessary with traffic insufficient to pay for them and to provide any dividends before 1819. Then the founding of the textile industry in the city of Lowell near the head of the canal at the falls on the Merrimac brought cargoes which had not been anticipated and some prosperity. Good times were not long for the Middlesex Canal, however, for it could not withstand the competition of the two railroads between Boston and Lowell. It was dead by 1846. As it passed out of existence, there was some talk of using parts of it to bring water into the city of Boston. Far superior means, however, were soon found for solving that municipal problem. Today, the course is hard to trace through the truck farms of Middlesex County and the woods which are springing up again everywhere on the idle lands of Massachusetts.

Another canal in Massachusetts deserves notice although it was only 2 miles long. It was blasted through rock much of the way in order to bypass the 50-foot fall in the Connecticut River at South Hadley and Holyoke. There were six locks in this little canal, replacing an earlier inclined plane called the Hampshire machine, which had been operated by two 16-foot water wheels. The hope of the canal builders in 1805 was that it would help to clear the water route for travel from tidewater at Hartford, Connecticut, to the Canadian border. Dutch bankers took shares in the project and lost heavily. Attempts to navigate the Connecticut with steamboats were discouraging. The river was not reliable; its water was often so low the passengers did a great part of their traveling on foot

Figure 8.4 A Morris Canal boat on an incline (From *Scientific American*, Supplement, Feb. 24, 1883)

along the banks while the steamer labored with the river. In 1847 the South Hadley Canal was taken over by the power company to which Holyoke owes its existence as an industrial city.

The Morris Canal was built across New Jersey's hills to join the Delaware River with tidewater in New York Harbor. As Gallatin had suggested in 1808, such a canal would provide a sheltered way for coastwise commerce which had to round Cape May in the open sea. Locally, it would carry anthracite coal from Pennsylvania near Easton to the iron mines and furnaces of northern New Jersey. While engineers elsewhere used locks for the most part, those who had to deal with the rolling terrain of New Jersey and Pennsylvania preferred inclined planes (Figure 8.4). These had been used for centuries on the Grand Canal of China to gain elevations of a few feet; a number had been installed in 1792 to raise 20-foot barges as much as 200 feet on the Shropshire Canal in the west of England.

Lake Hopatcong was enlarged to supply the water for the channel across the summit of the Morris Canal, which was 915 feet above the sea. To overcome such height, the designers, James Renwick (1790–1863) and David Bates Douglass (1790–1849), used 23 ordinary locks and 23 inclined planes, or boat railways, with a gauge of 12½ feet, along the 100-mile route. Each of these planes rose on an average 63 feet over lengths varying from 500 to 1,500 feet. Chains were used at first; they were later replaced by iron cables. A barge moved vertically 2,134 feet while passing through the canal; power to lift it came from cast-iron water wheels which produced in all 704 horsepower. These "Scotch turbines" operated on the principle of reaction from a head of water admitted from below so that the upward pressure balanced the weight of the wheel, the shaft and gear, and relieved the friction on the bearings. Acclaimed by its promoters "the boldest canal in existence," the Morris Canal was opened to traffic in 1831 and, despite a financial failure which angered William Willink of Amsterdam and other Dutch investors, withstood much better than most canals the competition of the railroads. Not until 75 years later was it virtually idle. Recently acquired by the state of New Jersey, the Morris Canal has now been appropriately marked, dismantled, and consigned to history. One turbine wheel has been preserved to show the engineering of a century past. By test under a head of 47 feet, its best speed was 87 revolutions per minute; the calculated actual efficiency was 60 per cent.

William Penn had proposed in 1690 a canal between his great city on the Schuylkill and the Susquehanna where he planned to establish his second. Communication by land, he said, was "as good as done already" and would "not be hard to do by water." However, it was not until 1772 that David Rittenhouse made some surveys for the Schuylkill and Susquehanna Canal. While the Lancaster Turnpike was being completed, William Weston directed the preliminary construction of the canal in 1793. Work then stopped as the company ran short of funds, and it was 1821 before the American engineer, the younger Loammi Baldwin (1790–1838), took over the project in Pennsylvania to put it in final form as the Union Canal. Baldwin, Harvard-trained in the law, may have early picked up some of his engineering on his father's Middlesex Canal, possibly under William Weston there. Baldwin resigned after three years because the directors would not approve the larger locks and deeper channel he advised and which later were necessary. Then Canvass White came down from the Erie Canal to finish the work, but he had to give up after a year on account of illness. The Union Canal, as originally planned, was to carry the produce of the inland farming country into Philadelphia rather than allow this traffic to descend the Susquehanna to Baltimore. The hope was that the canal would also attract the commerce of central New York into the market of Philadelphia. In the 1830s one could travel three-quarters of the way from Philadelphia to Pittsburgh on the state's Pennsylvania Canal. Leaving the railhead at Columbia the canal wound up the valleys of the Susquehanna and Juniata Rivers 172 miles to Hollidaysburg on the eastern slope of the Alleghenies. Here Pittsburgh passengers and freight had to cross the mountains to Johnstown in the

Conemaugh Valley, a 36-mile trip by means of the unique Allegheny Portage Railroad, which will be described in Chapter 9. From Johnstown the western division of the canal followed down the Conemaugh. Kiskiminetas, and Allegheny River Valleys, 104 miles to the Ohio at Pittsburgh.

Of all the new systems of transportation which came in the United States after the War of 1812, the Erie Canal, built by the State of New York from 1817 to 1825, combined engineering and geography most effectively to produce economic benefit. It had greater influence upon the development of this nation than any other route into the mid-continent prior to the construction of the railroads, and it was able even to compete with them. Its successor, the New York State Barge Canal, built from 1903 to 1925, is still transporting oil, grain, and other commodities of bulk, when it is not frozen.

The conception of the Erie Canal was bold by its very magnitude; its realization was an outstanding event. The construction itself was not extraordinarily difficult. The Chinese many centuries earlier had built the Grand Canal, 1,000 miles long. Other feats of engineering were more spectacular than the Erie Canal. Nature was not in general forbidding along its length of 363 miles from Albany on the Hudson River to Black Rock on Lake Erie. There was a climb of 420 feet from the Hudson to Utica, and the highest elevation above the sea, near Buffalo, was less than 600 feet. The Cayuga marshes caused some trouble; there were malaria and dysentery among the workmen. An aqueduct of red sandstone laid in cement was thrown 750 feet across the Genesee River at Rochester, with 12 piers and 11 arches, mostly 50-foot spans. There were other aqueducts of impressive size over the Mohawk, one at Little Falls with a span of 70 feet. The embankment carrying the channel across the Irondequoit Valley extended nearly a mile over three ridges, the highest rising 76 feet. There was a deep cut beyond Lockport where a series of five double locks (Figure 8.5) rose some 60 feet, but for the most part the terrain was fairly level and low, with few obstacles to the familiar routine of digging ditches and laying up masonry. No dams of any size or particular complication had to be constructed. The engineers had no great ridges to cross or to tunnel as on the Chesapeake and Ohio Canal. All the way there was abundant water from the lakes and streams of a glaciated countryside.

Some innovations were made on the Erie Canal in methods of excavation. Plows, root cutters, and scrapers, drawn either by oxen or by horses, were employed for types of work still being done at that time in

Figure 8.5 Erie Canal locks at Lockport (From *Harper's New Monthly Magazine,* 1881)

Britain by men with shovels and wheelbarrows. There was an ingenious stump puller on 16-foot wheels which applied the ancient principle of lever and wheel so effectively that one team and seven laborers could remove from 30 to 40 tree stumps a day. The organization of the whole construction was unusually well managed. Operations on the three divisions of the Erie as well as on the Champlain Canal went forward simultaneously. For political and practical reasons all were treated as part of a single task. Canvass White (1790–1834), who assisted Benjamin Wright (1770–1842) with the early surveys, made an extensive trip through Britain in 1817 examining canals, investigating cements, acquiring surveying instruments, and becoming expert on locks. The first hydraulic cement for the Erie Canal had to be imported at considerable expense from Britain, so White explored the neighborhood. He discovered a proper limestone in Madison County near Chittenango and from it made the first American hydraulic natural cement for which he obtained a patent in 1820.

Possibly no one among the American engineers who worked on the Erie Canal had ever seen a canal dug before. Some familiarized themselves with British practice, as did White, or studied the Middlesex Canal in Massachusetts; all learned in the training school of experience. From the Erie, many went to other canals and on into the new, though not too dissimilar, profession of railroad building. Their achievements on the Erie Canal, while worthy of note, should not be overstressed. The fact persists, despite legend, that the engineering of the Erie Canal, though effective, was neither especially difficult nor unique.

The first earth of the Erie Canal was turned at Rome near Lake Oneida on July 4, 1817. The locks were completed in the western division at Lockport so that boats could be passed through the canal into and from Lake Erie on October 26, 1825. Then came the parade across the state. Aboard the canal boat *Seneca Chief*, Governor Clinton received ovations in Rochester, Syracuse, Rome, Utica, Schenectady, perhaps tempered from place to place by local feelings, as in Schenectady where one editor spoke of the event as a "funeral." Governor Clinton was preceded by the news of his coming. Cannon spaced along the way sent the message from Lake Erie to New York City, 500 miles or so, in 100 minutes, and then relayed it back to Buffalo. Today's teletypists may be amused. At Albany, the steamer *Chancellor Livingston* took the *Seneca Chief* in tow for the final procession down the Hudson to New York Harbor. There, on November 4, 1825, DeWitt Clinton emptied into the ocean the much celebrated keg of water which he had brought from Lake Erie; how many other kegs made this happy trip from the Ohio country all the way to Manhattan was not recorded.

New York State's enterprise hastened the settlement of its western counties and increased migration beyond into northern Ohio, Michigan, Indiana, and Illinois. In 1816, Rochester was a village of 331 inhabitants; at the opening of the Erie Canal in 1825, there were nearly eight thousand and the city was becoming the flour-milling center of the United States. Buffalo, no more than a trading post, was on its way to possession of shipping, mills, meat-packing, and iron industries. Forty thousand persons had passed through Utica on the canal before the end of its first year. Before it was finished, its tolls were helping to reimburse the State of

New York for the initial cost to the people of $9,027,456.05, or some $24,869 per mile.

The effect of the Erie Canal upon transportation from West to East was especially marked. Men of Ohio thought of canals too and dug them across the portages of their state from Lake Erie to the Ohio River—continuations of the Erie Canal deep into the Mississippi Valley to bring eastward the grain and later the cotton which once had gone down the river to market in New Orleans. Cleveland, Ohio, quickly emerged as a distributing and manufacturing center. Lake vessels docked at the western terminus of the Erie Canal in Buffalo with cargoes from Detroit, Mackinaw and beyond, even from as far as Chicago. Primary Western industries producing such commodities as lumber, potash, pork, whisky, and flour were the first as a matter of course to be stimulated by the Erie Canal. Because of superior soil and growing conditions, these industries were certain to forge ahead of similar enterprises in the East. Rochester, New York, could not expect to retain the lead as the milling center of the United States. For a while, too, cargoes of bog iron from Cleveland came through the canal to the iron industries of the upper Hudson Valley. But great ironworks were soon growing with the settlements beyond the Appalachians nearer the sources of raw materials and fuel. The Erie Canal was not important as a carrier of ore from Western mines to Eastern mills. It was an effective agent in speeding the development of the communities in the Middle West, which later greatly stimulated iron and steel manufacturing within the region of the Ohio and the Great Lakes.

The 1825 ceremony off Sandy Hook marked the establishment of New York City rather than Philadelphia or Boston as the metropolis of the United States and the greatest port of North America. It also dramatized the economic and political union of the old Northwest Territory with New England and the Middle states of the Atlantic seaboard, and then came the railroad to accelerate Northern expansion at the expense of Southern supremacy in the West.

One of the later American canals, the Farmington or Northampton, deserves special mention because it was deliberately converted into a railroad. Extending north from New Haven, Connecticut, it was conceived by some of its projectors as one link in a waterway that would connect Long Island Sound with the St. Lawrence River. It was chartered in 1822 and was privately financed, in part by organizing a bank in New Haven. Started in 1825, the 56-mile ditch was completed in three years under the direction of engineers with previous experience on the Erie. It was later extended to Northampton, Massachusetts, on the Connecticut River. Compared with the Erie Canal, which had been finished just as work on the Farmington Canal began, the latter was a minor project. It did, however, provide leisurely and comfortable transportation to a large number of old towns. It also served dozens of small typical New England manufactories which hitherto had not been able to expand on account of difficulty in shipping their products over the rough country roads. The canal boats were horse-drawn, the largest with a capacity of 25 tons. Boats and horses were privately owned, and their owners paid toll to the canal company which operated with varying success for less than twenty years. It was during this time that the Hartford and New Haven Railroad was built, roughly paralleling the canal a few miles to the east. In 1847 it was proposed to build a railroad along the canal's towpath, both canal and railroad to operate, perhaps, simultaneously. In fact, for a time much of the material for the railroad was hauled through the canal. Finally, in the late 1840s the canal was abandoned and the railroad was built close to it but following the devious towpath only in certain places. The railroad was built by the financiers and engineers who had been actively associated with the canal company. It was one of the few instances in history in which a canal, instead of disappearing, submitted willingly to metamorphosis.

Stone Bridges

Old things often attain perfection after new devices have come to replace them. As the Erie Canal opened, the technique of building bridges with stone was entering upon its final development before the railroad brought radical changes in design and construction. Stone had been the best material for bridges, and the arch their requisite form since Roman days. Stone is still the ideal material for bridges if expenditures of time, effort, and money do not have to be taken into the accounting. No better substance has been found to withstand varying temperatures, the corrosive forces of weather, and decay. Engineers have yet to determine the exact limit beyond which they would not raise stone arches upon abutments sunk deep into the earth. The longest stone arch on record is 295 feet, erected in 1903 at Plauen in Germany. There are sites, like those of the Niagara Gorge, where the torrent makes difficult and hardly practicable, though not impossible, the centering or falsework that is necessary in laying up arches of masonry until their keystones are in place. But

Figure 8.6 Pont de la Concorde, soon after completion (From a contemporary engraving)

wherever wood, iron, and concrete have generally replaced stone, it has been not so much because they make better bridges; they are easier and less expensive to handle in certain forms and under particular conditions. Bridgebuilders in modern times have neither the labor supply of the Egyptian pyramid builders nor their seemingly boundless expense account.

It was the eminent French engineer Jean Rodolphe Perronet (1708–1794) who brought stone bridges to a remarkable degree of perfection at the end of the eighteenth century. Roman builders had raised theirs to majestic heights; the men of the Middle Ages and the Renaissance with great skill added variation and beauty. But in the low structures of all three periods the piers of their short spans had blocked a considerable fraction of the waterway, obstructing the stream as much as 65 per cent in the Roman bridge of A.D. 14 at Rimini and 50 per cent under the Pont Neuf of 1607 at Paris. Perronet understood that the intermediate piers would have only to support the deadweight of the arches and their superstructure if he could balance the horizontal thrusts of adjacent arches one against another between the abutments on either bank.

Using a hard stone to take the high compression in his extraordinarily flat arches, Perronet in 1787 made the piers of his Pont de la Concorde (Figure 8.6) in Paris so slender and so far apart that he reduced the obstruction in the waterway to 35 per cent. The deck of the bridge ran level from the quay walls. The arches rose but 13 feet over spans of 102 feet. Lateral thrusts were correspondingly great, but they were balanced and held in equilibrium at each of the central piers. To preserve the balance, the entire structure had to be built as a unit. No falsework could be removed until all arches were complete and the end thrusts were held by the abutments. It has been said that Perronet's achievements were all the more remarkable because he had no means of computing the thrusts of arches. It is hard to believe that one who was head of both the Corps d'Ingénieurs and the École des Ponts et Chaussées did not know and use the work of his great predecessor in France Pierre Varignon who had developed the "parallelogram of forces" from Stevin's work. In any case, the Pont de la Concorde still stands to prove that, if Perronet did not compute the thrusts of arches, he at least did know how to handle them.

More than traditions in bridgebuilding were reaching a climax before undergoing swift change as this man of eighty watched his assistant Chézy complete the masterpiece. The history of France itself had come to a crisis. Begun in 1787 under Louis XVI, the Pont de la Concorde was to have borne his name. However, stone from the Bastille entered into its construction before it was finished in 1791, and Louis rode through the streets of Paris to the guillotine in the Place de la Concorde in 1793. Perronet died in the following year. Before the century closed, Napoleon Bonaparte had taken command of the French Revolution and greatly enlarged its purpose to dominate Europe. Napoleon would break the hold of the British upon Egypt and the Near East, open the way to India and the Orient for Frenchmen, and if his luck were good, restore a vast French empire beyond the Mississippi in North America. But his luck was not good. It was not to be Frenchmen but Americans who built the bridges across the Mississippi, developed its steamboats, and laid the railroads through the Rockies to the Far West.

Trusses

Meanwhile builders with wood, progressing along rather different lines from Perronet's work with stone, were slowly approaching again the principles which Palladio had established in 1570. French engineers had been building wooden arches braced with a multitude of struts and ties crossed this way and that; they were handling their designing without clear-cut use of triangles. German carpenters, too, had constructed many truss bridges of relatively short span. The Swiss church and bridge

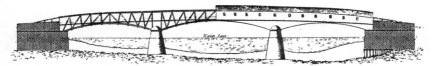

Figure 8.7   The Permanent Bridge spanning the Schuylkill (From D. Stevenson, *Sketch of the Civil Engineering of North America*, 1838)

builders developed this combination of wooden arch and irregular trussing to an even greater extent. The structure which Johannes and Hans Ulrich Grubenmann built at Wettingen near Zurich in 1764 had a span of 200 feet.[6] More credit perhaps would be given to their work had their bridges not been destroyed by Napoleon's army in 1799.

Experts tend to pass by the Grubenmanns as less capable than American builders of a few years later, far across the sea and apparently beyond European influence, but not all of the builders in America had learned their craft from native ingenuity and isolated experience as carpenters and shipwrights in the New World. Enoch Hale, Timothy Palmer, and Theodore Burr may have built without knowledge of European forms and practices, but one of the most expert Americans, Lewis Wernwag, had come at the age of seventeen from Germany in 1786. Enoch Hale (1733–1813) raised over the Connecticut River at Bellows Falls in 1785 the first wooden bridge with long spans in this country. Its two trusses, each 174 feet in length and some 50 feet above the stream, rested upon a framed wooden pier rising from an island; they were generously braced at the ends with heavy inclined struts. The record is not clear, but it is possible that the spans were slightly arched or cambered. With some rejuvenating, this structure served for half a century.

In 1805 Timothy Palmer (1751–1821) replaced the bridge of planks on floating logs which for years had crossed the Schuylkill at Philadelphia. His "permanent bridge" (Figure 8.7) to carry the Lancaster Turnpike stood 50 years before it was considerably altered. There were three spans, the center reaching 195 feet, the side spans each 150 feet. The western pier was remarkable in that it went down to bedrock nearly 42 feet below the surface of the river. Its cofferdam had been planned by the English engineer William Weston, who had come to America in 1793 to build neighboring canals. It stood another twenty years after alteration until destroyed in 1875 by fire.

Theodore Burr (1771–1822) raised wooden truss-and-arch bridges

[6] Jos. Killer, *Die Werke der Baumeister Grubenmann*, Verlag A. G. Gebr. Leemann & Co., Zurich, 1942, p. 40.

over the Hudson at Waterford, the Delaware at Trenton, and the Mohawk at Schenectady. In 1815 he dared to cross the main channel of the Susquehanna at McCall's Ferry, southwest of Lancaster, with a timber arch of 360 feet where the river was 100 feet deep, swift, and certain to fill with ice in the spring thaw. He built his arch in upright sections on floats along the shore ¼ mile below the abutment, "amidst tremendous storms and tempests," he said, "accompanied with floods and whirls and the bursting of waters." [7] Often in the night his men had to change the lashings, to brace off or haul in the floats as the winds tore at the structure, but they finished the arch and when the freeze came had it ready to drag upon the ice and swing into position.

The ice was not solid; it was floating chunks from ¼ inch to 2 inches thick which had formed upstream, broken upon the rocks, packed into the narrows at the bridge site, and then plunged beneath the surface until the mass of icy particles was from 60 to 80 feet deep. By digging into it 3 feet, Burr could thrust a pole down 60 feet into the mush ice below by the mere strength of his hands. Upon this uncertain floor, rising and falling a couple of feet in a day and moving imperceptibly but steadily downstream, Burr slid his arch in halves with capstans, set them upon the abutment and the pier, drove in the keys, and cut away the scaffolding on February 1, 1815. Standing back to view his work by the light of great fires, he enjoyed the "grandest spectacle" that he had ever seen—his arch "rising from the abutment and extending itself west out of sight." [8] Although liquor was "handed round in great abundance," only one man was hurt. He fell 54 feet, struck the braces twice and plunged into the water, but was back at work in a few days. Burr was very pleased with the quality of the workingmen from Lancaster and York counties, and he might have added, with their good luck. Two years later the ice destroyed Burr's bridge at McCall's Ferry on the Susquehanna. It was never replaced.

No experience so dramatic as Burr's came to Lewis Wernwag (1769–1843). Possibly it was because he did not build upon so difficult a site nor under such trying weather. His structures, nevertheless, advanced the science of bridgebuilding much further. It was he who erected the first cantilever bridges of modern times, at least in the Occident. There is evidence that Oriental bridgebuilders used the cantilever, as indicated by

---

[7] Theodore Burr, "McCall's Ferry Bridge," *Niles' Weekly Register*, vol. 9, p. 200, Nov. 18, 1815.
[8] *Ibid.*, p. 202.

Figure 8.8 Wandipore cantilever bridge, India, 1660 (From T. Pope, *Treatise on Bridge Architecture*, 1811)

the bridge of 1660 in India at Wandipore (Figure 8.8), though it looks more like simple corbeling with wood than the balancing of weights usually known as the cantilever. Wernwag's wooden cantilevers, built near Philadelphia in 1810, reached over 50 feet between piers. He said that he could extend the cantilevers to 160 feet. In his use of the truss, Wernwag anticipated the precision with that form which is generally credited to Ithiel Town, Squire Whipple, and other Americans who came after him. His Colossus at Philadelphia, finished in 1812, was an arch-and-truss bridge with a clear span of 340 feet, perhaps the longest ever built in wood or stone. Wernwag expertly combined five parallel arch ribs, braced and cross-braced, and used iron rods in every panel so that the joints could be adjusted if they worked loose. He fastened the wooden pieces with iron bolts and links without mortising so that any piece could be replaced with no danger to the superstructure. The arch rose 20 feet, but the roadway had less than half of that rise. Wernwag is credited with originating the type. The Colossus, like most wooden bridges, was covered to protect the timbers from the weather.

Wernwag's third bridge of particular significance was built at New Hope, Pennsylvania, in 1814 across the Delaware River. The innovation here was to connect the floor chords with the iron bedplates on which the arch rested, to keep the arch from shoving against the pier or abutment. The vertical timber posts with iron rods for diagonals suggested other forms of the truss which were soon to follow. During the next thirty years, Lewis Wernwag built some thirty bridges for highways and rail-

Figure 8.9 A typical Ithiel Town truss bridge (From his *Description*, 1821)

roads in Pennsylvania, Maryland, Virginia, Ohio, Kentucky, and Indiana and made significant contributions to other fields of engineering, particularly in improving the method of firing anthracite coal. Why he has not received more acclaim is rather difficult to understand. Possibly, like so many in his profession before and since, Wernwag had no flair for self-presentation. Possibly it was because his books, papers, plans, and reports were lost in 1870 during a flood in the Shenandoah Valley at Harper's Ferry where he had spent his later years and where he died in 1843.

Palladio had marked the way in 1570 with the first truss bridges of recorded history. Ithiel Town (1784–1844) of New Haven, Connecticut, now abandoned the arch altogether and patented in 1820 a truss bridge of uniform depth (Figure 8.9). It was made of wooden planks crossed in diamond pattern and fastened with wooden pins. There were vertical pieces only at the ends of the bridge. Town's lattice truss was popular on the highways and early railways of this country. The design was simple. Materials could be easily obtained, often by the side of the road, and construction was within the capacity of any good carpenter. Being a businessman as well as an engineer, Ithiel Town made a fortune from royalties of $1 to $2 per lineal foot for all bridges of the type built by anyone. The most notable of these bridges, 2,900 feet long, resting upon 18 piers,

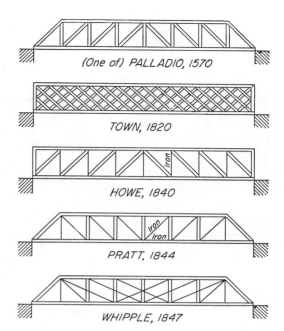

Figure 8.10
Typical early American
bridge trusses

(One of) PALLADIO, 1570

TOWN, 1820

HOWE, 1840

PRATT, 1844

WHIPPLE, 1847

spanned the James River at Richmond, Virginia. The Confederates destroyed it as they evacuated the city in 1865.

Once Town had reopened the way suggested by Palladio, many other forms of trusses appeared (Figure 8.10). One invented in 1840 by William Howe (1803–1852) of Spencer, Massachusetts, was first used on the Western Railroad, now part of the Boston & Albany. It was a composite truss; the verticals, under tension, were wrought-iron rods, but the diagonals, under compression, were made of wood. The end posts also were commonly wooden. This was the truss generally used on early American railroads. Four years later came another, designed also by New Englanders, Caleb Pratt and Thomas Willis Pratt, father and son. In their truss the diagonals sloped in the opposite direction from those of Howe's truss so that most of the verticals of the Pratt truss, whether of iron or wood, came under compression instead of tension. Its end posts were usually inclined. The Pratt truss is still in general use.

The first significant iron bridge in history had been erected by the third Abraham Darby (1750–1791) and John Wilkinson (1728–1808) in 1779 to replace the ferry across the Severn at Coalbrookdale, in west central England (Figure 8.11). They planned a semicircular arch of five

Figure 8.11 The world's first cast-iron bridge at Coalbrookdale

parallel cast-iron ribs with a 100-foot span, cross-braced to support a level roadway 24 feet wide and some 55 feet above the stream. Each rib was cast in halves 70 feet long and weighing about 38 tons. These were floated from the nearby foundry in barges, raised into place with block and tackle, and there joined at the center with cast-iron bolts. The weight of the banks subsequently tilted the abutments forward and pushed the arch up to a slight point at the center, but this in no way affected its strength. The bridge stands today, revealing the skill of its builders and the efficiency of iron.

Other cast-iron arches followed quickly in England, France, Germany, and later in America, although few if any of the bridges had the great castings of Darby and Wilkinson. It was easy to imitate stone construction in iron and to bolt together the "voussoirs" (tapered or wedge-shaped pieces). Thomas Telford constructed a number of such bridges. The first, built in 1796, lasted 110 years, and he drew plans in 1801 for a 600-foot arch of cast iron across the Thames at London. Parliament consulted authorities and deliberated upon the matter, but the bridge was not erected. In 1819, John Rennie built the Southwark Bridge over the Thames. It was impressive for its central span of 240 feet and for waste of metal. The plates were solid cast iron. The ribs of the arch would have been sufficiently strong if Rennie had used open-work voussoirs.

Engineers were learning to avoid cast iron for pieces under tension.

A cotton mill, seven stories high, had been erected in 1801 at Manchester, England, with its floors supported on brick arches resting upon inverted T-shaped cast-iron beams. These in turn were supported by cast-iron columns. Boulton and Watt had designed the structure as the first modern fireproof building. Others followed, but in spite of improved beams, the floors in such buildings too often collapsed, with the loss of many lives. Joseph Paxton took care to profit from such experience when he came to designing the famous Crystal Palace of 1851 in London (Figure 8.12). Its skeleton was made of both cast and wrought iron, but cast iron was used only for those pieces which were under compression. This huge structure enclosed 20 acres within its glass walls and roof. The design was controlled by the largest standard sheet of glass then manufactured, 4 feet long. As Sigfried Giedion has remarked, the astonishing thing was Paxton's "simple system of small prefabricated units." He completed the building within six months. Contemporaries praised his combination of iron and glass as revolutionary in architecture, but comparable work with the same materials was appearing in America at the same time. Removed from Hyde Park to Sydenham in 1854, the Crystal Palace stood until destroyed by fire in 1936.

By 1841 a native of Massachusetts had patented a truss bridge with cast-iron top chords which were always under compression but with wrought iron in the lower chords and in all other members under tension. It is not for this bridge, however, nor his own peculiar truss that Squire Whipple (1804–1888) deserves fame. It is for his thinking on the problem. As a boy in central New York, he had watched the engineers on the Erie Canal. He had attended Union College in Schenectady one year and graduated in 1830 at the age of twenty-six. He had worked in the field, building bridges and surveying for railroads, and like Smeaton and Watt, he had made instruments. And then in 1847, at the age of forty-three, Squire Whipple published his *Essay on Bridge Building*. In it he showed how one could compute the tensile or compressive stress in each member of a truss which was to carry a specific load. Whether or not he was aware of it, he was applying practically the knowledge which French mathematicians had already elaborated from the works of Stevin and Varignon.

Squire Whipple's achievement may have been quite independent of past studies in France and of contemporary thinking in Britain and America. His fame was not to be exclusive, however. At the very same time and

Figure 8.12 Interior of the transept, Crystal Palace, London, 1851 (From *Illustrated London News*, Jan. 25, 1851)

nearby, an American of German stock in Pennsylvania was bringing into form his own thoughts on bridges. Herman Haupt (1817–1905) had studied engineering in the tradition of the Frenchmen Sganzin and Crozet, at West Point. After graduating in 1835, Haupt spent three months in the Army but resigned to become an assistant engineer for the Norristown Railroad. In 1840 he had begun to study bridges. By 1845 he was professor of engineering at Pennsylvania, now Gettysburg, College. In 1851 Haupt published his *General Theory of Bridge Construction*, declaring in its preface that he had neither seen nor heard of any other work on the subject. Meanwhile in 1851 Robert H. Bow, an engineer in Edinburgh, also apparently unaware of Whipple's work, had written his *Treatise on Bracing*. And William Thomas Doyne and William Bindon Blood, British investigators, had made public in the *Minutes of Proceedings of the Institution of Civil Engineers* for 1851–1852 their analysis of the stresses in the diagonals of lattice trusses. The writings of Whipple, Haupt, Bow, Doyne, and Blood were the first in any language to give rational methods for calculating stresses in truss bridges. Thereafter, when designing bridges, no engineer had to rely entirely upon his intuitive sense or upon some rule of thumb from his experience. He could calculate in advance, by using mathematical analyses, the distribution of the stresses in the various members of the structure that he was about to raise.

Bridge designing by the middle of the nineteenth century was fixed for the most part within the truss form, and wood was giving way to iron. The Bavarian Carl Friedrich von Wiebeking (1762–1842) built wooden bridges with flat arches of bolted planks so effectively that fellow engineers credited him with establishing a system, but the truss and iron had come to dominance. The tensile strength of wrought iron, its resistance to fire, and other distinctive qualities made the metal a superior bridge material, especially for railroads.

As he laid the Stockton and Darlington Railway, George Stephenson had experimented with a little iron bridge at West Auckland in 1824. It was 50 feet long with four lenticular, or lens-shaped, spans in a combination of arch and suspension form, convex above and below. There were vertical stays, but no diagonal braces. In 1859 Isambard K. Brunel constructed his unique bridge over the Tamar River at Saltash (Figure 8.13) near Plymouth, England, along similar lines, though much enlarged and modified by the addition of diagonal tie rods. Ten years later another lenticular bridge was built over the Elbe River at Hamburg in Germany. But most builders were depending more and more upon diagonals to take

Figure 8.13 Brunel's Royal Albert Saltash Bridge—lenticular trusses (From Philip Phillips, *The Forth Bridge*, 1889)

advantage of the inherent stability of triangles, the only polygons that do not change shape under stress. All designers, whether they knew it or not, were building upon the principles which Stevin had made clear in 1586 with his "triangle of forces."

Early Suspension Bridges

There had been bridges without intermediate piers since man first swung himself on a vine across a chasm. The fiber-and-hide bridges of ancient China, Peru, and Chile were historic. There had been talk of suspension bridges made of iron chains and eyebars in Europe during the seventeenth century, but none was ever built. Nor is there evidence that the conceptions of Veranzio and Kircher had reached James Finley (ca. 1762–1828), local judge in Fayette County of southwestern Pennsylvania. He appears to have drawn upon his creative imagination for the ideas which produced the first modern suspension bridge with hangers of varying lengths supporting a level floor. This bridge Finley built over Jacob's Creek near Uniontown in 1801.[9]

"To find the proportions of the several parts of a bridge of one hundred and fifty feet span," wrote Finley, "set off on a board fence or partition one hundred and fifty inches for the length of the bridge, draw a horizontal line between these two points representing the underside of the lowest tier of joists—on this line mark off the spaces for the number of joists intended in the lower tier, and raise perpendiculars from each, and from the two extreme points, then fasten the ends of a strong thread at these two perpendiculars, twenty-three inches and one quarter above the horizontal line—the thread must be so slack that when loaded, the middle of it will sink to the horizontal line; then attach equal weights to the thread at each of the perpendiculars—and mark carefully where the line intersects each of them." [10] Thus James Finley had observed the final curve of his thread under loading, had determined the proper location for the level floor which he wished to have in his bridge, and had found the proper length of each hanger. "I know," he said, "the young mathematician, with mind half matured, would smile at my mode of testing the relative force and effect of the several ties and bracings of any piece of framing: but the well informed, will not so lightly treat any information obtained or supposed to be obtained by actual experiment." Finley's first bridge, built of wrought-iron chains with its level floor stiffened by wooden trusses, was only 70 feet long, but in his mind there was no doubt that when the subject was further understood his kind of bridge would be extended to 1,000 feet. The span across the Golden Gate of San Francisco Bay reaches 4,200 feet today. The well-informed have indeed not treated his information lightly.

By the time that James Finley received his patent in 1808, several of his bridges had been built or were under consideration. Albert Gallatin, Secretary of the Treasury, stated in his report of 1808 on the roads and canals of the United States that the "one lately thrown across the Potomac" (Figure 8.14) deserved notice for the "boldness of its construction and its comparative cheapness." "The principle of this new plan," wrote Gallatin, "derived from the tenacity of iron, seems applicable to all rapid streams of a moderate breadth." [11] Finley carefully explained

[9] James Finley, "A Description of the Patent Chain Bridge," *The Port Folio*, vol. 3, p. 442, June, 1810.
[10] *Ibid.*, pp. 445–446.
[11] Albert Gallatin, "Roads and Canals," U.S. Congress, *American State Papers*, Miscellaneous, vol. 1, pp. 738–739, 1834.

Figure 8.14   A Finley bridge spanned the Potomac above Georgetown in
1807. The location is still called Chain Bridge (From *The Family Magazine*,
1839)

his invention to the public in *The Port Folio* for June, 1810, one of the
earliest magazines in America. And Thomas Pope in the following year
criticized Finley's device in his *Treatise on Bridge Architecture*. It was
like the fiber bridge with horizontal floor which the British Captain
Samuel Turner had seen in India; it had, said Pope, the same weaknesses.
The whole bridge would fail if one link were to break. It could not be
of "long duration." He preferred the cantilever which Wernwag had
begun to develop at the time, and better still, his own quaint "flying
pendent lever" with which he dreamed of bridging the Hudson.

Thomas Pope's criticism, however, served to attract favorable attention to Finley's ideas as Templeman completed the chain bridge across
the Merrimac River above Newburyport, Massachusetts, according to
Finley's plan, and Finley himself constructed others. The British engineer
Thomas Telford, writing in 1814 about a proposed suspension bridge at
Runcorn near Liverpool, spoke of the possibility of a 1,000-foot span.
The very phrase reflected Finley's statement in *The Port Folio*. Telford
had Pope's book at hand as he made his report in 1818 and began his
suspension bridge with level floor across the Menai Straits on his great
road from London to Holyhead. When the French writers Cordier and
Navier published their accounts of suspension bridges in the 1820s,

Figure 8.15 Telford's Menai Straits Bridge in Wales (From his *Life*, 1838)

based in large part upon Gallatin's report and Pope's little book, it was James Finley who received credit for having first put a level floor in a suspension bridge.

As one of Britain's ablest engineers since Smeaton, Thomas Telford possessed the versatility and inquisitiveness to find out for himself how to put a rigid and level floor in a suspension bridge. It detracts nothing from his record to say that he had the advantage of knowing what Finley had done. Telford went at once to work upon the tenacity of materials. His plan for the suspension bridge over the Mersey at Runcorn near Liverpool, with a central span of 1,000 feet and two side spans of 500 each, was not dismissed, as had been his suggestion of a 600-foot arch of cast iron for London. The project for Liverpool was stopped primarily for want of funds to finance so great an undertaking. The bridge across the Menai Straits in Wales (Figure 8.15), on a smaller scale, seemed within financial possibilities, although it might not have been started had the promoters known that it was going to cost twice as much as their original estimate. But then, it was necessary to close the gap in Telford's post road from London to Holyhead, Isle of Anglesey, on the mail route to Dublin in Ireland.

For this famous structure, Telford thought first in 1811 of a cast-iron arch. Scaffolding over a deep navigable waterway with a tidal range of 20 feet and swift currents would have been impracticable, so he planned to build out the temporary supporting timbers or centering from each shore and to hold them in place with cables drawn back over curved towers to anchors in bedrock until the keys of the arch were set. This sort of temporary cantilever the American James B. Eads used later in constructing his great bridge at St. Louis. But by the time Telford had

Figure 8.16 The roadway of Telford's bridge after a century of use

finished his experiments with the tensile strength of wrought-iron bars, he had set aside the plan for a cast-iron arch and turned to the suspension principle.

Instead of chain links Telford substituted wrought-iron eyebars. These were tested up to 24,640 pounds to the square inch, then dipped in linseed oil, dried, and painted before shipment from the forge at Shrewsbury on the Severn. Each of the 16 main longitudinal members was made of five eyebars, each bar 1-inch thick by 3¼ inches wide and about 10 feet long, and assembled in three sections. One of these sections was hung from each tower; the third or central section was laid upon a narrow raft 400 feet long. It was then hauled up into position 102 feet above the waterway with ropes pulled by 150 men, "encouraged by music and liberal refreshment," and bolted to the end sections. The time taken for this operation was 1 hour and 35 minutes. The Menai suspension bridge was under construction 6½ years. As completed in 1826, it had two roadways, each of 12 feet width, and a center walk 4 feet wide (Figure 8.16). The central span was 579 feet. The final cost amounted to £120,000. The bridge suffered in gales and had to be restored and strengthened after a storm in 1839. Except for the towers, it was completely rebuilt in 1937 to 1939. The roadways were widened and two walks provided outside the cables. The new eyebars were made of steel.

Within two years after Telford had finished his celebrated suspension bridge over the Menai Straits, an Austrian engineer took a step beyond his contemporaries. The structure which Ignaz von Mitis (1771–1852) built over the Danube Canal at Vienna was not remarkable for its magnitude; it was narrow, its span was only 334 feet, little more than half that of Telford's bridge at Menai. But von Mitis made his eyebars of steel, a decarbonated cast iron from the furnaces of Styria, the first steel ever used in bridgebuilding. He was dissatisfied with his bridge, but not because it was weak or heavy or the material brittle; it had none of those faults, but it vibrated excessively, and it swung too much in the wind. Some thirty years later, in 1860, it was dismantled to give way to a much stronger railroad suspension bridge, the only one, it is said, on a main line in Europe. This bridge, too, was replaced after only twenty-five years of service. Suspension bridges are essentially flexible and not well adapted to the heavy, concentrated loads of railways.

Telford was experimenting with wire cables in place of eyebar chains even while he was at work on the Menai Bridge. Iron wire had been drawn by hand and by water power for varied uses in musical instruments, women's headdresses, and clothing as early as the thirteenth century. It had become increasingly important in the eighteenth for the carding machines of the expanding textile industry. It was now to compete in bridgebuilding with the eyebar. Chains of eyebars have to be raised into position. Wire can be spun back and forth across wide chasms innumerable times, its strands of relatively small diameter gathered and bound in cables as large as desired. Stronger and stronger wire cables have been developed as the loads they are expected to carry have increased.

The first use of wire in a suspension bridge had occurred at Philadelphia in 1816 where a structure not 2 feet wide provided a precarious toll path 408 feet long across the Schuylkill. It fell that winter under snow and ice. But once more the needs of this thickly settled community about the early American metropolis had made necessary a significant departure in engineering. France's engineers, soon alert to Finley's pioneering, adopted iron wire also for their bridges. Marc Seguin, whom we have observed developing multiple-tube boilers, wrote a book in 1824 on the subject of wire bridges and in 1825 built across the Rhone River south of Lyons the first wire suspension bridge for vehicles and general traffic. This Tain-Tournon Bridge was of two spans, each 275 feet long. It was restricted to foot traffic in 1847 as it was too light for vehicles.

In 1834 Seguin's associate, Joseph Chaley (ca. 1800–1870), completed at Fribourg, Switzerland, a single span of more than 870 feet. This he suspended from wire cables supported by towers 167 feet above the Sarine Valley. The four cables were 5½ inches in diameter, each made of

1,056 wires of $\frac{1}{12}$-inch diameter, bound together by wire wrapping. Wrought-iron bars gathered in chains, such as Telford's, had ultimate tensile strength estimated at between 40,000 and 50,000 pounds to the square inch of cross section. French wire like Chaley's has been tested to between 60,000 and 70,000 pounds to the square inch. Today wires of heat-treated alloy steel have ultimate strength up to 250,000 pounds per square inch.

The long suspension spans made it possible to cross rapid streams and deep gorges without intermediate piers or supports. The type became popular, and many appeared in the 1830s, especially in France. But like Telford's chain bridge over the Menai Straits, Chaley's wire bridge at Fribourg was too flexible. The Telford bridge is said to have heaved as much as 16 feet in storms. Chaley's had twice to be strengthened with more wires. After rebuilding in 1939, Telford's is still open to traffic. Chaley's was removed in 1922 to make way for a viaduct of reinforced concrete. Both of these famous suspension bridges survived as others did not. Nevertheless, stronger and more rigid structures were desirable to withstand the shock of storms, the concentrated loads of heavier vehicles, and moving loads of cattle herds and marching troops. The advent of the railway in the 1840s, with the great weight and the pounding wheels of its locomotives, made greater strength and stiffness imperative.

Tubular Bridges

Plans in 1843 for a railway from Chester, England, to Holyhead in Wales led to a remarkable type of bridge over the Menai Straits and to exceptional feats in construction. Although the type did not remain popular, Robert Stephenson's Britannia Bridge (Figure 8.17) will be remembered for its contributions to engineering. Permission was first sought to use one of the roadways on Telford's suspension bridge, but its shortcomings for a railway were apparent and permission was refused. Stephenson next considered two cast-iron arches, each similar to that of the little bridge at Coalbrookdale. Each would have had a span of at least 350 feet with a common pier midway in the Straits on Britannia Rock. The Admiralty denied this request because the arches would reduce the clearance over the waterway and hinder navigation.

Stephenson then returned to the suspension principle and asked William Fairbairn, experienced in iron shipbuilding, for advice on stiffening such a structure. They agreed at once that wrought-iron tubing might

Figure 8.17 Britannia tubular bridge showing Telford's bridge a mile north (From E. H. Knight, *American Mechanical Dictionary*, vol. 3, 1877)

prove effective, and Fairbairn undertook investigations. He called upon Eaton Hodgkinson, professor of mathematics, for assistance. Together the three men tested and calculated for months the relative strengths of circular, elliptical, and rectangular tubes until they became convinced that the rectangular design was the best. A 75-foot model, one-sixth the dimensions of the proposed bridge, with roof and floor stiffened by cellular bracing, proved so strong that they abandoned the supporting chains and thus fundamentally altered the nature of the bridge. It was no longer a suspension bridge but a tubular beam in rectangular form—a combination of vertical and horizontal girders.

Fairbairn had patented a solid-plate girder with cellular bracing in 1846. It is virtually the same as the form which Brunelleschi had used in 1436 for his double dome in the Cathedral of Florence and which became standard construction for domes. Cellular bracing later became general too in shipbuilding. But it has seldom been used in bridges once the popularity which Fairbairn and Stephenson gave to it declined; for the same strength and stiffness could be obtained with open girders and trusses that were lighter and much less wasteful of metal. Besides, the pent-up fumes and gases from the locomotives (Figure 8.18) passing through Stephenson's iron tunnels not only choked the passengers but tended to corrode the solid plates.

Figure 8.18 Britannia Bridge, Anglesey entrance (From Edwin Clark, *The Britannia and Conway Tubular Bridges*, 1850)

While deciding upon the rectangular form, Stephenson erected three towers and two abutments to carry the four tubes of each track 108 feet above high tide. Some of the stones in these towers weighed as much as 14 tons. The masonry was strengthened with nearly 400 tons of embedded cast-iron beams. The three central towers rose more than 200 feet upon bases 52 by 62 feet tapering to 45 by 55 feet at the top. They were built with vertical slots extending 6 feet into the tower face so that the tubes might be inserted and raised stage by stage as workmen filled the slots beneath them with solid masonry. So far events appear to have been routine and dull. The tubes for the end spans were constructed on false-work in position at the bridge level.

The scene became dramatic as the first of the central tubes floated upon pontoons with the tide and moved out to the towers into the current and wind of the Straits. The story of the resident engineer Edwin Clark recalls similar events in the lives of the Egyptian pyramid builders who loaded their huge stones upon barges in the Nile, and more particularly, that great moment for Domenico Fontana as he gave the signal on April 30, 1586, to raise the obelisk in the Circus of Nero. Then the onlookers were silent under the command of the Pope. Now in the early evening of June 20, 1849, a British crowd stood by the capstan on the Llanfair shore watching a hawser some 12 inches in circumference let the tube down slowly to the butt or temporary piling upon which it was to swing into the first slot and then be drawn into the second.

"The success of the operation," wrote Clark, "depended mainly on properly striking the 'butt.'" [12] This was to be done by means of the Llanfair capstan, but its line overrode itself and fouled so that it jerked the capstan from the platform, knocked the men down, and threw some of them into the water. "The tube was in imminent danger of being carried away by the stream or the pontoons crushed upon the rocks." "With great presence of mind," the man in charge called upon the crowd for help and handed out another hawser. Men, women, and children seized this "huge cable," ran with it up into the field, and held the tube until it was safely let down to the butt, "veered around," and drawn into position. As the tide subsided the pontoons deposited their cargo on the shelf at each end. It had taken about 1½ hours. Next morning Robert Stephenson was found sitting on the edge of the platform over the tube, swinging his feet and smoking a cigar. What Fontana or the Egyptians were doing the morning after, no one observed for the record.

All this was spectacular and noteworthy, but it was by no means the only engineering feat at the Britannia crossing of the Menai Straits. Huge objects had been floated before and transported considerable distances under trying conditions. The Romans had brought obelisks across the Mediterranean. The Russians moved a granite block of nearly 1,500 tons, measuring 42 by 21 by 17 feet, some 5 miles on rollers to the banks of the Neva and floated it down the river to serve as the base for the statue of Peter the Great in St. Petersburg. Robert Stephenson's greater achievement was the raising of the four central tubes into their final position in the towers more than 100 feet above the waterway. This he accomplished with two of the hydraulic presses or jacks (Figure 8.19) patented by Joseph Bramah (1748–1814) in 1795 upon the principle which Pascal had established about 1650 on the basis of Stevin's earlier work in hydrostatics.

Each of these central tubes was 472 feet long and weighed 1,400 long tons, or more than 3 million pounds. To raise them Stephenson assembled the hydraulic presses high in the towers, one in each, upon huge girders spanning the slots. The chains which hung from the presses down to the tubes were made of wrought-iron eyebars. Each bar was approximately 1 inch thick, 7 inches wide, 6 feet long, and rolled entirely in one piece; there was no welding to endanger its strength. As its length was the same

12 Edwin Clark, *The Britannia and Conway Tubular Bridges*, 3 vols., Day and Son, London, 1850, vol. 2, p. 683.

Figure 8.19
A jack that raised Britannia Bridge trusses (From *Illustrated London News,* Aug. 25, 1849)

as the lift of the presses, the fastenings were shifted with each stroke to the next eyebars. One lift may have taken a whole day, according to the conditions of the moment. The jacks were operated by hydraulic pressure furnished by steam pumps of 40 horsepower each.

With two steam engines and two hydraulic jacks, Robert Stephenson raised more than 3 million pounds 6 feet in each continuous lift. Domenico Fontana had to use the strength of 75 horses and 907 men at 40 capstans to lift 350 tons, and he could raise the obelisk no more than 2 or 3 inches in the operation. The contrast is striking and indicative of the development in engineering through the intervening 263 years. But one thing had not changed. Fontana and Stephenson had it in common; both were engineers with imagination and took every precaution they could imagine.

Fontana foresaw that his iron band about the obelisk might snap and was ready when it did. Stephenson kept timbers beneath his tube so that it might never be without support and insisted upon closely building up the masonry in the slot as the engines raised the tube foot by foot. When a jack actually did fail and the bottom of its cylinder, weighing a couple of tons, fell 70 or 80 feet upon the tube below, it crushed the timber packing and bent the plates, but the tube could fall only 8 or 9 inches to solid masonry instead of to the bed of the Straits. And Stephenson's assistant could thank God, as he did, that Stephenson had been so "obstinate."

There was death in the episode. A workman happened to be climbing a rope ladder from the tube to the press directly in the path of the block of the cylinder when it fell. There was delay from August 17 to October 1 until another cylinder could be installed, but by October 13 the first tube was in place at the top. The second span was floated on December 4 and its raising completed on January 7, 1850. The first run over the single track thus opened was made by three locomotives on March 5. Later the bridge was tested with a coal train of 27 wagons, weighing together 248 tons; a deflection of less than 1 inch was observed. On March 18, 1850, the first track of the Britannia Bridge was open to traffic, and by October 19 the second track was completed.

Riveting had been used before 1850 in boiler work and shipbuilding, but the Britannia and the smaller Conway Bridge, built on the same railway line at the same time, were the first bridges to have tubular construction. Fairbairn had invented a riveting machine, driven by steam, to do without the boilermakers who were on strike in the works at Manchester. One operated by two men and a boy could fix "as many rivets in one hour," he said, "as could be done with three men and a boy in a day of twelve hours on the old plan." [13] The boy was essential in both cases for attending the heating furnaces. A machine was used on the Conway Bridge when practicable and when it was not broken "purposely by workmen who objected to its introduction." It had a piston 48 inches in diameter working with a 9-inch stroke under 40 pounds of pressure, thus putting the rivet under 32 tons. However, the riveting on the Britannia Bridge was done by hand. The operation required two riveters swinging 7-pound hammers, another man to hold up the rivet with a heavy hammer fitted with a long handle, and two boys. As Clark described it, "the

13 William Pole, *The Life of Sir William Fairbairn*, Longmans, Green, and Co., London, 1877, p. 164.

rivets have frequently to be passed a considerable distance to the work. This is done by throwing them with a pair of pincers; and no spectacle has afforded so much amusement to spectators as the unerring precision with which boys only eleven or twelve years of age toss them to their destination. And when a large number of riveters were engaged on the top of the tubes by night, the constant succession of these red-hot meteors, ascending in graceful curves to a height of 30 or 40 feet, formed an interesting sight." [14]

Robert Stephenson gave the Britannia Bridge three other features of note in engineering. After raising the tubes, he had them continued through the towers and joined by riveted sections. In the central tower, these inserts were 32 feet long. This made for each track two continuous tubular beams just over 1,500 feet long. This continuous beam added stiffness and supporting capacity to the whole bridge. Stephenson also had the tubes cambered lengthwise. The roofs of the longer tubes were 7 feet higher at the center than at the ends, and the lower surface was given a camber of 9 to 10 inches to allow for sagging and still maintain a level floor. He also anticipated the effect that varying temperatures would have upon the tubes. To provide for their expansion and contraction, he set them upon cast-iron rollers at either end as the center was anchored. A range of 76°F in temperature was expected to cause a change of 9 inches over the whole distance. Allowance, therefore, was made for 12 inches, to be safe.

The tubular bridges of Robert Stephenson and William Fairbairn were decisive contributions to engineering history. They were successful in that they accomplished their immediate purpose, and they made advances in technique from which engineers have never turned back. None of them, however, was a good financial investment; their promoters lost money. The Britannia, though twice as wide and able to carry much heavier loads, cost something like five times as much as Telford's suspension bridge over the Menai Straits. More damaging to the future of tubular bridges was the fact that they wasted material. There was no need to build tunnels in the air; open trusses and girders could be made just as strong, even more lasting, and convenient. The evolution of railroad-bridge building, therefore, was away from the closed structure. First to go was the roof, leaving but two vertical girders parallel and cross-braced with a floor. With the use of steel were to come further departures.

[14] Clark, *op. cit.*, vol. 2, p. 629.

Bibliography

Gauthey, Emiland M.: *Oeuvres*, 3 vols., Firmin Didot, Paris, 1809–1816.

Jakkula, Arne A.: *A History of Suspension Bridges in Bibliographical Form*, Department of Commerce, Bureau of Public Roads, Washington, D.C., 1941.

Kirby, Richard S., and Philip G. Laurson: *The Early Years of Modern Civil Engineering*, Yale University Press, New Haven, Conn., 1932.

Labatut, Jean, and Wheaton J. Lane (eds.): *Highways in Our National Life*, Princeton University Press, Princeton, N.J., 1950.

McAdam, John L.: *Remarks on the Present System of Road Making*, 9th ed., printed for Longman, Hurst, Rees, Orme, and Brown, London, 1827.

Phillips, John: *A General History of Inland Navigation*, I. and J. Taylor, London, 1792.

Stevenson, David: *Sketch of the Civil Engineering of North America*, J. Weale, London, 1838.

Timoshenko, Stephen: *History* (see Chap. 6).

NINE

Steam Vessels and Locomotives

Horse-drawn vehicles and barges pulled by men or animals traveled on the roads and canals of the eighteenth century. The Industrial Revolution in England, with its requirement for increased amounts of fuels, raw materials, and finished goods to be moved, and the great distances to be traversed in America, created a demand for more rapid forms of transportation and greater carrying capacities. The most obvious advance to be made was the installation of the steam engine in wheeled vehicles and ships to propel them. The early road locomotives were not successful, particularly for hauling goods, and the advances made in steamboats for inland transportation were more rapid in the first part of the nineteenth century. The steamship was the first radical improvement in transportation following the fore-and-aft rigged ship developed during the Middle Ages.

The early evolution of the steamship was more rapid than that of the locomotive because there were fewer engineering problems to solve. A relatively large, low-pressure Watt engine could be fitted into vessels of several different types in which there was much more space for fuel than in a land vehicle. Moreover, the inland steamboats generally traveled on canals and rivers that were smooth and flat as compared to roads with steep gradients that were available for steam vehicles on land. To achieve success with locomotives, engineers first had to develop high-pressure engines and boilers, delivering many more horsepower per pound than the low-pressure Watt engines, and also a special roadway having iron rails and slight grades.

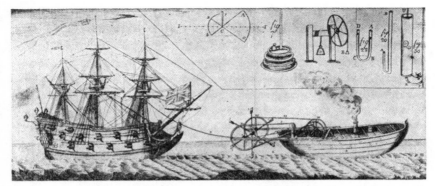

Figure 9.1 Jonathan Hulls's proposed steam tug (From his *Description*, 1737)

Steamboats

One Jonathan Hulls patented a steamboat in 1737 (Figure 9.1) after Newcomen's engine appeared. As the engine was to be single-acting and with no rotary motion, Hulls proposed to drive a stern wheel with ratchets, thirty-two years before Cugnot put his ratchet-driven tractor on a Paris street. Hulls's idea had been even less practical than Cugnot's. The Newcomen engine was too feeble; it would have been too cumbersome for its task. Moreover, the time was not yet ripe for the paddle wheel. There was a curious lag in comprehension of its effectiveness, although such a device had been in mind for a long time.

Book 11 of R. Valturio's *De re militari*, published in 1483, contains illustrations of boats with crank-driven paddle wheels (Figure 9.2); A. Ramelli's *Le Diverse et artificiose machine* of 1588 has a vivid drawing of armored amphibious military craft driven by paddle wheels while in the water (Figure 9.2). In both cases men supplied the power. Raphael in his famous 1514 painting of Galatea (Figure 9.3) shows her coming ashore in a shell drawn by dolphins but equipped with paddle wheels, an idea which Leonardo quite possibly suggested to his friend Raphael who was not mechanically minded. Among a number of other early references to paddle-wheel propulsion is a comment by Guido Pancirolli (1523–1599), written originally in Latin and published in English translation in 1715, entitled *The History of Many Memorable Things Lost*. He said, "I saw also the Pictures of some Ships, called . . . Liburnae, which had three *wheels* on both Sides without, touching the water, each consisting of eight Spokes, jetting out from the Wheel about an Hand's breadth, and *six Oxen within*, which by turning an Engine stirr'd the wheels, whose

Figure 9.2 Early suggestions for paddle-wheel boats: above, amphibious assault craft (From A. Ramelli, *Le Diverse et artificiose machine*, 1588); below, left, naval boats with rams (From R. Valturio, *De re militari*, 1483); below, right, a more complicated design (From H. Dircks, *Life, Times and Scientific Labors of the Second Marquis of Worcester*, 1865; original in Harleian Mss.)

Fellys driving the Water backward, moved the *Liburnians* with such a Force, that no *three oar'd* Gally was able to resist them."

    Henry Dircks in *Life, Times, and Scientific Labours of the Second Marquis of Worcester* (1865) mentions and reproduces a picture attributed to the fifteenth century from the Harleian manuscripts, *Italian Book of Sketches* parchment 3281, *Delinationes machinarum*, representing a small craft driven by paddle wheels turned by crank and gears (Figure

Figure 9.3 Raphael's "Galatea" of 1514, showing paddle wheels (Courtesy Yale School of the Fine Arts)

9.2). In the nineteenth century men were still arguing over the relative merits of paddles, mechanical oars, and similar forms even after propellers had been tried. Although he suggested the screw propeller in 1770, James Watt seems to have had no interest in promoting steam navigation beyond offering his patented engines to those whom the British Admiralty approved. Others experimented with considerable success and with decided influence upon Robert Fulton, from whose work on the Hudson River in 1807 the steamship as a commercial success dates.

In America, James Rumsey of Virginia (1743–1792) tried jet propulsion during the 1780s and put a vessel on the Potomac which actually made 4 miles an hour against the current. He did as well on the Thames in February, 1793, attracting the attention of Robert Fulton, native of Pennsylvania, in London at the time intent upon becoming an artist under the tutelage of Benjamin West. Rumsey died in the midst of his experimenting, and jet propulsion had to wait for the development of aeronautics

249

in the twentieth century. In Britain, Patrick Miller (1731–1815) and William Symington (1763–1831), poaching upon Watt's scheme for a condensing engine, constructed a steamboat with a paddle wheel driven by ratchets; with it they attained a speed of 5 miles an hour. They later increased the speed to 7 miles an hour, but the Admiralty was indifferent and Miller's capital was limited. Financed in 1802 to 1804 by Lord Dundas, Symington produced a steam tug on the Forth and Clyde Canal. Again Robert Fulton was observant. The canal's coproprietors, however, rejected the *Charlotte Dundas* because, it was said, its wash threatened the banks of the canal. The vessel had a horizontal engine, double-acting and driving directly to a paddle wheel recessed in the stern. Fulton did not copy this design, but designers returned to the idea fifty years later on the Mississippi.

Meanwhile, among other inquisitive Americans, two were coming close to success with their ideas on steamboats. John Fitch (1743–1798), a native of Connecticut, formed a stock company to obtain capital, engaged a mechanic, and built an engine with a chain sprocket to move six vertical paddles on each side of the hull. He demonstrated this vessel on the Delaware River, July 27, 1786. During the following summer, he displayed a better boat to the framers of the Constitution who were gathered in Philadelphia. His third, with paddles at the stern, steamed 20 miles up the Delaware River to Burlington. By 1790 he was making 8 miles an hour and maintaining a passenger and freight service between Philadelphia and Bordentown, 28 miles away. It has been estimated that Fitch must have run his steamboat more than 2,000 miles, but the company lost money. The machinery left too little space for cargoes. The vessel was laid up for the winter of 1790 and never used again because John Fitch had not really succeeded in putting an efficient power plant in a ship. He went to France in 1791, got no support there, returned to fail again in New York City, withdrew to his lands in Kentucky, and died in 1798 disgruntled and impoverished as he had always been.

Another American accomplished his purpose but did not win recognition as the first to operate a boat by means of a steam engine and make it pay commercially. Deeply stirred by the work of Rumsey and Fitch, John Stevens (1749–1838) of New Jersey, a lawyer by training and inventor by nature, was engaged for much of his life in promoting transportation by steam. Stevens experimented in 1796 with steam propulsion on the Passaic River. His collaborators were Robert Livingston, Nicholas Roosevelt. and Marc Isambard Brunel, French engineer who had fled from

France and who was soon to become prominent in British engineering. By 1802 Stevens had built a small steamboat with a screw propeller, water-tube boiler, high-pressure steam, and a crank engine giving rotary motion. It was as notable a piece of constructive engineering as had yet been done in America. The torque of the propeller, however, tended to move the vessel in circles, and so his next boat, the *Little Juliana*, in 1804, had two propellers turning in contrary directions. Stevens found that his engine was not as effective as he wished; the water tubes of the boiler filled with sediment, and the high pressure was likely to cause trouble. He turned to low pressure, Watt's reciprocating engine, and paddle wheels. The result was his *Phoenix*, begun in January, 1808, five months after Fulton's success. John Stevens sent the *Phoenix* in July, 1809, around Cape May into the Delaware River—the first ocean voyage by steam. It took thirteen days and much creeping between wind and weather, hardly a propitious beginning for ocean-going steamers, but Stevens's son Robert L. brought the vessel safely to Philadelphia where it served as a ferry. Robert L. Stevens (1787–1856) was later to make innumerable contributions to the development of steamboats, including the hog frame to stiffen their hulls lengthwise. He increased their speed to 15 miles an hour, but like his father, he has been more famous for his interest in railroads.

The contribution of Robert Fulton (1765–1815) to the steamboat was not the engine, for which he relied upon the British manufacturers Boulton & Watt. Nor did Fulton discover the laws of floating bodies and their propulsion. For these he studied the experiments of Mark Beaufoy in 1793 to 1798. Fulton's achievement was in coordinating a hull and a power plant so that the combination was economically successful. He might have succeeded with his earlier effort on the Seine in France if his hull had been strong enough to carry the weight of the engine. Artists and engineers have in common unusual powers of observation and analysis. However much they differed in other respects, Leonardo da Vinci and Robert Fulton were alike in possessing curiosity and persistence beyond the lot of most men. In appraising any achievement there are always the additional factors of good relationships with others and publicity. Few have known better than Fulton the value of favor in high places and how to go about getting it. Da Vinci had put his ideas at the service of the Duke of Milan and the King of France; Fulton dealt with Napoleon, the British Admiralty, and then with Chancellor Livingston of New York, who had access to the state legislature, where the power of granting monopolies lay at that time. Together, Robert R. Livingston and Robert

Figure 9.4 Fulton's *Clermont* after covering of paddle wheels (From *Transactions, Royal Society of Arts,* 1852)

Fulton had political favor and financial credit, scientific skill and imagination—not to stress unduly their common desire to make money.

Their steamboat (Figure 9.4) was a commercial success from the day of its first run on the Hudson, August 17, 1807. Fulton wrote exuberantly to his friend Joel Barlow, poet and politician, that he had overtaken many sailing vessels "beating to windward and parted with them as if they had been at anchor." [1] By October, from 60 to 90 passengers each trip were taking the steamer between Albany and New York, covering the 150 miles in 30 to 36 hours. There are no data on the consumption of wood for fuel or its cost. The vessel had cost $20,000, and it earned $1,000 in three months. Said Fulton to Livingston, "as this is the only method which I know of gaining 50 to 75 per cent, I am, on my part, determined not to dispose of any portion of my interest on the North River." [2] His arithmetic was more a matter of enthusiasm than of accuracy. The partners were soon planning larger boats for the Hudson and Long Island Sound and looking beyond the Appalachians to the Ohio, Mississippi, and Missouri. Perhaps they might establish as effective a monopoly at New Orleans as they seemed to have at New York City.

[1] Charles B. Todd, *Life and Letters of Joel Barlow*, G. P. Putnam's Sons, New York, 1886, p. 233.
[2] Henry W. Dickinson, *Robert Fulton, Engineer and Artist*, John Lane, London, 1913, p. 225.

As originally constructed, *The Steamboat* of Fulton and Livingston was 133 feet long, a flat-bottom scow, 13 feet broad and 7 feet deep, with a very shallow draught. The tonnage was 100. An attempt to duplicate this narrow design for the centennial celebration of 1907 indicated that it was unseaworthy, and the dimensions had to be changed in the reproduction. Fulton's engine (Figure 9.5) had a 24-inch cylinder and a piston stroke of 4 feet. There was a flywheel on a jackshaft geared to the crankshaft in a very complicated arrangement with a bell crank. The engine generated some 20 horsepower. The paddle wheels had 8 floats, or boards, 4 feet long and 2 feet wide. At first they were not in any way covered. As rebuilt in 1808 and eventually named the *Clermont*, the vessel's tonnage was increased to 182½, the length to 149 feet, and the breadth to 17.9 feet. Other dimensions remained the same. The horsepower was not changed, and the craft's speed has been recorded as 4.6 miles an hour.

It was not mere coincidence that while they were together in Paris Robert R. Livingston should have taken up with Robert Fulton and should have discerned the possibilities in his ideas when they returned home. Livingston had most to do with negotiating with France the Louisiana Purchase, a vast empire as large as the United States itself, reaching from the Mississippi to the Rockies. He had been experimenting with steamboats before he went as Minister to France for Jefferson in 1801. If Livingston and Fulton could develop a steamboat and get monopolies on both the Hudson and the Mississippi, what an opportunity

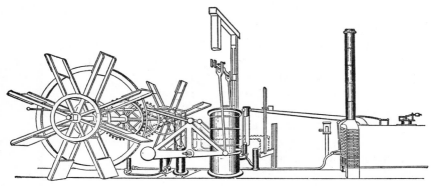

Figure 9.5 The machinery of Fulton's steamboat *Clermont,* 1807 (From E. H. Knight, *American Mechanical Dictionary,* vol. 2, 1876)

there would be, they fully perceived, to control the commerce in both great ports of the United States, New Orleans and New York! Livingston himself attended to the matter in the legislature of the State of New York. By February, 1804, his brother Edward was in New Orleans to make a new start for himself and to look after any interests of the family that might appear. He did this so well that by 1811 the territorial legislature of Louisiana had granted a monopoly to the steamboats of Livingston and Fulton.

Excluded from New York waters by Livingston and Fulton and hampered, too, by New Jersey's legislature, Nicholas J. Roosevelt made arrangement to participate in steamboating on the Ohio and the Mississippi. He built the *New Orleans,* the first steamer west of the Appalachians in 1811 at Pittsburgh. The vessel's rounded hull drew so much water that it slipped through the rapids below Louisville with only 5 inches to spare. How it survived the fury of the river during the great earthquake of 1811, which caused much damage, is a mystery. The engine of the *New Orleans,* however, was fully capable of stemming the normal currents of the lower Mississippi; it made the trip to Natchez regularly until wrecked on a snag at Baton Rouge in 1814.

The monopolists from the East soon had more boats on the Mississippi, but they could not hope to thwart forever the ambitions of others. Henry M. Shreve (1785–1851), a native of New Jersey, captained the *Enterprise* for Daniel French on the first trip upstream from New Orleans to Louisville, May 31, 1816. The *Enterprise,* built in 1814, had given aid to General Jackson in the Battle of New Orleans in 1815. Shreve won judgment in court against the territorial law of Louisiana granting a monopoly to the steamboats of Livingston and Fulton. Shreve obtained this finding in the United States District Court for Louisiana in 1819, five years before Chief Justice John Marshall gave the Supreme Court decision in *Gibbons v. Ogden* against the Livingston-Fulton combination on the Hudson River within the waters of the State of New York.

Like Roosevelt, Shreve appreciated that ocean-going hulls could not be expected to navigate the river except at high water. The *Enterprise* had reached Louisville only because of floods, and then it had ventured outside the regular channel. Its low-pressure engine had difficulty with the current. As French was not convinced by Shreve's report, Shreve made plans for a steamboat of his own which would move upstream against all currents and in normal water past the rapids of Louisville, to Pittsburgh. The keel was laid at Wheeling, Virginia, on September 10,

1815, for the *Washington,* first of the Mississippi steamboats which have contributed so much to American literature, from Mark Twain to *Showboat.*

Shreve's *Washington* was a stern-wheel flatboat powered by steam. He put the engine on the main deck and built another deck above it to carry the boilers. He abandoned the space-consuming condenser and exhausted the engine into the atmosphere. Labor was plentiful for the all too frequent job of cleaning mud from the boilers. He laid the two cylinders horizontal and made them stationary instead of oscillating as in French's engine. Shreve transmitted the power to the cranks with crossheads and connecting rods. The cranks were at right angles to one another so that the engine could not stick on dead center. The cylinders were 24 inches in diameter and the pistons had a 6-foot stroke. The four boilers also were horizontal, and like Oliver Evans, his contemporary, Shreve put flues in his boilers and operated them well above atmospheric pressure.

He risked explosions; in fact he had a severe one off Marietta, Ohio, on the first trip when the weight on the safety valve slid by accident to the end of the lever. Eight persons were killed outright and six fatally injured; Shreve himself was blown overboard and badly hurt, but he repaired the damage to the vessel and stubbornly retained high pressure, for he must have power and speed. As he thus rejected one of James Watt's ideas, Henry Shreve was adopting another whether he knew it or not; he installed a valve operated by a cam to shut off the steam early in the cycle. This arrangement took advantage of the expansive force of the steam within the cylinders, and the device saved much fuel.

Niles' Weekly Register carried on Saturday, July 20, 1816, a special item from St. Clairsville, Ohio, dated June 6, commenting upon Captain Shreve's steamboat. Gentlemen from New York said that its accommodations were better than any on the North River. Its main cabin extended 60 feet; it had three private rooms and a commodious bar. Its 100 horsepower was applied upon "an entirely new principle, exceedingly simple and light." There was no "balance wheel." The whole engine (the "invention of Captain Shreve") weighed "only nine thousand pounds." Accounts vary as to whether the *Washington* had side wheels or a stern wheel. The *Liberty Hall and Cincinnati Gazette* of September 23, 1816, observed, as the vessel was passing on the way to Louisville after it had been repaired, that there was "a single wheel placed at the stern." Pictures and other descriptions indicate that the *Washington* of 1816 had side paddle wheels, but these descriptions are all dated some years later when side-wheelers were prevalent and the *Washington* was no longer on the river. Henry Shreve's new *George Washington*, built at Cincinnati in 1824, was a side-wheeler. Its paddle wheels were connected separately to its engines so that one could be reversed while the other was going forward, and the vessel could be turned in less space than a stern-wheeler.

If ever an engineer accomplished his purpose, Henry Shreve did— and at once. The *Washington* demonstrated the superiority of its shallow draught and higher speed on its trip down the Ohio and the Mississippi to New Orleans in September, 1816. Shreve made the round trip between Louisville and New Orleans during the following spring in 41 days. He came upstream in 24 days, running over the rapids of the Ohio in normal water. The toastmaster at the dinner in his honor predicted that some would live to see this upstream trip done in ten days. As a matter of fact, before the railroad came and the Civil War, the voyage from New Orleans to Louisville had been made in less than five. A line of packet steamers, riverboats, and ocean-going vessels maintained regular schedules between Louisville and Liverpool by way of Havana, Cuba.

The effect of Fulton's success and Marshall's breaking of the monopoly on the Hudson River had been like that of raising a sluice gate. Steamboats appeared wherever there was quiet water to float their hulls and distance to make the voyage worthwhile. By 1824, as soon as the monopoly in New York State was broken, they were operating regularly on Long Island Sound to provide the best means of transportation for the river towns and seaports of southern New England. It was many years before the town-to-town companies merged into a system of communication; even then they supplemented the steamships on the Sound and along the coast to the east of Boston. The French engineer Jean-Baptiste Marestier (1782–1832), who came to observe for his government, had taken note of these steamers and others on the Delaware, Chesapeake, and St. Lawrence, and Lakes George, Champlain, and Ontario. Though sailing vessels continued long to carry freight on the Great Lakes, it was the steamboat that accelerated the growth of population and the rise of the lake ports of Buffalo, Cleveland, Detroit, Milwaukee, and Chicago. The industries of Chicago may be attributed in largest part to the railroad, but the steamboat carrying grain, iron ore, and general cargoes made a tremendous contribution, as it had done to the earlier development of Cincinnati on the Ohio River.

The need, the distances, the incentives were less in Europe than in the United States. There was, accordingly, notwithstanding Marestier's optimistic report, less haste in adopting the steamboat for the waterways of the British Isles and the European Continent. In 1811 Henry Bell (1767–1830) had produced the *Comet,* a 40-foot boat with a 4-horsepower engine, and had operated it three times a week on the Clyde from Glasgow to Greenock. It usually ran 27 miles in 3½ hours. It was not, however, a commercial success, and so Bell turned it into a "jaunting boat" to tour the coasts of England, Ireland, and Scotland, as he said, "to show the public the advantage of steamboat navigation over the other mode of sailing." [3] The engine is on display in the Science Museum, South Kensington, London.

In 1815, after Fulton's death, a project for the Russian government led Charles Baird to put an engine in a barge on the Neva and in 1817 to construct another steamboat. The English Channel was crossed by steam from Brighton to Le Havre in 1816. John Rubie built the *Prinzessin Charlotte* for the Elbe River in 1816. The paddle wheel of this vessel was placed in the center and covered, and its engine developed 14 horsepower. However, for steamboats the larger rivers like the Rhine, Danube, and Rhone were too swift until more powerful and less clumsy engines could be produced. It was not until 1830 that the Englishman J. Pritchard succeeded in conquering the rapids of Floresdorf in the Danube with his steamer *Francis the First.* He received exclusive rights to operate steamboats on the Danube for fifteen years.

Improvements on river boats came from every quarter. So many persons had shares in them that it is futile to try to distinguish any one contributor from another. There were so many improvements in details that all cannot be cited here. A feathering paddle wheel increased driving efficiency. Coal gave better heat than wood. Trusses or hog frames, usually credited to Robert L. Stevens, although the ancient Egyptians used the principle, stiffened the long shallow hulls which otherwise would have been too flexible. A skeleton diamond-shaped iron walking beam, first used in 1823, applied the laws of stress to maintain the strength of the beam while reducing its weight as compared with solid wooden beams. Boiler pressures increased with the introduction of compound engines of Hornblower's type.

Steamships were slower in getting upon the high seas because of conservative devotion to sailing ships, although they had still to reach

[3] Robert Meikleham (pseud. Robert Stuart), *Historical and Descriptive Anecdotes of Steam-engines,* 2 vols., Wightman and Cramp, London, 1929, vol. 2, p. 526.

their glory with the clippers of mid-century. Uncertainties in the engines, especially, and the space required for fuel held in check full reliance upon steam for ocean-going vessels. If their fuel gave out, they could not drop down with the current to the next woodpile or coal bin. So unreliable was steam power on the seas that it was nearly 1900 before even the great liners abandoned auxiliary sailing rig for emergencies.

The first transatlantic voyage with steam has been credited to the *Savannah* (Figure 9.6), a full-rigged ship of 350 tons. Carrying no

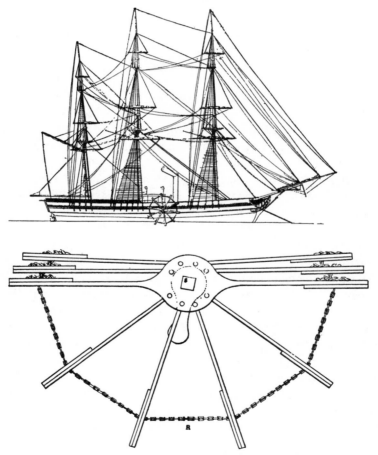

Figure 9.6 The American steamship *Savannah*, 1819, and folding paddle wheel (From Marestier, *Mémoire*, 1824)

passengers or cargo, it left its home port of Savannah, Georgia, on May 26, 1819, for St. Petersburg, now Leningrad, at the far end of the Baltic Sea. Its owners hoped to capitalize Russian interest in American steamships and sell the craft there. The *Savannah*'s engine greatly interested the French observer Marestier, and he published drawings of it. The single cylinder had a diameter of 4 inches and the piston a 6-foot stroke. It is said to have generated 72 actual horsepower and to have been capable of driving the ship at 5 to 6 knots. The iron side wheels could be folded like fans and stowed on deck when not in use. The ship carried 75 tons of coal and 25 cords of pine for fuel. No coal was left by the time it reached Ireland. More had to be taken aboard there for the run to Liverpool and presumably still more before the *Savannah* reached St. Petersburg, 50 days after leaving the United States. Significant as this voyage is in the history of navigation, it is altogether too much to say that the *Savannah* made the first transatlantic crossing under steam power. The log shows that her engine was in use only parts of 18 out of the 29½ days to Liverpool, a total of but 3½ days, and then for the most part in leaving and entering port. The owners could not sell in Russia, and the return voyage was not historic. The ship spent the rest of her life as a sailing vessel on the American coast.

It was not wholly the assurance of a popularizer which trapped the widely known lecturer on science, the Reverend Dionysius Lardner (1793–1859), F.R.S., when he declared in 1835 that a voyage by steam directly from New York to Liverpool was "perfectly chimerical"; they might as well "talk of making a voyage from New York to the moon." [4] He criticized, too, among other things, Samuel Hall's (1781–1863) new surface condenser for salvaging condensed steam after it had passed from the exhaust. The purpose was to retain the pure water to feed the boilers again and again on a long voyage. Lardner was a very articulate scientist, on this occasion doubtless too articulate. He had good reason to question whether vessels could hold the fuel required by the steam engines of the time on a transatlantic voyage from New York to Liverpool and have any room left for profitable cargoes. Several steamships crossed the Atlantic in the 1820s, using sail part of the way. The *Royal William* had made three short trips in 1831 from Montreal to Halifax and in August, 1833, from Pictou across the Atlantic to Gravesend, under steam much of the way but with auxiliary sails. The coal for its 200-horsepower engine

4 William S. Lindsay, *History of Merchant Shipping and Ancient Commerce*, 4 vols., Sampson Low, Marston, Low, and Searle, London, 1876, vol. 4, pp. 168-170.

Figure 9.7 The *Sirius* after its arrival in New York, 1838 (Courtesy New York Public Library)

had filled most of the hold. The cargo which it had taken across the Atlantic amounted to "254 chaldrons of coal, a box of stuffed birds, and six spars . . . one box and one trunk, household furniture and a harp . . . and seven passengers."[5]

Two groups in Britain were less interested in the pronouncements of Dionysius Lardner than in the *Royal William*'s achievement. The Queen Company had for some time been running the steamer *Sirius* of 700 tons between London and Cork in Ireland, a distance of approximately 600 miles. They knew how far a ton of coal would carry the *Sirius* (Figure 9.7). They were aware also that a larger vessel, nearing completion, was expected to cross the Atlantic, and so they loaded the *Sirius* with more than 450 tons of coal and sent it out with a few passengers from Cork for New York on April 4, 1838. Eighteen days later it appeared off Sandy Hook and anchored the next day at the Battery with its fuel entirely consumed. However, a rival, the *Great Western* (Figure 9.8), which had set out four days later, arrived on the same day. There was much in the newspapers about the race across the Atlantic. The *Great Western*, in fact, might have been the first to leave for America

[5] Robert Ker, "The Pioneer of Atlantic Steamships," *The Canadian Magazine of Politics, Science, Art and Literature*, vol. 29, p. 13, May, 1907.

Figure 9.8 The *Great Western* of 1838 (From *Transactions, Newcomen Society*, vol. 10; original in Science Museum, Kensington; courtesy Newcomen Society)

and to arrive there had it not been delayed by a fire and by grounding at Gravesend near London.

Thinking of New York as the next stop on the line after Bristol, the directors of the Great Western Railway had formed in 1835 a subsidiary steamship company and authorized its engineer, Isambard K. Brunel, to construct a vessel for the transatlantic service. Brunel and his associates made the *Great Western* 235 feet long with a beam of 35¼ feet and a displacement of 1,320 tons. Its side-lever engines with 73½-inch cylinders seem to have developed as much as 750 horsepower. The paddle wheels were 28 feet in diameter, and the ship attained a speed of 8 knots. It consumed 655 tons of coal in crossing the Atlantic from Bristol to New York in fifteen days. Before it was sold in 1847, the *Great Western* had made records between New York and Bristol of twelve days and eighteen hours westbound and twelve days and seven hours eastbound. Whether this pioneering steamer met every expectation of the directors of the Great Western Railway is doubtful. There is no question that it marked, with the *Sirius*, the beginning of transatlantic travel by steamship. In 1851 Dionysius Lardner denied that he had made his "perfectly chimerical" statement which had been reported in the *Liverpool Albion* for December 14, 1835.

Figure 9.9 The *Great Britain*, first iron steamship, 1843 (From Claxton's *History and Description of the S.S. Great Britain*, 1845)

Cunarders began to appear in 1840. The first *Britannia*, a wooden side-wheeler, left Liverpool for America on Friday, July 4, 1840, defying superstition in favor of publicity, and arriving in Boston on the 18th. The ship displaced 1,154 tons. Its four boilers operated under 9 pounds pressure. Its single cylinder engine burned coal at the rate of 38 tons a day, generated 740 horsepower, and drove the ship at 8½ knots. By 1843 the Great Western Company had launched an iron ship for the transatlantic service. John Wilkinson, William Fairbairn, and others for some time had been constructing iron hulls. The advantages of iron over wood were many: iron hulls were lighter, stronger, more durable, and less susceptible to decay. The ship's capacity was greater; it could be made longer and larger without danger of buckling. Iron met far better than wood the demands of the "big ship" men. The *Great Britain* (Figure 9.9) was so large that a special shipyard had been constructed for it, and it could not be berthed at any of the docks on the Mersey. It displaced 3,443 tons.

A short time before this, John Ericsson (1803–1889), Swedish-born inventor famous later for his *Monitor* in the American Civil War, and Francis P. Smith, his English collaborator, had demonstrated screw propellers. They were superior to paddle wheels in rough or icy water and freezing weather; they would remain submerged most of the time, and they would not clog with ice. Correctly pitched, screws did not "bore holes" in the water. They delivered power more effectively than did paddles. Brunel's new iron ship, the *Great Britain*, though planned in 1839 to be a side-wheeler, was changed over to screw propulsion. The paddle-wheel shaft, for the forging of which James Nasmyth (1808–1890) had made his steam hammer, was never made; in its place the screw shaft was produced which was rotated by four flat endless chains over a drum 18¼ inches in diameter with a 36-inch face. The drum was mounted on a crankshaft which was driven by four cylinders; each cylinder was 88 inches in diameter and had a piston with a 6-foot stroke. The pulley on the propeller shaft was 6 feet in diameter, so that the propeller shaft rotated about one-quarter as fast as the crankshaft. Steam came from three boilers at a pressure of 5 pounds, generating 1,500 horsepower. The propeller was 15 feet 6 inches in diameter; its four blades were later increased to six.

With some fifty or sixty passengers aboard and 600 tons of cargo, the *Great Britain*, largest ocean-going ship up to that time, steamed out of Liverpool Harbor on July 26, 1845, and arrived in New York 14 days and 21 hours later. It had averaged 9½ knots. Although this was not remarkable speed, the *Great Britain* had set standards for years to come. Old things were to reach perfection after new devices had arrived to replace them. The full-rigged clipper ships of the 1850s were still to "crack on the dimity" and race through rioting seas, delightful for their spirit and their beauty. But they would idle in the doldrums, too, as steamers came droning by, bound on schedule to port.

Steam Engines in Ships

While Henry Bessemer demonstrated his steelmaking to the ironmasters of Britain, Isambard Kingdom Brunel (1806–1859) was planning the world's biggest ship. It was to exceed beyond comparison, in tonnage and power, anything afloat. His *Great Britain* of 1845 displaced 3,443 tons and had 1,500 nominal horsepower; the *Great Eastern* (Figure 9.10) was to have a gross tonnage of 27,060 and to generate more than 11,000 nominal horsepower. The *Great Britain* measured 322 feet long by 50 feet 6 inches beam. The *Great Eastern* was to reach 692 feet with a beam of 83 feet inside the paddle boxes and 120 feet outside. There would be accommodations for 800 first-class passengers, 2,000 second-class, 1,200 third-class, a total of 4,000, and a crew of 400. The passengers were to be quartered midships over the machinery, an arrangement which was

Figure 9.10 The *Great Eastern* on the ocean (From *Pictorial History of the Great Eastern Steam-ship*, 1860)

considered a startling innovation nearly twenty years later when adopted by the White Star Line. There was space for 6,000 to 8,000 tons of cargo, and the bunkers would hold 12,000 tons of coal for the voyage to India, China, or Australia and return, around either the Cape of Good Hope or Cape Horn. The *Great Eastern* was to take command of the seas.

Construction came too early for the ship to have in it any steel. It was launched on January 31, 1858. The wrought-iron hull nevertheless embodied principles which were to govern shipbuilding with steel to this day. It had longitudinal framing or webbing and a cellular or tubular design remindful of Brunelleschi's double dome and the Fairbairn truss in Stephenson's Britannia Bridge. The distance between the inner and outer skins of the *Great Eastern* was 2 feet 10 inches. The hull was subdivided by 10 transverse bulkheads, spaced 60 feet apart, and by two longitudinal bulkheads 36 feet apart and 350 feet long. This design gave not only strength throughout but also safety, by dividing the interior into cells, in case any compartment were pierced and flooded from the sea.

The vessel was driven by a combination of paddle wheels and a screw propeller. The engine to drive the paddle wheels had four oscillating cylinders, each 74 inches in diameter, thrusting obliquely upward under 25 pounds pressure with pistons of 14-foot stroke. It developed a maximum of 5,000 horsepower at 16 revolutions per minute. The engine which drove the screw propeller had four opposed horizontal cylinders, each 84 inches in diameter with pistons of 4-foot stroke and generating 6,500 horsepower at 55 revolutions per minute. These cylinders also worked under 25 pounds pressure. Both engines were simple or single-expansion engines; that is, the steam was used but once in the cylinders. The paddle wheels were 56 feet in diameter with paddles 13 feet long by 3 feet wide and 30 in each wheel. The screw propeller was 24 feet in diameter and had 4 blades.

There were 10 wrought-iron boilers, rectangular in shape, 6 for the screw and 4 to serve the paddle-wheel engines. They were equipped with brass fire-tubes and were some 18 feet wide by 14 to 18 feet high. Each boiler had 10 furnaces in it, to be stoked day and night by hand in suffocating heat. Brunel did not adopt Hall's new surface condenser for salvaging spent steam and reusing it as fresh feedwater. Instead he kept the old method of water jets in the condensing chambers; sea water still entered the boilers as feedwater with inevitable damage to them from dissolved solids and with loss of efficiency and necessity for frequent change by "blowing down" or emptying the boiler periodically. Even so, they delivered power enough to drive the *Great Eastern* at 14 to 15 knots.

The fastest clipper ships, sailing greyhounds of the 1850s, could make 18 knots with very strong favoring winds. The *Great Eastern*, fairly independent of the wind and free to set a straight course, would outrun the best of them from port to port—if nothing happened to its engines. The ship was also equipped with six masts. All were rigged with fore-and-aft sails, the second, third, and fourth also with yards, and the masts were 130 to 170 feet tall. Five of the masts were made of wrought iron rolled in tubular form, 2 feet 9 inches to 3 feet 6 inches in diameter. The mizzen mast was wood with hemp stays to reduce the magnetic interference of iron with the ship's compass, which was mounted high on the mast.

With thirty-eight paying passengers and eight guests, the *Great Eastern* set out across the Atlantic on June 17, 1860, for the United States and arrived at New York on June 28, in eleven days. There had been no trouble on the voyage except some anxiety over the bar at Sandy Hook where the ship had "only 2 feet to spare." The trip was a clear prophecy of things to come in transatlantic navigation, but the vessel was clumsy, oversize, and expensive to operate. It was too large for the docks and shore facilities of the time. Nearly forty years were to pass before the *Oceanic* exceeded the *Great Eastern* in length, and its tonnage was not exceeded until the *Celtic* was built in 1901.

The opinion appears to be general that the *Great Eastern* came too

soon in the history of steam navigation for its own success. The heavy
cost led to the failure of the company and to sale of the vessel to another company that was not interested in the Oriental traffic for which
the ship had been planned. Brunel's death has been attributed in part
to worry about the great ship. The transatlantic run was too short to take
proper advantage of the size and range of endurance which the ship
possessed. It served in laying the Atlantic cable in July, 1866, but it failed
to be of any profit on the sea and ended its days in the Mersey as a
"floating billboard and amusement hall." It was broken up in 1889 and
sold as old iron for £16,000, about one-sixtieth of its original cost.

The big ship was planned before work started on the Suez Canal. If
the ship had been able to establish its size and its speed on the long sea
lanes in the early sixties before the Suez Canal was opened in 1869, the ship
might have had something to do with determining the dimensions of the
canal; if it had, the subsequent history of transoceanic navigation would
certainly have been different. The Suez Canal, once in operation, made
clear that all vessels seeking the Oriental trade should be small enough to
pass through its waterway rather than large enough for the long run
around the Cape of Good Hope and across the Indian Ocean. Speculation on what might have happened is interesting, but the historical fact
remains. The significance of the *Great Eastern* lies not in failure or
success but in its effect upon the ultimate practices of marine engineering. William H. White's address as president of the Institution of Civil
Engineers in 1903 summed up the achievements as, among others, ample
structural strength with a minimum of weight, safety by watertight subdivision and cellular double bottom, economy of coal and endurance for
long-distance steaming, dimensions to minimize resistance and to favor
good performance at sea. Brunel, said White, had displayed a knowledge
of principles such as no other ship designer of the time seemed to have
possessed; his courage had been justified in the experience of subsequent
years. Above all, Brunel had proved that size in a ship was an advantage
and, as he said, was "limited only by the extent of demand for freight,
and by the circumstances of the ports to be frequented." [6]

The emphasis for some time after the *Great Eastern*, however, was
upon smaller ships with increased efficiency and speed. Transition from
iron to steel came with the growth of the Bessemer and Siemens-Martin
processes of making steel, bringing lighter weight and greater strength

[6] Isambard Brunel, *The Life of Isambard Kingdom Brunel*, Longmans Green and
Co., London, 1870, p. 292.

for a given section; by 1891, 80 per cent of all steamships under construction were of steel. Single-screw propellers replaced paddle wheels on ocean-going vessels. Compound triple- and quadruple-expansion engines operating under higher pressures reduced the consumption of fuel as much as 50 per cent. There was thus more room for pay cargoes. Steam power was now clearly in profitable competition with sail on the transatlantic run, as had not been the case in the first days of the *Great Western.*

Locomotives

Following the medieval inventions of the horse collar and horseshoe, enabling the horse to pull heavy wagons, nearly a thousand years passed before the next important advance in land transport was achieved. The vehicle that superseded the horse-drawn wagon for long-distance hauling was the steam locomotive—the iron horse. Although the early road locomotives were not successful, they have historical importance because they were the progenitors of the railway locomotive.

A Flemish Jesuit abbé in far-off China appears to have been the first to produce a vehicle which moved itself. Using Hero's aeolipile, mechanical knowledge that he had acquired in his education at home, his own ingenuity, and spare time from his labors among the heathen of Pekin, Ferdinand Verbiest (1623–1688) built a toy cart of light wood 2 feet long with the axle of its front wheels geared to a steam turbine, or aeolipile, fed by a vessel full of live coals. Verbiest made his tiny vehicle move in a circle "with a motion and not slow for an hour or more, for as long a time, be it understood, as the steam continued to be forcefully expelled from the aeolipile." [7] Verbiest seems to have made his little cart some time between 1665 and 1681. An account of it first appeared in Europe at Dillingen, Bavaria, in 1687, in a chapter on "Pneumatica" in the abbé's book *Astronomia Europaea.*

The French scientist Denis Papin (1647–1712) made the second model of a self-propelling vehicle about 1698 and used in it his idea of cylinder and piston. It was to be for military purposes. The French military engineer Nicholas Joseph Cugnot (1725–1804), however, in 1769 was the first to put a self-propelled vehicle on a road (Figure 9.11), using a steam engine developed quite independently from those of Newcomen and later of Watt. Cugnot used Papin's well-established ideas regarding piston, cylinder, and steam to operate above atmospheric pressure without

[7] Ferdinand Verbiest, *Astronomia Europaea,* Joannem Federle, Dillingen, 1687, p. 88.

Figure 9.11   Voiture à vapeur de l'Ingénieur Cugnot (1770) (Courtesy New York Public Library)

the condensing feature of the British engines. As improved the next year, Cugnot's vehicle was intended to be a gun tractor mounted on three wheels, with the power delivered straight to the single wheel in front. Thus he avoided the problem of the differential gear or other arrangement necessary to allow moving one wheel more slowly than the other when making a sharp turn. However, he suspended his boiler upon the front guide wheel instead of loading it within the carriage, which, of course, made the thing clumsy, top-heavy, and likely to tip over at the first obstacle, as it actually did. Besides, his two cylinders were not equal to their task although they were 13 inches in diameter. They were expected to work under relatively high pressure and to deliver their power through pawls to ratchets on the front wheel—one pushing while the other was withdrawing. This arrangement rapidly exhausted the pressure, and the tractor had to stop frequently to get up more steam. When running, it could move no faster than a man could walk, 4 miles an hour at best. It had no capacity for hauling a trailer; artillery horses had not been replaced. Cugnot's steam vehicle retired to the Conservatoire des Arts et Métiers in Paris where it may still be seen.

There was further experimenting in Britain in 1784 at Redruth, Cornwall. James Watt's brilliant assistant, Scottish-born William Murdock (1754–1839), constructed a working model a foot high with a spirit lamp to generate steam and tried it out on the path leading to the church. The fiery, hissing little monster came suddenly upon the parson, like the "Evil One," he said. Watt considered making an engine which would move itself and had included the idea among his patents, but he pooh-poohed Murdock's device so much, as was his habit regarding every high-pressure engine, that his young assistant seems to have been talked out of attempting to make it practical. Not so the Cornish strong man Richard Trevithick. For him, whatever James Watt believed was either to be refuted or excelled.

Trevithick built a model in 1798 and developed it by 1801 into a full-sized steam locomotive engine with which he planned to carry passengers. Obtaining a patent on March 24, 1802, he shipped this engine by sea from Plymouth to London in 1803 and actually ran it in the streets. It was an adaptation of his noncondensing or puffing engine with a cylinder 5½ inches in diameter and a piston stroke of 2½ feet. It operated under 30 pounds pressure, making 50 strokes per minute. The three wheels were of wood, their tires wrought iron. There is no information regarding the speed of this engine on a smooth road, but a significant event in the history of transportation was at hand if only Trevithick were to strive to bring it about. His imagination as an engineer, however, was more forceful than his persistence. When his engine twisted its frame, he put it to stationary work, lost interest in steam locomotion on the highways, and turned his genius to developing a locomotive engine for the rails. Here too he pioneered, but turned away to other ideas which fascinated him before he had made his excellent plan practical.

In the meantime, Oliver Evans, Trevithick's American rival in creating the high-pressure engine, had also thought of his "steam waggons," but like Trevithick, he was to accomplish far more with stationary power plants. The legislators of Pennsylvania were quite as much to blame as Evans for his having ceased work on road vehicles. They thought him insane when he asked for the rights of steam transportation on the roads of the state and they ignored his request. He worked with self-propulsion long enough, however, to put one of his high-pressure engines into a wooden dredging scow on wheels (Figure 9.12). This he named Orukter Amphibolos, literally, the digger which works both ways. It trundled its 34,000 or more pounds 1½ miles up Market Street and several times around Centre Square, Philadelphia, in July, 1805, before launching itself upon the Schuylkill. There its paddle wheel took over and pushed it 16 miles to its dock on the Delaware.

The Orukter Amphibolos was expected to dredge around the docks along the water front for the city's Board of Health, but after three years of experimenting, the whole project was abandoned. The board had spent $4,800 on the "mud machine." From the ease with which this strange device had moved up Market Street and circled the square, Evans drew the conclusion that "the engine was able to use any ascent allowed by law on turnpike roads, which is not more than 4 degrees," [8] which would be just over 7 per cent. Nobody else seems to have thought so. He estimated the maximum speed at 15 miles an hour. It is not easy to think of the contrivance as anything more than a historical curiosity, a forerunner, but in no sense an ancestor, of the automobile.

With these developments and the contemporary success of the steam locomotive on rails, many sought in the next twenty years to build steam-propelled vehicles for the public highways, especially in Britain where Telford, McAdam, and others were laying hard-surfaced roads through a thickly populated countryside—the best roads for general traffic until recent times. Outstanding among these engineers who put self-driven

[8] Oliver Evans, *The Young Steam Engineer's Guide*, H. C. Carey and I. Lea, Philadelphia, 1805, p. 50.

Figure 9.12 Evans's Orukter Amphibolos (From *The Mechanic*, July, 1834)

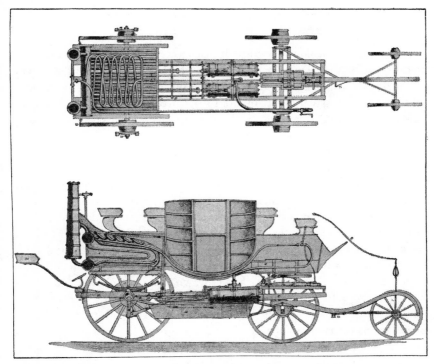

Figure 9.13 Gurney's steam carriage of about 1832 (From *Histoire de la loco-motion terrestre*, editée par *L'Illustration*, 1936; courtesy *France Illustration*)

machines on the roads of Britain were Goldsworthy Gurney (Figure 9.13), Timothy Burstall, John Scott Russell, and Walter Hancock. Russell later became the builder of the steamship *Great Eastern*. Burstall competed with Stephenson in the trials of locomotive engines on rails at Rainhill. Gurney was noted for his research in chemistry, and his lectures on the subject had some influence upon Michael Faraday.

As a child, Gurney (1793–1875) had been deeply impressed in 1804 by the Trevithick engine on wheels. He declared in a public lecture in 1822 that "elementary power" could be used to propel carriages along the common roads. His views, as he recalled ten years later, were not then accepted either by practical engineers or by the scientists, who were sceptical, although their eminent colleague Humphry Davy (1778–1829) was experimenting even then with the mechanical use of compressed carbon dioxide in place of steam. Gurney set out to prove them all mistaken and succeeded. By 1832 his steam coach was running on schedule four times daily over the 6 or 7 miles between Gloucester and Cheltenham. Gurney's engine weighed 2 tons. It had a water-tube boiler fired with coke to prevent smoke. It operated under pressures up to 70 pounds and attained a speed of 24 miles an hour. The wheels were 4 feet in diameter with iron tires nearly 4 inches broad. In 396 trips, it carried three thousand passengers without a serious accident, in less time than the horse-drawn stages and at one-half their fares.

Similar carriages averaged between 12 and 20 miles an hour. Nathaniel Ogle testified before a committee of Parliament that he was able to reach speeds between 30 and 35 miles an hour under fairly normal conditions of weather and road. Another operator claimed that his steam vehicle had run a mile in 70 seconds. Closer reading of his testimony reveals, however, that his course had been downhill, his brake had failed to hold, the engine had run away with itself and its load, which was hardly demonstrating the power of steam. Interests vested in the old method of travel opposed the new. Owners of horse-drawn stages made the outcry to have been expected of them. For apparently it was correct, as claimed by one operator, that the steam carriage could run 100 miles a day and replace as many as fifty horses. Farmers who raised oats joined in the clamor. Those who in all times are forever looking, and possibly hoping, for a good moan over catastrophe raised their voices too. These contrivances were not only uncomfortable, they were absolutely dangerous, they were sure to explode and take everyone in them straight from this world into the next.

The facts were that there were frequent boiler failures with sudden loss of power. Injury from them, however, was remarkably slight because the boilers contained so little water. Sometimes the passengers did not even know of the failure until the vehicle had come to a stop. Explosions were few from the beginning, and loss of life attributable to them was small. Parliament, nevertheless, listened to the complaints, held hearings upon the dangers to peace and quiet, property rights, and damage to the public roads, passed legislation levying heavy taxes on road locomotives and making other restrictions. Finally it was required that a man on foot with a red flag had to precede each self-propelled vehicle on the public highways. The restrictions were not removed in Britain until 1896.

It was not, however, primarily the jealousy of horsemen and farmers that drove steam carriages from the roads of Britain in the 1840s and kept them from the highways of America. There were differences between the tolls charged for steam vehicles and those levied upon horse-drawn coaches, but these differences were hardly enough to force the steam carriages off the roads. It was the railway which brought misfortune to the predecessors of automobiles. In commenting upon the operation of steam carriages in Britain the *American Railroad Journal* of New York, June 30, 1832, had remarked, "It would be a little singular if *steam carriages on common roads* should treat Railroads with as little ceremony as Railroads are treating Canals." It was not to be so. The rail companies. like the farmers, horse owners, and turnpike companies, threw their weight into Parliament to exert pressure upon the lawmakers. More determining than restrictive legislation, however, was the fact that the railways developed high-speed service for both passengers and freight so well that the self-propelled carriers on the roads could not compete for the traffic with the carriers on rails. This triumph of railroading was before the coming of gasoline, production on assembly lines, reinforced concrete, and rubber.

Locomotives on Rails

As was mentioned in the introduction to this chapter, the success of the locomotive depended principally on the development of compact, high-pressure engines and boilers and the use of iron rails on roadbeds with slight gradients. The initial advances in the solution of these problems were made in England, where the steam engine had been invented and improved and where large-scale production of iron had first been achieved. Engineers had improved high-pressure boilers and engines for the road locomotives described in the previous section, but the solution to some of the fundamental problems of the railways was not attained until about the time that a ship first crossed the Atlantic entirely under steam power. Not until the end of the first quarter of the nineteenth century was the iron industry able to supply adequate iron rails. By this time engineers realized that only an almost level roadbed would enable a locomotive to haul a trailing load heavy enough to be profitable. Within a relatively few years after the construction of the first railways, steam locomotives began to replace steamboats in inland transportation. Although there are still many inland steam vessels, the number of ton-miles they now haul is but a small percentage of that carried by railroads in most parts of the world.

Richard Trevithick had made a trial with steam locomotion in February, 1804, on the rails of a Welsh coal road, probably the very first

Figure 9.14 Trevithick's railway exhibited in Euston Square, London, in 1808 (From *Transactions, Newcomen Society,* vol. 1; courtesy Newcomen Society)

time a steam locomotive engine ever ran on rails. His engine hauled 10 tons of iron, 70 men, and 5 trailing cars, 9½ miles at 5 miles an hour, except when stopped to remove obstacles from the track. Later, in March, Trevithick wrote that he had tried the carriage with 25 tons of iron and had sent the steam from the exhaust into the stack above the damper as a matter of convenience. He took note, however, that it made the draft much stronger "by going up the chimney." When this engine went off the track, he put it to stationary work, as he had done with the steam carriage which he had demonstrated on the streets of London. Trevithick shipped a steam locomotive engine of 8 tons to London in 1808 and exhibited it for an admission fee of 5 shillings behind a fence on a circular track (Figure 9.14). This was the engine acclaimed as the "racing steam horse" and as "the catch-me-who-can." A passenger timed it at 12 miles an hour, and Trevithick asserted that it could do 20 on a straightaway. Its cylinder was 14½ inches in diameter and the piston stroke 4 feet. The engine ran off the track and overturned, and since there had been few shillings spent to see it, it was not put back on the rails.

Richard Trevithick has been credited by many with being the first to show that smooth wheels would adhere sufficiently to smooth rails on ordinary gradients, the first who used the steam jet in the smokestack for increased draft, and the first to couple the four wheels of an engine so that all of them were driving wheels. He did this coupling by means of gears. As the engine had but one cylinder, however, there was the danger of its sticking on dead center when it stopped. Trevithick had also used the return flue in his boilers. All of these devices had to be coordinated and much more accomplished before the steam locomotive engine could be considered practical.

It is useless to say that Richard Trevithick would have succeeded if he had tried. He turned to steam dredging and tunneling the Thames, improved his stationary engines, and in 1816 went to the mines of Peru where he remained until ousted by the revolution for independence from Spain. There is a story of his planning a railway in Central America from the Atlantic to the Pacific. During 1827 in Venezuela he met Robert Stephenson (1803–1859), who also was mining in South America on behalf of himself and Britain. Those who know the political history of Britain in liberating the republics of Latin America will catch here a glimpse of more than antipathy for the misrule of the Spanish monarchy. As Trevithick was destitute, Stephenson supplied him with funds. Both returned home shortly, Stephenson to fame in railway and bridge building, Trevithick to scant attention and penury when he died in 1833, although he had been one of Britain's ablest inventors.

John Blenkinsop (1783–1831) had constructed in 1811 a locomotive engine with cogwheels on a rack. It could haul 94 tons at 3½ miles an hour on a level and climb a grade of over 5 per cent. As proposed by Matthew Murray of Leeds, this engine had two cylinders driving cranks at right angles so that it could not stop on a dead center—an arrangement used in subsequent locomotive designing. Blenkinsop's engine was capable of operating at a profit, but it was unsteady; the strain upon the rack at one side pulled the rails apart, and the boiler finally blew up. Robert Stephenson's father George (1781–1848), self-educated engineman of the Killingworth Colliery in northeastern England, had observed the weaknesses in Blenkinsop's engine. Stephenson decided that he could build a better "Travelling Engine," and before he finished he had made the coordinations and adjustments that were necessary to the technical efficiency and economic success of steam locomotion on rails.

Stephenson returned to the use of a smooth wheel and rail. He named his first engine the *Blücher*, in honor of the Prussian general then campaigning with Wellington against Napoleon, and tried it out at Killingworth on July 25, 1814. It had two vertical cylinders set into the boiler, one for each pair of wheels. The cylinders were 8 inches in diameter with a 2-foot stroke. The pistons drove upward to crossheads from which the power was sent downward on both sides of the boiler through connecting rods to cranks fastened outside the wheels. The cranks were set at 90 degrees from one another to prevent the engine from stopping on dead center. The axles were connected by chains on sprockets so that all wheels, like Trevithick's, operated together. Stephenson made his cylindrical boiler of wrought iron, 8 feet long and 2 feet 10 inches in diameter, with a single flue of 20 inches diameter running through it. As Trevithick had done earlier, he carried the exhaust steam to a nozzle in the smokestack where it induced a draft for the fire. With this locomotive, Stephenson hauled trains of thirty "waggons" 4 miles an hour at the colliery. But it cost too much to operate. He had still to make a better "Travelling Engine."

Having built a second in 1815 without the success of which he was so confident, Stephenson by 1818 came to the conclusion that he needed to know more about the resistances his locomotive had to overcome. They were mainly three: first, the friction of the axles; second, the "rolling resistance" or friction between the wheel and the rail; and third, the force of gravity on grades. He constructed an instrument to measure the drag of the train against the pull of his engine and, with the assistance of Nicholas Wood (1795–1865), made a most significant finding for railway builders. Friction was constant at all speeds. The resistance of a train of cars on straight-level track amounted to approximately 10 pounds per long ton, but so slight a rise as 1 foot in 100 required an additional pull of 22.4 pounds per long ton to overcome the force of gravity. In other words, a gradient of 1 per cent would add more than twice as much resistance as on level track and thus require of the locomotive more than three times the tractive effort. Stephenson made it obvious that if a railway locomotive were to be successful it should operate upon grades as low as possible. This essential factor was not applied to the first railway which Stephenson began in 1822 to connect the coal field of Darlington with the port of Stockton-on-Tees. Neither the funds nor the experience of the promoters made a nearly level road possible. There were several inclined planes with stationary engines on the route, and in some sections horses were still used. The two locomotives, however, were marked improvements over his first. On the opening day, October 27, 1825, *Locomotion No. 1* hauled a load of 90 tons made up of 36 small cars, about 600 passengers, and some freight at a speed of 12 miles an hour.

Earlier in the year, Nicholas Wood, as he produced the first edition of his book *A Practical Treatise on Rail-roads,* had committed himself to this judgment: "It is far from my wish to promulgate to the world that the ridiculous expectations, or rather professions, of the enthusiastic speculist will be realized, and that we shall see them travelling at the rate of 12, 16, 18 or 20 miles an hour: nothing could do more harm towards their adoption, or general improvement, than the promulgation of such nonsense." When he published the second edition of his book in 1831, he had to omit his pronouncement.

The Stockton and Darlington Railway had drawn the attention of the world. Plans which had been germinating for some time in many places had taken form. A depression checked most of these in Britain but only for the moment. The economic value of a better connection between the textile city of Manchester and the port of Liverpool than Brindley's Bridgewater Canal furnished was evident; a survey for a horse tramway had been made in 1797, another in 1822, and now a third more direct route was selected in 1826. George Stephenson became the chief engineer; his son Robert built the locomotive and practically had charge of the construction of the roadbed. The young engineers whom he gathered about him at this time were to go on to other successes and fame of their own.

The Liverpool and Manchester Railway was a decisive success, which its contemporary historian understood. "Speed—dispatch—distance—are still relative terms," he wrote, "but their meaning has been totally changed within a few months." [9] George Stephenson was able to disregard the cost and apply his principle that grades should be as low as possible. At the time of the opening, costs had mounted to twice original estimates, and later expenses increased the outlay considerably more. In one cut 2 miles long through rock at Olive Mount near Liverpool, the track in places was down 80 feet. Through the 1¼-mile tunnel near Liverpool the gradient was as much as 2.1 per cent, and at first the trains were hauled

[9] Henry Booth, *An Account of the Liverpool and Manchester Railway,* Wales and Baines, Liverpool, 1830, p. 89.

up this incline by a stationary engine, which was removed by 1832. Another feature of Stephenson's engineering, which added to costs but deflated critics, was the layer of "hurdles, thickly interwoven with twisted heather, in a double layer with their ends overlapping, thus forming a floating road" [10] over the bog at Chat Moss where he carried the track straight across for nearly 5 miles. An arched viaduct became famous as the Sankey Viaduct, but it was the locomotive which George Stephenson and his son Robert produced for the Liverpool and Manchester Railway that put an end to all speculation as to whether a "Travelling Engine" on rails could replace horses.

George Stephenson had persuaded the promoters to hold public trials and offer a prize of £500 for the best locomotive which should meet stipulated conditions with regard to weight, speed, and tractive power. Of the four engines entered in the contest, held during the historic week of October 6 to 14, 1829, two met with mechanical difficulties and the third was far too weak. Stephenson's *Rocket* (Figure 9.15) triumphed, meeting all of the stipulated conditions and on its second run drawing a carriage with 30 passengers from 25 to 30 miles an hour. In the *Rocket* were combined the features that are most essential to an efficient steam locomotive. Stephenson's boiler was multitubular, equipped with 25 copper tubes, 3 inches in diameter, through which flowed the hot gases to the stack. It does not matter whether Stephenson learned this tubular arrangement from the French Marc Seguin (1786–1875), or Seguin acquired the idea from him, or both developed it independently, as John Stevens in America seems to have done. It does matter that Stephenson used it in his *Rocket* and thus increased the thermal efficiency of his engine and more especially its capacity for making steam. The firebox at the rear end of the boiler was surrounded by water compartments, or water legs, to conserve the heat. This arrangement has not been essentially improved since Stephenson's time. He sent the exhaust into the stack and brought the blast under control by a variable opening. According to their early biographer Samuel Smiles, the Stephensons in this way doubled the power of their engine. They still kept the cylinders inclined at an angle but connected the pistons directly with the pair of driving wheels. The *Rocket* with its tender weighed about 7¼ tons. Some twentieth-century steam locomotives with tenders weigh as much as 500 tons.

After the victory of the *Rocket*, the Stephensons modified the design

[10] Chapman F. D. Marshall, *Centenary History of the Liverpool and Manchester Railway*, Locomotive Publishing Co., London, 1930, p. 22.

Figure 9.15 Stephenson's *Rocket* (From Smiles, *Lives of the Engineers*, vol. 3, 1862)

of their engines somewhat. They placed the cylinders horizontal at the front end of the locomotive so that they drove directly backward to the main axle. The engineman's control station was put at the rear of the boiler; both he and the fireman could tend the fire and the controls. For many years, European locomotives were inside-connected; that is, the connecting rods were fastened to cranks in the axles between the wheels. American engine makers started early to build outside-connected locomotives, with their connecting rods outside the driving wheels. Other changes followed, but it is fair to say that subsequent locomotives were for the most part merely improvements in detail upon the *Rocket* of George and Robert Stephenson.

The improvement of the railway followed rapidly after the trials at Rainhill had demonstrated that locomotive engines were quite practical. Rails of a sort had been used underground in medieval mines, perhaps earlier. The first illustration of a four-wheeled cart on a track (Figure 9.16), published about 1519, has been mentioned with the story of mining in Europe. It appears likely that German miners brought the idea to the

Figure 9.16 An early sixteenth-century four-wheeled cart on rails (From S. Münster, *Cosmographia*, 1628; first printed in 1550 edition)

collieries of Northumberland in Britain about 1600—not the last time that imported labor has made significant change in a country. Little imagination was required to extend the rails from the pit to the wharf on the nearest waterway, and none at all to load the cars with materials besides ore and coal. Horses dragged or pushed the "waggons." The next improvement was to fasten iron strips to the wooden rails, and the next,

Figure 9.17
The Quincy Granite Road, one of the first United States railroads, 1826 (From C. H. Snow, *A Geography of Boston*, 1830)

to put iron tires on the wheels. These two improvements reduced the rolling resistance and prolonged the life of both cars and rails. Thus originated the private tramways of the nineteenth century. Among them was the 3-mile railway in Quincy, Massachusetts, built in 1826 (Figure 9.17) which carried granite from the quarry to the wharf for the Bunker Hill Monument, Minot's Ledge Lighthouse, the Custom House of Boston, and other notable structures. This road was the best known, if not the first of such tramways, in the United States.

The first railway chartered for the use of the public was built from the Thames River at London southward 8 miles to Croydon, today the site of London's first great airport. This was the Surrey Iron Road, opened in 1803, later extended and maintained until 1846. Its users paid tolls and owned or hired their conveyances. It amounted to little more than a plank road, although it followed the pattern of the tramways. It had rails set on crossties rather than close-laid boards upon parallel beams or stringers. Plank roads of the latter construction were popular in early American and Canadian communities where the forests provided cheap timber and stone surfacing, for roads had not yet arrived. Though these plank roads were privately owned, they too were open to the public upon payment of tolls. Users provided their own conveyances.

The problem of keeping the cars on the rails was solved before the locomotive engine made it necessary to do so. The device at first was a simple prong below the car, caught in a trough or groove between the rails. This device can be seen in Agricola's *De re metallica* upon a miner's truck of 1556. The system has been used as late as the 1870s for a horse-tramway at Geneva, Switzerland. Just when the flanged wheel came is not known, but the idea had been fully matured by 1730. Ralph Allen's car to haul stone from his quarries down to the River Avon had deep flanges on its cast-iron wheels. They fitted closely upon edge rails of wood. Flanges on the rails themselves were tried, both on the inside and on the outside of the rail, but their control of the car was manifestly inferior to the contact and restraint which were to be had with rail and flanged wheel.

The origin of the crosstie is no better known than that of the wheel and axle. Certainly it is not as old, but experience with cross timbering in bridges, trestles, and wooden construction generally, since the days of Caesar, at least, was common. The earliest railways had wooden crosspieces. The use of stone blocks to support rails without crossties, which came into vogue with the first locomotive engines, was in reality an ill-advised departure from a proven device. Ties not only were better to keep the rails from spreading; they also were superior for the stability which the stone posts were intended to ensure. By 1838, Nicholas Wood, Stephenson's associate, was stressing the necessity of crossties where the ground was likely to settle. Rigid stone, steel, or concrete supports are often used today in terminals and at loading platforms where permanence is desired and the speed of trains is not to be considered, but such supports are too unyielding for high-speed railways. The strongest track today must have some elasticity, or rolling stock under any speed at all will be injured by the pounding. Trains, like horses, require some spring to the path upon which they travel.

The question as to what should be the width of a railway caused in the 1840s and 1850s the "battle of the gauges" in Britain between the Stephensons and Isambard Kingdom Brunel (1806–1859), builder of the Great Western Railway to the port of Bristol. The Stephensons thought that eventually all railways in Britain would become one system. Brunel visualized regional groups, such as did come later, which might vary with local conditions. He conceived of wider tracks, more powerful engines, heavier trains for the transport of the future. The Stephensons built for Brunel the locomotives with the 7-foot gauge which he desired, but they preferred to keep the approximate width of 4 feet 8½ inches between the railheads, based on the traditional width between the wheels of carts. More than any other influence, their engines made this gauge standard. Many gauges have been tried, from 2 feet to 7. The narrower gauges had many advocates on the score of economy; the wider gauges on the score of capacity. Hundreds of miles of both were built, but most of them have been changed to standard gauge. The Erie Railroad in the United States, for instance, used a 6-foot gauge until 1878, but the Erie had more influence upon the fine art of manipulating the finances of railroads than upon their engineering. The Stephensons' superior engines of standard gauge won the battle of the gauges in America. The desirability of a uniform gauge became obvious as the town-to-town companies merged into interstate and international railroad systems. Only local interests such as the proprietors of transfer stage companies and lunch counters cared to have one gauge entering town and a different one leaving it. As soon as the Stephenson gauge became dominant, it was too late for any other to persist. The tread width of ancient carts of about 5 feet survived as the standard width of successors, which would have astounded the owners of those carts in every other respect.

There are several unlikely theories as to the origins of the 4-foot 8½-inch gauge, and the battle of the gauges still goes on in various parts of the world as the Spaniards keep their 5-foot 6-inch gauge, Ireland its 5-foot 3-inch gauge, and the Russians move the rails in satellite countries to meet their own requirement of 5 feet. There are many narrow gauges as well as broad gauges throughout the world.

The rail itself responded to the quickening demands of the steam locomotive. Wooden tramways had been overlaid with iron strips. Although likely to curl and break, they continued in use, especially in America, years after iron rails set on edge had been adopted in Britain. Rails like beams were made of cast iron at first in 1789 but given a deeper section in the middle of their span between ties in order to sustain heavier loads. From their appearance they were known as fish-belly rails. George Stephenson advised the use of edge rails on the Stockton and Darlington Railway in 1821. "According to Gregory's Mechanics," he wrote, " 'in rectangular beams the lateral strengths are conjointly as the breadths and squares of the depths.' Hence the substance in edge Rails is disposed of in the most advantageous manner, viz. by increasing the depth." [11] This was a fundamental observation of the greater strength to be had by taking advantage of the distribution of metal and using it in the height rather than the width of the rail. The next improvement was to eliminate as much as possible of the excess metal near the center, or neutral axis, of the cross section and to concentrate it at the top and the bottom, where the stresses of compression and tension are the greatest. This form came to be known generally as the T section. But Stephenson's edge rails for the Stockton and Darlington Railway were not so designed; they were rectangular in cross section, of wrought iron, 12 to 15 feet long, and weighed 28 pounds per yard. They rested in cast-iron supports, or chairs, which were pegged into stone blocks.

[11] Randall Davies, *The Railway Centenary*, London & North Eastern Railway Co., London, 1925, p. 10.

The next improvement was made by the American Robert L. Stevens (1787–1856), son of the John Stevens (1749–1838) who had been prominent in the development of the steamboat. Robert Stevens, as president of the Camden and Amboy Railroad—on the original line of travel by rail between New York and Philadelphia and now part of the Pennsylvania Railroad—had been instructed by his directors to study and report upon railroading in Britain. On the voyage abroad in 1830 Stevens seems to have given much thought to the materials and types of rails. He decided against using the British iron edge rail set upon expensive chairs. When he arrived in England he placed with a firm in South Wales the first order for rolling T-section rails. They were wrought iron, 18 feet long, and weighed 36 pounds to the yard.

Experimenting with miscellaneous cross sections on steam railroads ceased in America during the 1840s as the mills for rolling T rails began production. To Robert Stevens, therefore, goes the credit for establishing the principle of the rail section virtually as it is today on American railroads. Its dimensions have been enlarged; some rails stand today $7\frac{1}{2}$ inches high upon a base $6\frac{1}{2}$ inches broad, and their weight has been increased to as much as 155 pounds to the yard. The material has been changed to steel carefully rolled and heat-treated to make it tough and long-lived. The British type designed by one of Stephenson's engineers, Charles B. Vignoles, is double-headed or symmetrical top and bottom and supported in cast-iron chairs.

Stevens's T rails were set upon stone blocks. But wooden crossties, or sleepers, quickly proved superior and fixed the type of railroad track which has been laid since then. Practice in spacing the ties had varied, but it has become customary in the United States to place them 20 to 22 inches apart. The first ballasting amounted only to leveling and tamping the crossties with earth or sand, gravel, broken stone, or cinders, almost any material at hand along the way. The popular notion that broken-stone ballasting was made necessary by the greater weight and speed of modern trains is true as far as it goes, but it does not explain the whole problem of the roadbed. Ballast is necessary for holding the track in place and to check slipping and sliding; it is indispensable to prevent settling. The shock of engine and train is carried through the network of rail and tie to the subsurface. As McAdam discovered in making his roads, a railroad has to be kept dry if it is to be firm. Modern track, therefore, is laid upon carefully graded and drained subsurfaces. Stone ballasting serves primarily to facilitate draining water away from the ties.

If railroad builders had their choice, every line would be straight as well as level, in accordance with George Stephenson's analysis. Each curve on the early railroads was simply an arc of a circle. It was soon realized, however, that for smooth riding at any speed there should be at each end of each curve an easement or transition curve of gradually changing radius. Trains would then, with a minimum of jar, pass from the curve where the outer rail is superelevated above the inner in order to balance the centrifugal force to straight stretches, or tangents, where there is no superelevation. It was an English inventor who first observed the need to flex the wheel base of a vehicle on rails. William Chapman (1749–1832) patented in 1812 what is called a bogie by the British. The wheels of his swivel truck moved around the rail or circumference of the curve as the vehicle above it took the chord of the arc. The long granite columns for the new post office of Boston were hauled underslung between two four-wheel trucks in 1826. The principle was the same as that in Chapman's patent. Two-truck carriages may have been used for hauling timbers and other long objects in earlier times.

British engineers did not at once take up Chapman's idea, whether old or new, but American engineers did. John Bloomfield Jervis (1795–1885), chief engineer of the Delaware and Hudson Canal Company, who had received his early training on the Erie Canal, and his young assistant Horatio Allen (1802–1889), who had come to engineering from the law, were among the first engineers in America to turn from canals to railroads. Their plan was to build a 17-mile railroad from the Carbondale, Pennsylvania, coal mines to Honesdale on the canal. Jervis drew up the specifications for the construction in England of the *Stourbridge Lion*, which Allen went abroad to purchase. As he piloted the newly imported engine on its 6-mile trial run, August 8, 1829, some two months before the famous Rainhill trials, Allen was the first in America to drive a locomotive. Its 8 tons, however, were too much for the wooden track with iron-capped rails; its wheels were too rigid for the curves, one of which spanned Lackawaxen Creek on a frail 30-foot-high wooden trestle. Allen got the *Lion* safely across at something less than 5 miles an hour, but the locomotive, after one more attempt on September 9, was never operated again.

The next year Horatio Allen became chief engineer of the South Carolina Railroad and supervised at the West Point Foundry in New York City the construction of its locomotive, *Best Friend*. During the summer of 1830 and 1831 he visited in Albany with Jervis, who had taken charge of building the Mohawk and Hudson Railroad to compete with the Erie Canal, and together they discussed how they might design a locomotive which would have its weight distributed over more wheels without increasing its "action" upon the railway. Forty years later Jervis recalled that Allen's idea had been to put two engines together and suspend one boiler between them on a "transom and pin." How Allen would have assembled the cylinders, Jervis did not make clear. Both men knew that two cars had been connected and used to carry timbers at moderate speed. They knew also from Nicholas Wood's writings that there had been experimenting with a similar engine having two four-wheeled frames with its boiler resting upon one of them. Jervis himself insisted that they should keep the driving wheels as well as the boiler on one truck. He placed his second truck, or swivel, under the front end simply to support and guide the engine into the curves. The result of their discussion and experimenting was that the system of two pairs of driving wheels and a two-axle, forward swivel truck, known as the American type (Figure 12.1), became common throughout the United States.

Railways advanced generally after George Stephenson's demonstration at Rainhill. Construction got under way with gradually settling political conditions in several European countries, even in France where Adolphe Thiers advised his King, Louis Philippe, that, however good for Britain, railways were not for France. Marc Seguin, as always, pressed to the fore and introduced steam locomotives on his Lyons–St. Etienne Railway in 1831. By 1842 the French government was planning nine lines, seven to radiate from Paris. Swift mobilization of military forces was the major factor in determining their location, but there was some thought of travel and business in other countries. The longest road was built to connect Paris with Rouen for the cross-channel trip to Britain.

By 1835 Belgium appears to have had a railroad system on a small scale, built, owned, and run by the government. It was in shape an irregular cross centered upon Malines to provide railheads at the borders of neighboring countries, including Britain across the Channel. German lines began with the Nürnberg-Fürth road in Bavaria, opened in 1835. Others connected Leipzig and Dresden, Munich and Augsburg, Berlin and Potsdam. They were designed for passengers rather than freight; some were private enterprises, others were built by the states. There was no linking of the German railroads into a national system until after Bismarck's time. Holland joined Amsterdam and Haarlem in 1839, with an eventual plan to import German coal and transship Spanish iron ore to German industries on the Rhine; King Willem II financed and built the railway at his personal expense. Russia, Austria, Italy, and Spain also laid plans. Most of the early equipment and the contractors for these European projects were British, but American engineers were soon in the competition. The American George Washington Whistler, father of a more famous artist son, went to Russia in 1842 at the request of Czar Nicholas I to build the St. Petersburg-Moscow line.

Like the Liverpool and Manchester in Britain, the Baltimore & Ohio Railroad was the experimental workshop, the experimental laboratory, for railroad building in America. Here Peter Cooper's engine *Tom Thumb*, in August, 1830, raced a horse, lost to the horse, but demonstrated the superiority of steam power. Here the use of privately owned conveyances was proved impracticable because of the lack of the company's control over the vehicles. Here the railway was first extended from the usual river valleys, where grades were lowest, into mountainous country. Better traction was obtained by increasing weights upon driving wheels, and pusher locomotives overcame gradients as high as 4 per cent. Grade, rail, and tie designs were coordinated through experience to become the modern roadbed. By 1853, twenty-five years after ground was broken at Baltimore for the little horse-drawn railway it started out to be, the Baltimore & Ohio Railroad, under the direction of two American engineers, first Jonathan Knight (1787–1858) and later the younger Benjamin Henry Latrobe (1806–1878), had crossed the Appalachians to its then western terminus at Wheeling, Virginia, on the Ohio River. The European counterpart of the Baltimore & Ohio was the Semmeringbahn, opened in 1853, on the main line from Vienna to its seaport Trieste on the Adriatic. This bolder project crossed the Alps almost 3,000 feet above the sea through 15 tunnels and over 16 viaducts in a stretch of 33 miles.

Other railroads in the United States raced the Baltimore & Ohio toward the West. The Mohawk and Hudson and town-to-town companies in central New York paralleled the Erie Canal to complete a railroad line from New York City to Buffalo in 1851. We have already referred in Chapter 7 to the canal system that formed in the 1830s a large proportion of the through route between Philadelphia and Pittsburgh, roughly following the line of the present Pennsylvania Railroad. The easterly 82 miles of this route, the Philadelphia and Columbia Railroad finished in 1834, was, like the canal, built by the state of Pennsylvania. The most spectacular feature of the Pennsylvania Canal was the 36-mile Portage

Railroad which carried its freight and passengers over the Allegheny Mountains through Blair's Gap, 2,326 feet above the sea.

David Stevenson (1815–1886), Scottish engineer, viewed the achievement of Pennsylvania with the eye of a contemporary expert. His account, published in 1838, praised the "mountain railway" for boldness in design and difficulty of execution; not even "the passes of the Simplon and Mont Cenis," he wrote, struck him as "more wonderful" works of engineering than the Allegheny Portage Road. The inclined planes appealed to Stevenson as the "most remarkable works which occur on this line." [12] He notes also the 900-foot tunnel and the several bold viaducts. The railroad consisted of 11 level or nearly level stretches whose aggregate length was 32 miles, alternating with 10 inclined planes, totaling 4 miles. The total rise from Hollidaysburg to the summit level at Blair's Gap was 1,400 feet; from this point there was a descent of 1,171 feet down five planes and along five gentle gradients to the canal at Johnstown. The 10 inclined planes varied in length from 1,500 to 3,100 feet with gradients from 7 to 10 per cent. The line was double-tracked, and on the nearly level stretches some of the tiny cars were pulled mainly by steam locomotives. The cars were handled on the inclined planes by endless hemp ropes 6 or 7 inches in circumference, passing around grooved wheels which rotated horizontally, controlled by brakes and moved by stationary steam engines of approximately 25 horsepower each. There were two of these engines at the head of each plane so that one could always be in use. Four cars, each with a 3½-ton load, could be drawn up, and four let down at the same time.

David Stevenson took note that in 1836 from April to October the Allegheny Portage Road carried 14,300 cars—about 100 a day. It could have transported more if it had been used twenty-four hours a day. The total freight handled during that period was 37,081 tons, and there had been 19,171 passengers. No comparison, of course, is to be made with traffic today on the Pennsylvania's Horseshoe Curve. Stevenson's journey from Hollidaysburg over the Portage Road to Johnstown lasted from nine in the morning to five in the evening, with an hour's stop at the summit for dinner. The average speed seems thus to have been 5 miles an hour for the 36 miles. Stevenson's whole "extraordinary journey" from Philadelphia to Pittsburgh took 91 hours of traveling time at 4.34 miles an hour. Of the total 395 miles, 118 were by railway and 277 on the

[12] David Stevenson, *Sketch of the Civil Engineering of North America*, John Weale, London, 1838, p. 268.

canal; the journey cost Stevenson £3. Today the Pennsylvania's passenger trains make the trip in six hours or less. The fare, however, appears to be much the same.

Charles Dickens should have come earlier to America for experience in perspective before making his very British commentary on the American scene. As he journeyed over Pennsylvania's mountain railroad in 1842, "It was very pretty," he said, "traveling thus at a rapid pace along the heights of the mountain in a keen wind, to look down into a valley full of light and softness, catching glimpses . . . of scattered cabins; children running to the doors; dogs bursting out to bark . . . terrified pigs scampering homeward; families sitting out in their rude gardens, cows gazing upward with stupid indifference, men in their shirt sleeves looking on at their unfinished houses, planning out to-morrow's work; and we riding onward high above them, like a whirlwind." [13] Drivers on the high-speed motor road from Harrisburg to Pittsburgh today might enjoy their Dickens more if they could stop long enough for a detour of 25 miles and a look at the grades, curves, and 900-foot tunnel on the old Portage Road over which he passed.

The Charleston and Hamburg, or South Carolina, for a brief moment the longest railroad in the world, started toward the Ohio Valley under the driving enthusiasm of Robert Y. Hayne. He died in 1839 and more conservative Southerners took possession. The project made Atlanta a railroad center and moved on to Chattanooga, but veered then toward Memphis, Tennessee, on the Mississippi to become an all-Southern route that could never compete with the railroads from New York and Pennsylvania for the traffic of the Great Lakes and the Northwest.

Chicago, cluster of houses around Fort Dearborn near the portage from Lake Michigan to the Illinois and Mississippi Rivers in 1830, had but 28,620 residents twenty years later. Enthusiasts, however, had built its first railroad westward to the lead mines of Galena in 1848. The Michigan Central and the Michigan Southern came into Chicago from the East in 1852. The Illinois Central began its service southward through the state to the Ohio River in the same year. By 1860, as the Pennsylvania system arrived from Pittsburgh and Cincinnati, the population of Chicago had leaped to 100,000. In ten years more it was 330,000.

Northern interests there were far stronger than Southern on the eve of the Civil War. Main streams of inland commerce had shifted from

[13] Charles Dickens, *American Notes for General Circulation,* D. Appleton and Co., New York, 1868, p. 65.

north and south to east and west. The steamboats of the Mississippi and
the Ohio, instruments of Southern penetration into the Northwest after
1811, had lost the bulk of the produce to Northern railroads and the Erie
Canal during the 1850s. The movement of wheat and flour alone is proof
enough. The traffic had been evenly balanced in 1839 between ports on
the Gulf of Mexico and those of the Atlantic seaboard. In 1852, Northern
lines of transit to the East had an advantage of five to two over the
Mississippi riverway to New Orleans. By 1861 this Northern advantage
had increased to 15 to 2. Without a doubt railroads had much to do with
the outcome of the Civil War.

As the war approached, New Orleans remained a great port, almost
equal to New York. In fact there was an absolute increase in traffic as
cotton, an all-Southern product, swelled the total shipment from the
lower Mississippi, but the relative loss to the Atlantic ports was heavy,
and Southerners knew why. Northern enterprise had "rolled back the
mighty tide of the Mississippi and its ten thousand tributary streams until
their mouth, practically and commercially, is more at New York and
Boston than at New Orleans." [14] The city of steamboats must have its
railroads, too, radiating into markets and sources of supply like those networks which were spreading about the Northern cities of New York,
Boston, Cincinnati, and Chicago.

Bibliography

Brown, William H.: *The History of the First Locomotives in America*, rev. ed., D. Appleton and Co., New York, 1874.

Chatterton, Edward K.: *Steamships and Their Story*, Cassell and Co., London, 1910.

Dugan, James T.: *The Great Iron Ship*, Harper & Brothers, New York, 1953.

Flexner, James T.: *Steamboats Come True*, The Viking Press, Inc., New York, 1944.

Gibson, Charles E.: *The Story of the Ship*, Abelard-Schuman, Inc., Publishers, New York, 1948.

Gilfillan, S. Colum: *Inventing the Ship*, Follett Publishing Co., Chicago, 1935.

Tredgold, Thomas: *A Practical Treatise on Rail-roads and Carriages*, J. Taylor, London, 1825.

Tyler, David B.: *Steam Conquers the Atlantic*, Appleton-Century-Crofts, Inc., New York, 1939.

[14] James D. DeBow, *The Industrial Resources, etc., of the Southern and Western States*, 3 vols., DeBow's Review, New Orleans, 1853, vol. 2, p. 484.

TEN

Iron and Steel

Swiftly expanding railroad systems stimulated the rise of the steel industry soon after the middle of the nineteenth century. The later demand for steel for steamships and tall buildings further influenced the increase in steel production. A need for a new product often directly motivates men to development of it, but for two of the three most influential inventors of new steel processes such motivation was conspicuously lacking. Among the many men who contributed to the evolution of efficient and economical methods for the manufacture of steel on a large scale, the three most outstanding were the Englishman Henry Bessemer (1813–1898), the German-born British-naturalized William Siemens (1823–1883), and a Londoner, Sidney Gilchrist Thomas (1850–1885). Bessemer invented his converter in 1856, Siemens's brother Frederick patented a regenerative furnace in the same year, and Thomas patented his basic, or alkali, process in 1878. It is interesting to note that none of these men was in the iron industry.

The general need for steel was not the immediate stimulus that led Bessemer and Siemens to make their important inventions. During the Crimean War, Bessemer had developed a heavy, elongated projectile which would rotate around its axis in flight, but the British War Department was not interested in it. Subsequently the French tested Bessemer's projectile at Vincennes, and after a successful trial the commandant of the fortress remarked that the real question was, "Could any guns be made to stand such a heavy projectile?" In his *Autobiography* Bessemer wrote, "This simple observation was the spark which kindled one of the greatest industrial revolutions that the present century has to record, for it instantly forced on my attention the real difficulty of the situation, viz.:

How were we to make a gun that would be strong enough to throw with safety these heavy elongated projectiles? I well remember how, on my lonely journey back to Paris that cold December night, I inwardly resolved, if possible, to complete the work so satisfactorily begun, by producing a superior description of cast-iron that would stand the heavy strains which the increased weight of the projectiles rendered necessary." [1] In less than two years, Bessemer had invented and perfected his converter.

Similarly in 1847, William Siemens had no immediate interest in the production of steel for railroads when he invented his metallic respirator to conserve the heat in steam exhausted from a steam engine. The respirator was not successful, but William and his brother Frederick attempted to apply the principle of returning lost heat to various other processes. They invented the regenerative furnace in 1856 and subsequently substituted preheated gas for solid fuel. Although the first practical application of the Siemens furnace was for melting steel in 1857, its principal uses during the next decade were in glass manufacture and the heating of air for blast furnaces. Not until about 1865 was the regenerative furnace first used to produce steel.

Once Bessemer steel began to be available and reliable, the demands of the railroads for steel rails to replace wrought iron caused a rapid expansion of the new industry. The first Bessemer steel rails were rolled during either 1857 or 1858. Although small Bessemer steel steamboats were built in 1858 and 1859, shipbuilders did not use steel extensively in ocean-going steamships until about 1880, despite Bessemer's repeated urgings. The first Bessemer rolled beams were in the earliest of the skyscrapers, Chicago's Home Insurance Building of 1884. Although apparently the principal demands for steel were not the factors that led Bessemer and Siemens to initiate their important developments, there can be little doubt that these demands stimulated Thomas to work out his basic, or alkali, process in the 1870s.

Steel is a solid solution of iron and carbon containing up to 1.7 per cent carbon, in contrast to pig iron or cast iron with its 2.5 to 4 per cent carbon. Wrought iron has a carbon content of normally less than 0.1 per cent, but it also contains 1 or 2 per cent of slag, which differentiates it from very soft steel. The ancients appear to have known of steel, as did the medieval smiths who hammered, reheated, and hammered again until

---

[1] Henry Bessemer, *An Autobiography*, Offices of "Engineering," London, 1905, pp. 135–136.

they produced the supple flashing swords of Toledo and Damascus and the tough steel of Sheffield. The medieval smiths made steel from wrought iron by what is known as the cementation process. They heated the wrought iron, packed in contact with powdered charcoal, to a cherry-red heat for about ten days or two weeks. They next rolled or hammered, reheated, and again rolled or hammered the metal to bring about the absorption by the iron of a desired proportional content of carbon. Obviously, highly skilled workers were required to make steel in this way, and they could make only small quantities. Such steel was available only for special purposes at high cost.

In addition to cementation, prior to Bessemer's invention of his converter, there was one other principal steel-manufacturing technique. Working near Sheffield, England, Benjamin Huntsman developed the crucible process about 1740; he kept it secret for a number of years. Using blister steel, produced by cementation and having a high percentage of carbon on the surface and much less at the center, Huntsman melted his charge of about 35 pounds in a covered crucible for several hours. The resulting steel, with a relatively high but evenly distributed carbon content, was exceptionally hard. Because he cast it in molds, Huntsman called it cast steel. Obviously it was even more restricted in quantity and higher in cost than cementation steel, one form of which was the charge that went into the crucible. A few manufacturers still make crucible steel for special purposes, but the charge today usually consists of wrought iron and a carbonaceous material such as pellets of charcoal. Crucible steel is still expensive, selling sometimes for well over a dollar per pound.

The Industry

The prime significance of the Bessemer process was that it industrialized steel production. The cementation and crucible processes require highly skilled operators upon whose judgment rests the success of the processes. As Henry Bessemer put it when he made his first ingot, "we had as much metal as could be produced by two puddlers and their two assistants, working arduously for hours with an expenditure of much fuel. We had obtained a pure, homogeneous 10-in. ingot as the result of thirty minutes 'blowing,' wholly unaccompanied by skilled labour or the employment of fuel."[2] Using phosphorus-free Swedish pig iron that cost him

[2] *Ibid.,* p 153.

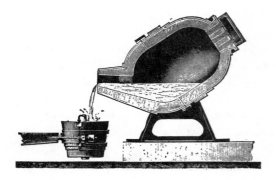

Figure 10.1
A simple Bessemer converter (From E. W. Byrn, *The Progress of Invention in the Nineteenth Century*, 1900; courtesy *Scientific American*)

only £7 per ton, Bessemer produced cast steel equivalent to steel currently selling for £50 to £60 per ton, by melting it and then blowing air through the molten mass.

Bessemer was not the first to produce steel by blowing air through molten pig iron. Apparently trying to overcome the difficulties of a shortage of charcoal at his iron furnace near Eddyville, Kentucky, William Kelly (1811–1888) first achieved success about 1847 in refining pig iron into steel by blowing air through it nine years before Bessemer made his announcement. Nevertheless, it was Bessemer who created the machinery for effective production. His converter (Figure 10.1) consisted of a huge cylinder lined with brick and having an open conical top. Its double bottom made possible the holes through which jets of air could be forced into the molten pig iron to oxidize or burn and thus remove most of the carbon. The converter was mounted on trunnions so that the operators could tip it like a huge teakettle to pour the steel into molds. The resulting ingots could be rolled, forged, or pressed into any desired shape. During a single "blow" the converter could produce some 20 tons of steel from cast iron in as many minutes.

Bessemer's process, however, was so rapid that it was not reliable for producing steel except with a very low carbon content, the residual amount left after the blow has ended. Such steel is very soft, and it was difficult to leave just enough carbon in the charge to make a steel with greater tensile strength. Moreover, too much air would cause a "burnt" steel, overoxidized and brittle. Bessemer's converter also could not remove such impurities as the sulfur and phosphorus that are present in many ores. In little more than a month after Bessemer announced his new process, in August, 1856, Robert Forester Mushet (1811–1891), English metallurgist and a competitor in steelmaking and invention, overcame the difficulty of producing steels containing the right amount of carbon. He added a specific amount of spiegeleisen, a pig iron rich in manganese, to the molten charge after the completion of the blow. The manganese removed the excess oxygen, and the carbon in the spiegeleisen recarburized the steel to the desired percentage. Mushet produced high-carbon steel which could be hardened and was especially suitable for machine tools.

Although William Siemens patented a process in 1868 for manufacturing steel from iron ore and pig iron in his regenerative furnace (the so-called Siemens process, which was never much used), the French brothers Émile and Pierre Martin patented in 1865 a more important advance by producing steel in a Siemens furnace from steel scrap and pig iron. This Siemens-Martin process now produces about twenty times as much steel in the United States as does the Bessemer process. The essential feature of the Siemens-Martin procedure is William Siemens's regenerative furnace patented in 1856. This furnace consisted of two heating chambers made of brick grids through which, alternately, Siemens sent hot waste gases to raise the brick to a high temperature. He then preheated the incoming air for the furnace by sending it through one of these chambers while the other was heating and so returned to the furnace large amounts of heat which otherwise would have been wasted. As used by the Martins, this additional heat further raised the temperature of the charge of pig iron, scrap steel, and limestone in the shallow container, or open hearth.

During the eight to twelve hours required for the process, each furnace could produce upward of a hundred tons of steel from known quantities of materials; the operators could make steel having accurate predetermined proportions of ingredients as indicated by samples taken from the furnace and analyzed from time to time, for, as is not the case with the Bessemer method. the reaction can be stopped at any point. However, neither the Bessemer process as improved by Mushet nor the original Siemens-Martin process could remove the phosphates in pig iron manufactured from the highly phosphoric ores which are plentiful.

A London police-court clerk, Sidney Gilchrist Thomas (1850–1885), developed a process to remove phosphorus from pig iron in the Bessemer converter; the Thomas process was later adapted to the Siemens-Martin open-hearth furnace. While taking evening courses at London's Birkbeck College, Thomas heard one of the lecturers say that whoever should discover how to manufacture steel from a phosphoric pig iron would make a fortune. In his spare time, Thomas immediately began reading technical and chemical literature and experimenting in a cellar laboratory. During the years 1871 to 1875 he devised the Thomas process which in its essentials consists of a furnace lining of basic or alkaline materials, chiefly dolomite. This material, a carbonate of lime and magnesium, absorbs phosphates from the charge. A flux is also added to form a slag which absorbs phosphates and other impurities and can be poured off. Thomas's cousin, Percy Gilchrist, a chemist in an ironworks, tested the process. Thomas patented his method in 1878, but steelmakers took little notice of it until in 1879 one of the Cleveland, England, ironworks using the highly phosphoric ores of that district demonstrated the benefits of the Thomas process.

A considerable amount of the steel manufactured in the United States is alloy steel which owes its properties to the presence of one or more elements in addition to the iron and carbon. Of the many alloys manufacturers produce, nickel steel, chrome steel, and chrome-nickel steel make up the largest total tonnage. Julius Baur, of Brooklyn, New York, obtained his first patent for chrome steel in 1865. An abstract of his patent in the *Scientific American* for September 2, 1865, stated that he could combine iron with metallic chromium. He made his alloy either in crucibles or "by the pneumatic process"; he could make "a triple compound of iron, carbon and chromium." Some have discounted Baur's claim because he gave too much credit to chromium and too little to carbon for the excellence of the alloy. His work, however, had an important effect on the industry, for it attracted the attention [3] of the French manufacturer H. A. Brustlein, who is usually credited with the development of chrome steel for engineering purposes. By 1874 chrome steel was going

[3] H. A. Brustlein, "On Chrome Pig Iron and Steel," *Journal of the Iron and Steel Institute*, pt. 2, p. 770, 1886.

into the arches of the Eads bridge across the Mississippi at St. Louis. A decade later, Marbeau introduced nickel steel.

At this time heat treatment of finished carbon steel was relatively simple. If a hard steel was desired for a cutting edge, the manufacturer heated the piece to a red heat and quenched it in a bath of water, oil, or some other material. However, this technique for producing a hard cutting edge also made the carbon steel brittle. If a more malleable product was required, the operator annealed the piece by allowing it to cool slowly after heating, and thereby he sacrificed hardness. Combined procedures could produce intermediate characteristics, but in general hardness was achieved at the expense of malleability and vice versa. In the twentieth century, metallurgists removed these limitations.

As Siemens was building his experimental plant in England at Birmingham in 1865, Alexander Lyman Holley (1832–1882) was manufacturing Bessemer steel in Troy, New York, and by 1867 steel rails were being rolled on commercial orders at the Cambria works in Johnstown, Pennsylvania. Abram S. Hewitt attended the Paris Exposition that year as United States Commissioner and saw the Martin open-hearth process. Recognizing at once its significance, Hewitt obtained the rights to use it in the United States and built the first American open-hearth furnace at Trenton, New Jersey. Andrew Carnegie organized his company in 1870 at Pittsburgh. By 1873 the Bethlehem Iron Company was erecting a Bessemer steel plant with Holley as consulting engineer, who was urging all steelmakers to employ chemical analyses as well as physical tests of their products. Following the experiments of Thomas and Gilchrist in 1877, the manufacture of basic steel spread rapidly, especially in Germany where ores containing phosphorus were plentiful. By 1887 the Krupp steel works had become the largest in the world, even though Krupp was unsuccessful in trying to evade Thomas's patent.

By 1865 the insistent demand for a material stronger than wrought iron, yet somewhat elastic, had brought about the production of 225,000 tons of steel in Great Britain, 98,000 tons in Germany, 41,000 in France, and 14,000 in the United States. Bessemer's first steel at £42 per ton was hardly cheap steel, but he apparently could have made a profit at £20. Prices dropped fairly rapidly, and by 1881 the British were selling railroad rails for £6/10 per ton. This price fell irregularly to a low of £3/15 in 1895. In this same year Great Britain produced nearly 3½ million tons, Germany 4 million, France nearly 1 million, and the United States over

6 million. In the thirty years, production had been multiplied by more than 37. This rapid rise of the steel industry was one of the most important aspects of the Industrial Revolution in the nineteenth century. Moreover, without steel and its alloys the engineering advances described in subsequent chapters could not have occurred.

Bridges of Steel

Among the great ferrous bridges built with the spread of railroads during the last half of the nineteenth century, three were outstanding. In conception and design the three structures were distinct. James B. Eads chose the arch for his bridge, opened in 1874, over the Mississippi River at St. Louis. For their famous Brooklyn Bridge, which was opened in 1883, John A. Roebling and his son Washington A. Roebling held to the principle of suspension with which they had been successful at Niagara. John Fowler and Benjamin Baker used cantilevers in their bridge of 1890 across the Firth of Forth in Scotland. All had experiences in common, especially with their underwater work, and all used steel. Moreover, all three bridges are still in use.

One of the first to appreciate this newest form of building material was James Buchanan Eads (1820–1887), born of Maryland stock in Lawrenceburg, Indiana, and builder of iron-clad gunboats for the Union government during the Civil War. Eads had steel with chrome in it tested by the British expert David Kirkaldy (1820–1897) even before Julius Baur had improved production methods. Eads was relentless with the manufacturers of the steel for his bridge until they delivered a product that met his requirements. As a salvager of sunken river craft before the war, Eads had walked the bed of the Mississippi in a diving bell of his own construction 65 feet under floodwaters and had felt the surging sands of the river about his knees. He had seen ice plunge beneath the hard-packed surface and scour gorges to great depths. He was sure that to withstand undermining the bridge piers had to go all the way down to bedrock.

The people of St. Louis were eager to build Eads's bridge. They were proud of their heritage from the fur-trading post of the French near the thoroughfares of the Missouri, upper Mississippi, and Ohio; jealous for the future of St. Louis as the industrial city on the great river and the center of trade between Eastern states and the West as far as San Francisco. They were spurred too by the certainty that Chicago, across the state

Figure 10.2 Eads's St. Louis Bridge (From Philip Phillips, *The Forth Bridge,* 1889)

of Illinois at the head of Lake Michigan and in touch there with the traffic of the Great Lakes, would be the center of a network of railroads. These might readily bridge the Mississippi above the Missouri and bypass St. Louis into the West. One, in fact, had already done so at Rock Island in 1855.

The Mississippi had served St. Louis in the past and had brought it wealth from every quarter. The river must not now be allowed to block the city's progress. The Mississippi was a barrier when its surface froze too thickly for riverboats and ferries yet remained too thin for wagons to cross. The ice jams were fearful when the rising waters of the Ohio or the Missouri or both swept into the valley before it had cleared with the spring thaw. The naturalist Audubon and his partner Rozier had been delayed more than a month opposite Cape Girardeau before they could work their boat through the ice to Ste. Geneviève. The approach of the railroads across the streams and prairies of Ohio, Indiana, and Illinois, therefore, could mean but one thing to the people of St. Louis. The railroads must come straight over the Mississippi into their city.

It was nonetheless the river that dictated the terms upon which James B. Eads was permitted to construct the bridge (Figure 10.2). The swirling flood rose and fell with the seasons. Less usual conditions brought torrents as much as 40 feet above normal low water, increasing the speed of flow from 3 to over 9 miles an hour and producing capricious currents in the channel that made a drawbridge inadvisable even if the river men had not opposed it. Spans would have to be long enough to give ample room for passing through without danger, and the clearance above high water great enough for the stacks of the river boats. Eads in 1868 planned his arches to rise about 50 feet above the flood level of 1844, the highest on record to that time. They were also to exceed in length of span any previous arched structure. For Eads was acquainted with the judgment of Thomas Telford that a cast-iron arch of 600 feet could be thrown across the Thames and was counting upon the much greater strength and toughness of steel to make secure his own spans of 500 feet. There are steel arches today with spans of more than 1,600 feet.

Steamboat men could not be satisfied. They wanted no bridge at all. It was bound to ruin the ferries and to create competition by rail with the river trade. They appealed to the army engineers. There were boats on the river, said the steamboat men, with stacks exceeding 100 feet in height; there might be need for even higher stacks. Moreover, Eads's arches were too long; his piers were so far apart that pilots could not use them as guides. Eads demonstrated that top sections of the stacks could be lowered and that properly designed stacks not only offered less resistance to the wind than older types but obtained better draft. For the charge that his piers were too far apart, he had only the amused contempt which it deserved. As the bridge was nearing completion in 1873, a board of army engineers agreed with the steamboat men that it should come down. But Secretary Belknap heard from higher up. President Grant had lived in St. Louis, and he knew James B. Eads. The bridge was completed.

Anticipating the heaviest traffic for this bridge across the Mississippi, Eads designed a double-deck bridge to support as many people, he said, as could "stand together upon the carriage-way and foot-paths from end to end of the Bridge, and at the same time have each railway track below covered from end to end with locomotives." [4] Even then the bridge would not be taxed to one-sixth of its strength; it would be capable of sustaining 28,972 tons uniformly distributed and of withstanding flood, ice, and tornado with a safety factor of 6. To these ends, Eads made two decisions and clung to them against every discouragement, the obstructions of envious rivals, ignorant or acquisitive associates, the assaults of storm and river. He chose to build a bridge of three steel arches sprung from granite piers set more than 500 feet apart. Authorities said that such

[4] Calvin M. Woodward, *A History of the St. Louis Bridge*, G. I. Jones and Co., St. Louis, 1881, p. 48.

arches never had been built and therefore could not be built. He determined to set the foundations upon bedrock at depths which never before had been attempted. The results were revolutionary in bridgebuilding.

Eads's first intention had been to excavate within cofferdams. The abutment on the western shore was so constructed, going down 30 feet through sunken boats, wreckage, and general debris to bedrock. However, as the plan for the east pier in the stream was taking shape, Eads suffered from a severe cough and had to leave on a voyage to rest. While recuperating in France he thoroughly investigated the use of compressed air to keep water out of closed caissons, a technique European engineers had been employing for some time. He decided that their method, adapted to conditions in the Mississippi, would speed his work down through the sands and would prove less expensive. He had to learn from experience what would happen to his men under increased pressures at such great depths.

The revised method was to float an iron-shod wooden caisson (Figure 10.3) of great strength into position behind icebreakers and protective piling and there to build the granite pier upon it while workmen within the caisson dug away the sand so that its edges cut down to bedrock. Eads invented a high-pressure water pump to force the sand, gravel, and small stones up to barges on the surface. He placed his air locks at the bottom of the shaft to the caisson and within the air chamber, so that those who supplied the men within the chamber could get close to them without being themselves under increased air pressure. The caisson, or wooden box, provided for the construction of the east pier was in position October 17, 1869, and by November 17 it was on the sand. With crews working day and night, it had reached bedrock by February 28, 122½ feet below the water level, a depth comparable to the height of a 10-story building. Meanwhile, operations had begun on the west pier, and by April 1 its caisson was down 86 feet to bedrock. But all had not gone well; twelve men on the east pier and one on the west pier had lost their lives.

From previous experience with compressed air, the engineers had expected that too rapid release of the pressure would cause distress in the ears, pain particularly in the joints, and sometimes death. They knew that men working under compressed air felt the exhilaration of the additional oxygen, sweat profusely, and grew tired sooner than under normal conditions. They knew that the gasoline lamps and candles they had to work by burned fast under compressed air and that danger from fire was great, but they were not prepared for the muscular paralysis which made work-

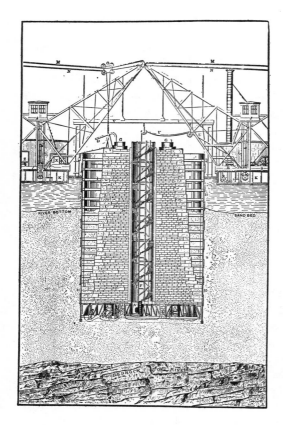

Figure 10.3
Cross-sectional view of a caisson, Eads's St. Louis Bridge (From E. H. Knight, *American Mechanical Dictionary, vol.* 1, 1874)

men stumble and stoop when the depth reached 65 feet where the pressure had to be raised to nearly 30 pounds to the square inch. The men resorted to galvanic arm and waist bands, rubbing oil externally, and doubtless alcohol internally. Eads and his family physician Dr. Jaminet were anxious.

The working hours had been reduced when a foreman became ill. The shifts had been made 4 hours each with 8 hours of rest between. At 65 feet the working time was further shortened to a six-hour day, three shifts of two hours each, with rest periods of two hours. At the bedrock of the east pier, then under more than 90 feet of water where the pressure was 44 pounds above atmospheric pressure, the hard work of concreting began. With it came severe cases of the bends, or caisson disease, and in quick succession, three deaths. Dr. Jaminet went down into the chamber to investigate, fell ill, climbed the long flight of stairs in great pain, became badly paralyzed and even deprived of his speech for a while, but he had learned much.

Decompression was slowed down so that the workmen could not hurry through the air lock. They were compelled to rest thirty minutes after coming out of the chamber. Their great loss of body heat with subsiding pressure was counteracted in some degree by warm food and covering, and alcohol was strictly forbidden. The working period was reduced to one hour. Eads took further precautions when they came to building the abutment on the eastern shore, descending more than 135 feet below high water. Every workman had to rest a full hour after leaving the chamber. At 100 feet under, two working periods of forty-five minutes each were a day's labor. There was but a single death on this abutment. That man had forgotten his lunch and had rushed home to eat; moreover, he had stopped on the way back for a drink of beer. Eads and Dr. Jaminet had achieved a notable success in engineering. Compressed-air safety laws of today have been drawn in large part from their conclusions. Their victory, however, was not complete enough at that time to save Washington A. Roebling from similar difficulties and personal injury in the construction of the Brooklyn Bridge.

As significant as his underwater work were Eads's steel arches leaping the Mississippi in three great spans, each more than 500 feet. They were built as continuous arches, without hinges, four abreast, and held transversely by wrought-iron braces. The ribs or arch members were made of two 18-inch cylindrical tubes, one 12 feet above the other. Eads's first assumption that the stresses in a parabolic arch would prove easier to calculate than those in the segment of a circular one did not survive the checking of his German-born assistant engineer Charles Pfeifer, and his mathematical adviser William Chauvenet (1820–1870), Chancellor of Washington University in St. Louis, but the difference in favor of the circle was slight. The curves of the finished bridge are arcs of circles that vary from parabolas by not more than 6 inches at any point. With Chauvenet's expert assistance, Eads and his associate engineers had also calculated in advance the amount of expansion to be expected in the arches with temperatures ranging from 20°F below to 141°F above zero. The crown of the center arch, 520 feet in length, would rise 8 inches. The roadways of the Niagara Suspension Bridge, a span of 821 feet, which John A. Roebling had built in 1851 to 1855, had been found to rise and fall in cold and hot weather, 2¼ feet at the center of the span, with no injury whatever to the bridge.

Figure 10.4 Method of erection by cantilevers, St. Louis Bridge arches (From *Scientific American*, Nov. 15, 1873)

To obtain his cylindrical tubes, Eads banded six chrome-steel staves in an envelope, like the staves of a barrel, 18 inches in diameter and 12 feet long. There were 6,000 of these staves, carefully machined, each with ultimate strength of 120,000 pounds to the square inch and elastic limit of 50,000 pounds. The tubes were fitted with wrought-iron couplings, tested to 40,000 pounds to the square inch. Both cast steel and "semisteel" failed in the testing machines. Eads was so exacting that when his contractor, Andrew Carnegie, came to write his autobiography, he remembered the engineer as "an original genius *minus* scientific knowledge to guide his erratic ideas of things mechanical." The editor of *Engineering* for October 10, 1873, had a different view. The alliance between the theorist and the practical man was complete, he said; James B. Eads had used the "highest powers of modern analysis" to determine the stresses; had taxed the resources of the manufacturer to the utmost in producing material and perfection of workmanship and had used the ingenuity of the builder to put the "unprecedented mass" into place.

Once the materials for the tubes and couplings were past the testing machines and in production, the arches went up easily. As falsework in the river was out of the question, Eads used the system of temporary cantilevers (Figure 10.4) which Thomas Telford had thought of using over the Menai Straits. Stays across towers on the piers and abutments held the double tubes of the arch ribs as they grew, section by section, first the upper and then the lower tube, simultaneously at each pier, for balance, with the cross-tying completed as they rose from skewbacks, or inclined surfaces of cast iron, set in the piers and abutments. The stays were held by jacks which were raised or lowered with changing temperatures. When complete the arches came under even stress and there was none upon the couplings except that of direct compression. The building out of the arches, though spectacular to watch from the shore, was relatively simple as engineering problems go, but there was a tense moment when the first arch rib arrived at the center. The last tubular section was just too long for the last space. The engineer on the job telegraphed to Eads for advice and strove meantime to solve the problem in his own way. He packed the tubes with tons of ice and hoped that the weather god would join his staff, but Eads's reply solved the problem. He had devised a last section which could be screwed into place and lengthened to make the joint tight. The news that the arch was closed, it is said, encouraged the American banking house of Morgan in London to obtain more British pounds sterling for investment in the St. Louis Bridge.

There was a tense moment when the first teamster crossed the roadway, another when General Sherman drove the last railroad spike, a third when 14 locomotives crossed two abreast and then in single line. Finally came the Fourth of July, 1874, with President Grant to honor Captain James B. Eads, who stood abashed before a high medallion of himself, with his bald head and his beard, hearing acclaim as one of the great engineers of all time.

John August Roebling (1806–1869) did not follow the reasoning which governed James B. Eads. When Roebling and his son Washington A. Roebling (1837–1926) came to build the Brooklyn Bridge (Figure 10.5) in 1869, they thought of using steel as Eads was doing, but they conceded no superiority to the arch over the principle of suspension. John Roebling had been reasonably successful with a suspension bridge for the road and railroad across the gorge at Niagara. He was triumphing again with cables of wrought iron imported from Britain, supporting a highway span of more than 1,000 feet across the Ohio River at Cincinnati. His skill with the suspended, or catenary, structure was established. There was in his mind no reason why they could not do as well with a span half again as long, more than 1,595 feet, from lower Manhattan to Long Island, high above the arm of the sea. The suspended roadway could be made rigid and held steady under every kind of traffic; there would be a maximum

Figure 10.5 The Brooklyn, or East River, Bridge in 1883 (From *Appleton's Annual Cyclopaedia and Register*, 1883)

clearance over the waterway for the masts of the ships, coastwise and ocean-going vessels constantly passing or docking along the East River. For Roebling, as for Eads, there was no difficulty with size except that of expense.

The Roeblings paid little attention to Eads's design for arches spanning the Mississippi where John Roebling had once proposed a suspension bridge, but Washington Roebling inspected Eads's underwater work with caissons, since the problem of setting foundations in the East River was similar. While the Roeblings did not have to take into account the turbulence of the Mississippi and its habit of plunging beneath its sands to bedrock, they did have to beware of tides that swept through from the Narrows to Hell Gate. They had to go deep for the footing of the towers for so great a span. Their caissons and piers were larger in every way, but the basic requirements of the construction varied little from those at St. Louis.

John A. Roebling lost his life as the work began in 1869. He was standing at the edge of the dock sighting across the river to line up the locations of the towers just as a ferryboat came in. It bumped the piling where he stood and crushed his foot. Tetanus developed and he died two weeks later on July 22. Washington Roebling took charge to complete his father's greatest project. Within a year and a half he too fell victim to the bridge. When fire in the Brooklyn caisson threatened the whole structure, he stayed in the air chamber from ten at night to five in the morning. Another experience with the bends in the New York caisson forced him in 1872 to give up active work altogether. At the age of thirty-five he was confined to a chair in the window where he watched the structure rise, unable to confer with his assistants but sending them instructions by his wife.

After the experience in the Brooklyn caisson, Washington Roebling and his assistants made the wooden top of the New York caisson as fire-resistant as possible with a sheathing of thin boiler plate. Lighting was a major problem. All open flames were especially dangerous in the compressed air of the caisson. Gaslights raised the temperature in the chamber unduly; the compressed air deadened the sense of smell and made the odor of escaping gas hard to detect. Oil lamps were too smoky. Calcium lights gave fair general lighting but were expensive. Candles remained the best for close work. The sand hogs in the caissons of the 1870s had to do without electric lamps, although arc lighting was to be ready for the roadway when the bridge was finished in 1883. The caisson on the Brooklyn shore came to rest some 44 feet below high water upon a conglomerate of clay, sand, and boulders that resisted picks, crow bars, and explosives, like solid rock. The caisson on Manhattan Island, however, encountered quicksands and had to go down more than 78 feet to the underlying ledge where crags were leveled and crevices blocked with concrete so that the sands could not flow. Roebling had cautiously experimented with pistol shots and small explosions under the compressed air before venturing to use normal blasts.

The towers of the Brooklyn Bridge rise 271½ feet above mean high water. They were built of limestone and granite brought from Kingston on the Hudson, Lake Champlain, and Maine. At 266 feet, four saddle plates were set in each tower and upon these were placed wrought-iron rollers to support the grooved saddles which carried the cables over the towers. Thus there was to be no wear upon the cables as they crossed the towers and no lateral stress from the moving loads of the roadway transmitted through the cables to the saddles and so to the towers. Eyebars carried through the towers below the saddle plates were to hold the stays which reached downward and outward to the roadway, where they were attached at successive intervals of 15 feet for a distance of 150 feet on either side of the towers. These counterstays, with the aid of the trussing of the roadway, were expected to sustain the roadway which rose in a slight arc. Together with these supports, the cables were to carry the weight of the bridge, the snow or ice load, and any traffic which might cross it. Roebling's design of his roadway trusses and supporting cables, together with his massive masonry towers, prevented the excessive vibrations due to aerodynamic instability which developed in some later bridges.

Distinctive features of the anchorage which Roebling patented were its wooden floors, anchor plates, eyebars, and the curves of the anchor chains. The yellow pine floors were to be kept wet so that they would be resistant to decay. The elliptical anchor plates were cast iron, weighing 23 tons each and measuring 16 by 17½ feet, with radiating arms 2½ feet from the center. Upon these rested the huge pile of masonry to offset any pull from the cables. The wrought-iron eyebars of the anchorage inclined upward through the masonry toward the cable ends in a curve which was designed so that the bars would bear downward upon the masonry and thus divert some of the stress from the straight pull upon the anchor plates.

In each of the four cables of the Brooklyn Bridge there are 5,296 wires, a total of 21,184 wires. They were made of crucible steel about ⅛ inch in diameter, drawn at the Haigh works in Brooklyn according to John Roebling's specifications. Their tensile strength was to be 160,000 pounds to the square inch, several times as great as the load which Roebling expected ever to come upon the wire. The wires were laid or spun straight and bound closely; they were not subject to torsion, shear, or any appreciable stress except tension. For splicing, the ends were threaded right-hand and left-hand, with corresponding threads in a coupling which drew the ends together. The joint, galvanized with melted zinc to prevent rusting in the threads, tested to better than 95 per cent of the tensile strength of the wire.

To guard further against rust, three coats of oil mixed with resin and lead oxide were applied to the wire as it came from the factory. Then it was oiled again as it ran upon drums through a sheepskin held in a man's hand. The two coats of oil could probably have been omitted since they did not stick well to galvanized wires; most of it rubbed off before the wires were finally compressed into the cables. The precaution, nevertheless, was taken, and another coating of white lead and oil was applied to the wrapping wire. Constant watch has been kept ever since, and a staff of painters works continuously on the bridge. The Roebling engineers had the satisfaction and reassurance of finding the wrought-iron wires of the suspension bridge at Niagara in fair condition after forty-two years. That bridge was replaced by a steel arch, not because the cables failed, but because the bridge was not strong enough to carry the increasing loads of heavier locomotives and trains.

With a footbridge laid on special cables and stayed against air currents and storms and with cradles hung crosswise from the footbridge on other wire ropes to regulate the cable laying, the long process went forward winter and summer in full view of ships and ferries beneath. No scaffolding blocked the waterway. A traveling sheave or wheel, 5 feet in diameter, passed from Brooklyn to New York, rolling out a loop of wire as another sheave returned empty to Brooklyn on the carrier. The first running time of 13 minutes eventually declined to about 10 minutes. The 278 lengths composing a strand were actually one continuous wire looped at both anchorages to make fastening with the anchor bars easy. Men in the cradles on stagings above the river and on the land spans brought each wire precisely into line with a guide wire which had been measured to the length of the finished cable.

As the wires were properly sagged they were clamped in a hand vise and gathered with lashings at 28-inch intervals into bundles of 278 wires each. New wrappings every 10 inches made these bundles or strands more compact, and then they were gathered into cables. There were 19 bundles, or strands, in each of the four main cables. As the lashings on the strands were removed and replaced by clamps, squeezers consolidated the more than five thousand wires, and wrapping machines bound them tightly with galvanized steel wire. A coat of white lead and oil immediately covered the completed cable; the individual wires had already been protected as has been indicated. Twenty feet a day was good progress with the wrapping machine operated by three men. On sunny days the work was speeded as much as 10 feet more. Each cable of the Brooklyn Bridge measures 15¾ inches in diameter. The combined ultimate strength of the four has been estimated at 18 million pounds. In comparison, Chaley's four 5¼-inch cables in the bridge at Freiburg, built in 1834, had a strength of 1,500,000 pounds. The four 36-inch cables of the George Washington Bridge of 1931 have a strength of 180 million pounds.

Builders before the Roeblings had used cables of wire for suspension bridges, but none had succeeded so well in reducing vibration and sway. John Roebling feared the winds and the impact of unbalance in the driving wheels of locomotives and the ambling rhythm of marching men even more. His able contemporary Charles Ellet (1810–1862) had lost three bridges because they had been too free to sway. For additional stiffness,

Figure 10.6 Brooklyn Bridge, original walkway, tower, and cables (From *Scientific American*, 1883)

the Roeblings drew their cables laterally out of line. The inside cables of the Brooklyn Bridge spread outward as they descend to the floor level. The outer cables draw inward toward the center of the span from the towers, and there are diagonal braces underneath the roadway, diverging from the towers to the sides of the roadway opposite those from which they spring (Figure 10.6).

Injuries and deaths marred the construction of the Brooklyn Bridge. Lines fouled, derricks fell, loose boards and falling objects accounted for many accidents. Disregard of rules caused cases of bends which took the lives of three men. Many of the men would not obey the rules about food, rest, and sleep, or report their illnesses, or avoid alcohol. The working day under compressed air was maintained at eight hours since the depths were not as great as at St. Louis and the men were never under pressures more than 34 pounds above atmosphere.

The Roeblings had begun the Brooklyn Bridge during the rule of the notorious Tweed Ring which managed the affairs of New York City with a skill in graft that has seldom been excelled. The original plan was for the cities of Brooklyn and New York to buy stock in the company along with private investors. The opportunity for misuse of funds seemed great. Work was delayed for months as John Kelly, Comptroller of New York, refused to pay its share, and trustees for the company had to take their suit to the court of appeals. By special act of the state legislature, June 5, 1874, the two cities assumed full control of the corporation and the bridge at last progressed to completion.

On May 24, 1883, President Arthur and his cabinet were present for the formal opening of the Brooklyn Bridge as were Grover Cleveland, Governor of the State of New York, and other dignitaries. Warming to the occasion, the orator, Abram S. Hewitt, declared that "the faith of the saint and the courage of the hero have been combined in the conception, the design, and the execution of this work." Hewitt took note also that cities in the Middle Ages had "walled each other out"; now men were "breaking down the barriers established by nature or created by man." [5] It could no longer be claimed that one traveled by rail from Albany to New York in less time than it took during the winter to cross the ice-filled East River from Brooklyn to Manhattan, and the bridge had a very large share in consolidating and expanding the two cities into the metropolis of America. With its railways now replaced by lanes for motor cars and buses but no trucks, the Brooklyn Bridge carries a type of traffic very different from that planned for by John A. Roebling. He had set the pattern for the huge suspension bridges of today.

Britain's engineers were aware of the success which the Roeblings were having in America with suspension bridges, because the communication of engineering advances had been vastly improved. The *Scientific American* had carried British correspondence from the 1840s; Britain's learned societies were recording addresses and papers of members from other countries. The London journal *Engineering*, founded in 1866, had become a clearinghouse for the profession throughout the world. Its special articles were reappearing on the Continent and in the United States. Reports from military and civilian imperialists overseas, on Chinese, Tibetan, Indian, and early Egyptian building, had long since been supplemented by news from America. It was well known that Finley, Pope,

[5] Abram S. Hewitt, "The Meaning of Brooklyn Bridge," in *Selected Writings of Abram S. Hewitt*, Columbia University Press, New York, 1937, p. 300.

Wernwag, and others there, as well as the Red Indians in the British Dominion of Canada, had ideas about suspension bridges and cantilevering. Thomas Telford had profited from the work of engineers abroad, and his successors kept informed.

Many engineers prefer cantilever to suspension cables for railway bridges because of greater rigidity. None of the several railway suspension bridges built had been altogether satisfactory. Telford's achievement at the Menai Straits in Wales had been overshadowed by the nearby tubular structure of Stephenson and Fairbairn. With all its shortcomings, the Britannia Bridge of 1850 had demonstrated the superiority of rigid structures in comparison with suspension bridges for railways. Prior to the construction of the Forth Bridge, one of the largest cantilever bridges was that which Charles Shaler Smith (1836–1886) had built over the deep gorge of the Kentucky River where John A. Roebling had raised towers for a suspension bridge. Speaking before the Institution of Civil Engineers in 1878 on the subject of long-span iron railroad bridges, the American engineer Thomas C. Clarke had declared that Smith's bridge was one of the most bold and original in the United States and had placed it high among the engineering structures of the world. Smith had thrown three 375-foot spans across a canyon 1,200 feet wide and 275 feet deep. He had intended originally to build a continuous Whipple truss but had altered his plan to the extent of inserting hinges in the outer spans. The result was a combination of balanced trusses, or cantilevers.

Following the tragic failure in 1879, during a December gale, of a large part of the iron railway truss bridge across the Firth of Tay near Dundee in Scotland, the engineering firm of Fowler and Baker took over a site on the Firth of Forth near Edinburgh where work on a suspension bridge had already begun. John Fowler (1817–1898) had distrusted the 2-mile-long Tay Bridge of 85 spans, then the longest bridge in the world, which had been opened for general traffic in 1878, and had forbidden his family to go upon it. He remarked to James Nasmyth that the bridge might have stood if the designer had used for its tall piers the "straddle of Henry VIII" in Holbein's portrait. Fowler and Baker determined in their proposed bridge for the Firth of Forth to make use of such a straddle, together with the principle of cantilevering which they had studied for several years (Figure 10.7). The adaptability of the principle to railway bridges of long span became more apparent as Bessemer made cheaper steel available that had the rigid strength to carry the rolling loads

*Figure 10.*7   The Forth Bridge (From Philip Phillips, *The Forth Bridge,* 1889)

of the trains which were to cross the bridge at full speed and to cope with the winds of the firth.

The Forth Bridge engineers designed approaches of traditional short spans uniform in depth. There were 15 of these, counting both sides of the firth, approximately 168 feet each in addition to masonry arches. Then came the innovation. The two great spans reached 1,710 feet on either side of the Isle of Inchgarvie and weighed some 16,000 tons apiece. The anchor spans at the shoreward ends were each 675 feet long. The cantilevers with their central girders or suspended spans, assembled as a whole, constituted a giant's stride in engineering (Figure 10.8). The towers of the Forth Bridge, 330 feet high, were given a uniform lateral batter or inward slope so that they were 120 feet wide at the bottom and only 33 at the top to provide the straddle with which Fowler expected to reduce the stresses due to the wind upon the structure. The same batter was carried into the bays as they were also narrowed laterally. The central spans, simple trusses with curved upper chords joining the cantilevers, were 350 feet long and, like the rest of the bridge, wide enough to carry a double-track railway. Benjamin Baker (1840–1907) operated upon the assumption that stresses caused by the winds were greater dangers to large bridges than the heaviest trains which could run on them. He experimented with models to determine the action of wind upon flat surfaces, curved sections, and cubes, and set gauges to take the record of the winds in the Firth of Forth. These came for the most part from the southwest and not from the North Sea as one might expect. The easterly gales scored from 15 to 20 pounds per square foot; the westerlies up to more than 40 pounds, with the greatest pressures coming in gusts and squalls. A maximum of 56 pounds was chosen for purposes of design.

Figure 10.8 The Forth Bridge under construction (From Philip Phillips, *The Forth Bridge*, 1889)

The sun also had to be taken into account. At the sliding ends of the suspended spans an expansion of 2 feet was anticipated, and the ends of the cantilevers were certain to move a foot. The only breaks were to be at these junctions of the cantilevers with the suspended spans, and so the whole structure would expand and contract with the cantilevers. The engineers solved the problem with an ingenious combination of ball and socket in a rocking post. What to do about the complex distortions occurring with direct sun on one side of the bridge and shade on the other was a different question. Fowler and Baker were confident that if they made provision for a margin of safety (or ignorance) in the strength of the materials over and above the requirements then known, they would have the strongest and stiffest as well as the biggest railway bridge in the world. Its margin of safety has been ample to this day. But an enormous number of unknown factors remain. Engineering still is in some respects empirical.

All tension members in the Forth Bridge are of open-lattice construction. But the great steel columns under compression were built of cylindrical tubes, 1¼ inches thick and as much as 12 feet in diameter. Baker had given much time to investigating their construction and seems to have ruled out the rectangular tubes of Stephenson and Fairbairn. He had then as precedents only the elliptical tubes in Brunel's Saltash Bridge of 1859 and the cylindrical tubes in the Eads bridge of 1874. Baker, like Eads, had realized the obvious fact that the hollow cylindrical form was the strongest, inch for inch and pound for pound. The steel was made by the Siemens-Martin, or open-hearth, process at Glasgow and at Swansea near Cardiff, Wales. Fowler and Baker expected the rolling load on the bridge to be 5 per cent of the dead weight of each of the long spans, or some 800 tons. To prevent tilting in case two trains happened upon one span at the same moment with no trains on the next span, they weighted the fixed ends of the cantilevers nearer the shores with counterpoises equal to the maximum train load plus one-half of the weight of the suspended span.

The river bottom, consisting of whinstone or basaltic traprock and a hard boulder clay just as stubborn, was excellent for the foundations of the heavy structure. On the Fife shore, piers could be built within cofferdams offering few difficulties that were new to engineers. On the Isle of Inchgarvie and the southern shore, the foundations had to go down so far that pneumatic caissons were as necessary as at St. Louis and Brooklyn. Instead of the rectangular wooden boxes of Eads and Roebling, the Forth engineers used wrought-iron cylinders for their caissons. It was kept in mind that Washington Roebling had come near to disaster with fire in his caisson on the Brooklyn shore and that cylindrical caissons are more efficient under compression than any other shape.

Once divers had leveled off the bottom, the caissons went down with few mishaps to their final locations, the deepest being 89 feet below high water. Their double skins were filled with concrete to give greater weight to the cutting edge and resistance to the water pressure. The mud encountered was diluted with water, forced down one pipe into the air chamber and blown out through another pipe by the compressed air. The caisson was entirely filled with concrete after it reached its final location. The pressures in the air chambers were raised and lowered to balance the weight of the water as the tide rose and fell, but in no instance were the men obliged to work under more than 35 pounds to the square inch above atmosphere. There was temporary paralysis on occasion but no death from the bends as there had been at Eads's St. Louis Bridge, possibly because the depths were not so great. It is possible, too, that the physicians in charge took better care of their men because of the experiences of Eads and Roebling. The workmen of the Forth had a great advantage over the sand hogs under the Mississippi and the East River. Incandescent lamps were a "great and lasting boon" in the air chambers.

Even the divers were supplied with watertight lighting equipment, and there were arc lamps with as much as 1,500 to 2,000 candlepower in the shops. The construction company installed steam engines and generators to supply its own electricity.

Laborers came for the most part from neighboring Scottish villages and towns, though some were imported from Belgium and France. The problems of housing and transporting the workmen were much the same as anywhere else. There were special trains and paddle steamers to take them to and from their homes and so keep them out of nearby public houses. In all some 3,500 men were employed on the Forth Bridge, 900 working on the Isle of Inchgarvie at one time.

As the Forth Bridge came into operation in 1890 John Fowler extolled the advances that had been made in comparison with the Britannia Bridge of but forty years past. Thanks to the achievements of Bessemer, Siemens, Thomas, and others, steel could now be produced at less cost than iron. If built of iron, said Fowler, the Forth Bridge would have both weighed and cost twice as much. Plates in the Britannia Bridge measured 12 by 2 feet; many in the Forth Bridge were as large as 30 by 5 feet. Fowler quoted the engineer of the Menai tubular bridge as having calculated that with the iron of 1850 a bridge of the span of 1,710 feet "might be constructed" but "not an ounce of weight" must be put upon it or "a breath of air allowed to impinge against it." Fowler added, "Each span of the Forth Bridge happens to be 1,710 feet. When tested by the Board of Trade inspectors, 1,830 tons were put on the 1,710 feet, but 4,000 tons might have been put on it without injuriously affecting the structure." [6]

As for aesthetic qualities, the critic William Morris, given to comments on the art of living, declared that the Forth Bridge was the "supremest specimen of all ugliness." However, Morris also stated that every improvement in machinery was uglier than the last and that there never would be an architecture in iron. Benjamin Baker replied to William Morris that beauty was relative to function; the duties of a structure should be considered when determining its place in art. In a lecture before the Edinburgh Literary Institution on November 27, 1889, Baker explained why the underside of the Forth Bridge had not been made a true arc instead of a series of chords. To have done so, he said, would have "materialized a falsehood." The Forth Bridge was not an arch and it said

[6] Thomas Mackay, *The Life of Sir John Fowler*, John Murray, London, 1900, pp. 308–309.

so for itself.[7] This statement may not have been a satisfactory answer to William Morris but it certainly was for the supporters of the new movement for "morality in architecture."

In spite of the experience with the St. Louis and the Brooklyn Bridges, the great bridge across the Firth of Forth marked in a way a transition in engineering practice in the use of steel in place of iron, particularly for structural purposes. Twenty years after steel had become available in quantity, builders had clung to the use of wrought iron. The lag was due in largest part, no doubt, to the fact that railways absorbed the increasing supply of steel; iron rails were so inferior that steel quickly became a necessity. But even when steel was cheaper and prevalent, many engineers still preferred structural iron for its uniformity and proven capacity. By 1890, however, this conservatism had given way; manufacturers were listing steel shapes for bridges and buildings in their handbooks, to the neglect of iron. Steel today is in general and virtually exclusive use for rails, girders, beams, and every sort of structural form, although wrought iron is preferred by some builders in such particular uses as pipes. For these, it is still thought by some to be the better material because it is less subject to corrosion than carbon steel.

After Eads's initial success with steel at St. Louis in 1874, his friend William Sooy Smith (1830–1916) completed an all-steel railroad bridge for the Chicago & Alton across the Missouri River at Glasgow, Missouri, in 1879. It was notable for its five spans of Whipple trusses, each 311 feet long, but although it remained in service for twenty-two years until replaced by a heavier structure, it had started no trend toward steel trusses in bridge construction. Charles Shaler Smith attracted more attention in 1877 with his wrought-iron cantilever spans 375 feet long over the Kentucky River.

French engineers were to gain even wider notice with their high wrought-iron viaducts. One of these, built for the Paris and Orleans Railway in the 1860s, was a single-track bridge extending over 1,000 feet in five trussed spans which were set upon clusters of cast-iron columns rising more than 181 feet above the Cere River. The Garabit Viaduct, designed by Alexandre Gustave Eiffel (1832–1923) and completed in 1884, was still more impressive with its parabolic central span of latticed construction, reaching 541 feet. Preceding the work of Fowler and Baker on the Forth, it was spread laterally to 66 feet at the spring line though but 20

[7] *Ibid.*, pp. 314–315.

Figure 10.9 Eiffel's Garabit Viaduct (Courtesy French Government Tourist Office)

at the top to give greater resistance to winds of the valley. The Garabit Viaduct soars over 400 feet above the Truyère River (Figure 10.9).

Eiffel, noted also for the wrought-iron skeleton of the Statue of Liberty in New York Harbor, went on to determine how high metallic frames could be raised with safety. There had been serious talk of a 1,000-foot tower at the Centennial Exhibition of 1876 in Philadelphia; Eiffel made it a reality at Paris in 1889 (Figure 10.10). Although Fowler and Baker were building in Scotland with steel, Eiffel preferred to use iron despite the fact that more metal would be required. Like Baker, Eiffel calculated the wind pressures that would come upon the tower at all heights and made certain that its design would be more than strong enough to withstand those pressures. The tower has a height of 984 feet; with lightning conductor it rises above 1,000 feet.

The most distinctive feature of the Eiffel Tower is its careful design. The structural members numbered 12,000; each required a special drawing, and all were finished at the shop and delivered ready to place. Of some 2,500,000 rivets, 800,000 were set by hand. The work, including the foundations, was completed in twenty-six months with so much precision from design to construction that no corrections were necessary. The *Scientific American* of June 15, 1889, applauded the achievement as "without error, without accident, and without delay."

The contrasts with another famous edifice in Paris have passing interest. Napoleon's Arc de Triomphe, mimicking Roman tradition and the architecture of the past, proclaims a warrior's success. The Eiffel Tower, heralding the buildings of the future, rises above the Champ de Mars; the laboratory high in the lantern has from its origin been devoted to

Figure 10.10
Eiffel Tower (From
Engineering News,
Supplement, June 8, 1889)

scientific research, first in astronomy, more recently in radio and radar. Neither of the two structures was planned to give economic return, but the Eiffel Tower has paid for itself in sightseers' fees.

High Buildings

Men have always yearned for the sky. Though varying their structures in form from ziggurat and pyramid to cathedral and skyscraper, they have expressed continuing ambition. The towering hotel and office buildings had begun to take shape before Gustave Eiffel raised his iron column. To achieve tall buildings there were combined two developments that were seemingly unrelated to one another. Until elevators were devised, the useful height of a structure was limited to the five stories or so that an average person could climb on stairs and still have breath left to speak of the business on which he had come. Also there was the very practical difficulty of raising walls beyond a certain level without having them fill the lower stories with inordinately thick masonry. A structural steel framework, supporting walls as well as floors, eliminated the need for bulky masonry walls.

There had been hoisting machines since the time of Archimedes or before, but the Romans and their successors were not interested in improving the mechanism for daily use. They felt no great need, although the occupants of the Roman *insulae*, or apartment houses, might have enjoyed the service. It was not until railways hastened the congestion of urban centers that engineers put their minds really at work upon vertical lifts and made crowding in the cities worse. Freight elevators of a crude type were running in some New York buildings by the 1850s. Their power was quite generally hydraulic, based on Pascal's hydraulic principle. Capacities and speeds were low. James Bogardus (1800–1874), enthusiast in the use of cast iron, proposed to build a 300-foot tower for the New York World's Fair of 1853 that would be firmer the higher it rose and would have a steam elevator to take observers to the top. It was his fellow townsman Elisha Graves Otis (1811–1861) who put into operation the first passenger elevator. Otis demonstrated his safety appliance at the New York fair of 1853 and in 1857 installed his elevator in a five-story store at Broadway and Broome Street. An appliance to check the fall of the car came into play if the rope broke or the mechanism failed. By November, 1866, the Franklin Institute of Philadelphia had taken note of an 11-story house in Paris that would have "a platform ascending noiselessly every minute, and raised by hydraulic power." Sigfried Giedion states that the first passenger elevator in Paris was at the Exposition of 1867. Both quietness and speed seem to have appealed, indicating factors besides safety that were essential.

The first hydraulic elevators were of the simple-plunger type. Water pressure was applied by a pump to a vertical cylinder from the top of which extended through a stuffing box a vertical steel column or plunger which raised the elevator. Releasing the pressure allowed the elevator to descend. There was counterbalancing by means of cables or ropes so that the water pressure was not required to raise the entire weight of the elevator and its load. The plunger made necessary, however, a pit beneath the building as deep as the building was tall. With increasing stories, a different type was developed, the rope-geared hydraulic elevator which had multiple-grooved pulleys; the motion of the car with respect to the piston was increased proportionally to the number of turns of the rope

Figure 10.11
The Pulitzer (New
York World) Building
(World Wide Photos)

around the pulley. Gustave Eiffel's hydraulic elevators of 1889, like his
Tower, attracted the attention of the world. They were of three designs;
one by the American Otis; the others were French. Together they formed
a system that raised the sightseeing crowds to the top of the platform in
three stages. The whole ascent took seven minutes; some 2,350 passengers
could be carried to the summit and returned to the ground in an hour.
Hydraulic elevators, while smooth-running and sure, are relatively slow.

Close upon the invention of elevators came the idea that reversed the practice of the builders of tall buildings and changed the primary function of walls. From earliest times walls had supported the framework and roofs of buildings, although medieval engineers had let the walls support only themselves as they enclosed the cathedrals. Now the frame was to carry the walls; builders of tall structures were to rest their walls independently, a story at a time, upon a comprehensive framework, or skeleton, which was self-sustaining. The walls could be built story by story from the top of the building downward if desired, for they no longer supported the structure, they merely enclosed it. This reversal in practice was not instantaneous. In the 1880s many buildings equipped with elevators were still carried up far beyond five stories without a skeleton. Among the most notable of these buildings still standing in 1955 is the Pulitzer (New York World) Building (Figure 10.11), completed in 1890, with 14 stories. It represents the last of the type whose walls carry their own weight. At their base the exterior walls of the Pulitzer Building are more than 9 feet thick.

In the 1850s James Bogardus of New York had come close to the skeleton or comprehensive framework without, however, seeming to realize it. His cast-iron frames, 70 feet high in some cases, had gained publicity in the *Illustrated London News* of April 12, 1851 (Figure 10.12), before William Fairbairn had published his book on building with iron. Bogardus himself boasted that the greater part of his ironwork might be removed or destroyed and still the frame would remain firm. Together with his contemporary in Britain, Joseph Paxton, designer of the Crystal Palace of 1851 (Figure 8.12), Bogardus is to be credited with an architectural achievement in combining glass effectively with iron. His building in 1854 for the publishers Harper & Brothers of New York had a façade that was little but framed-glass windows. The glass was supported independently at each story, but the walls continued to support only themselves between cast-iron columns at each floor. Their weight was not carried down to the foundations by any skeleton. It cannot be said that Bogardus had changed the function of the wall from support to enclosure.

The French builder Jules Saulnier took the further step of constructing a skeleton frame building with curtain walls. The structure which he raised in 1871–1872 for the Menier chocolate works (Figure 10.13) at Noisiel-sur-Marne straddled the river on four stone piers to make use of the waterpower. The site may also have influenced Saulnier to adopt prac-

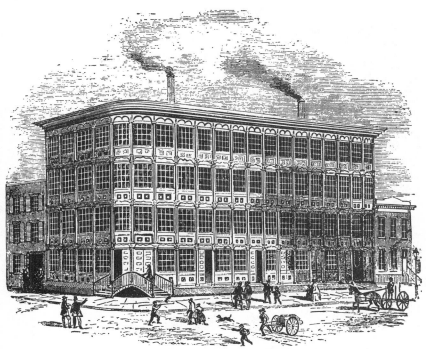

Figure 10.12 A Bogardus cast-iron building, Centre and Duane Streets, New York (From *Illustrated London News*, Apr. 12, 1851)

tices that were already familiar in bridgebuilding, both with wood and with iron; his diagonal bracing suggests this possibility. In any case the result was an iron skeleton which carried down to the piers the whole weight of the structure. The walls of hollow brick were no more than curtains. Though fully described in French journals, this new procedure in building was apparently not known in America at the time. The presumption has been that the discovery was independently made and applied to high structures in the United States. William LeBaron Jenney (1832–1907) is entitled to most of the credit for producing the first modern skyscraper, which he designed and built for the Home Insurance Company in Chicago in 1883 to 1885 (Figure 14.1). Carnegie, Phipps & Company asked permission to supply Bessemer steel girders instead of iron as the Home Insurance Building reached its sixth floor. The steel girders subsequently used were the first delivery ever made of structural steel for buildings, in the modern commercial sense of the term. The third essential

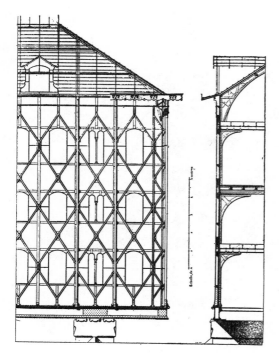

Figure 10.13
Jules Saulnier's Menier chocolate factory, earliest skeleton frame (From *Encyclopédie d'architecture,* 1877)

factor had come into the development of the skyscraper—steel. With the elevator and the comprehensive framework or skeleton, there was now available this relatively light, strong, and cheap building material.

A native of Fairhaven, Massachusetts, trained in Phillips Academy and the Lawrence Scientific School of Harvard, Jenney was graduated from the École Centrale des Arts et Manufactures of Paris in 1856 with high honors. In 1868 Jenney established himself in Chicago; there Louis Sullivan, later famed for his architecture and his personality, worked as a young assistant in Jenney's office and found him no architect, nor even "really, in his heart, an engineer at all," but a "*bon vivant*, a gourmet," who "knew his vintages, every one, and his sauces, every one." [8] The Empire State Building, the Chrysler Building, and other skyscrapers of today, whether in New York, Chicago, Dallas, Seattle, or Wichita, trace their ancestry to Jenney's 10-story building of the Home Insurance Company of 1883–1885, if not to Saulnier's Menier chocolate works of 1871–1872.

[8] Louis H. Sullivan, *An Autobiography of an Idea,* American Institute of Architects, New York, 1924, pp. 203-204.

Until Saulnier constructed the chocolate works, the primary function of the walls of large buildings, as has been said, had been to support the building or at least themselves. For centuries, however, there had been wooden-frame houses in which the frame supported the house and the walls which were hung on the frame. Prior to George Washington Snow's (1797–1870) invention of the balloon frame for houses, the wooden frame with its heavy posts, girts, beams, and braces was held together with mortise-and-tenon joints. Snow revolutionized frame construction by substituting for the heavy posts and beams with expensive and weakening joints thin wooden plates, or boards and studs held together by nails. Snow invented the balloon frame in Chicago, and the first building to have this type of construction was St. Mary's Church put up in Chicago in 1833. Balloon framing, used extensively in Chicago after 1833, spread rapidly throughout the Middle West and West. The successful development of the balloon frame depended on the mass production of machine-made nails and improvements in sawmill machinery. In turn the manufacture of wire nails and cut nails of steel and iron depended on improvements of iron and steel metallurgy.

The balloon frame industrialized house construction and made it possible for unskilled labor to replace the skilled carpenter. A balloon-frame house costs only 60 per cent as much as a mortise-and-tenon frame. Moreover, because skilled labor was not required, many more houses could be put up. Writing in the *New York Tribune* of January 18, 1855, Solon Robinson stated, "If it had not been for the knowledge of the balloon-frame, Chicago and San Francisco could never have arisen, as they did, from little villages to great cities in a single year." [9] After the middle of the century the balloon frame was widely known as "Chicago construction."

In the 1890s, following Jenney's construction of the Home Insurance Building, many skyscrapers with steel frames supporting them were built in Chicago. In fact it was the "Chicago school" which perfected skyscraper construction. By the end of the century, Chicago construction meant a steel-frame skyscraper and not a balloon-frame house. It is significant that this term should apply to two types of framing. It seems quite possible that the balloon frame had some effect on the development of the steel frame.

[9] Quoted in Sigfried Giedion, *Space, Time and Architecture*, Harvard University Press, Cambridge, Mass., 1944. p. 273.

Bibliography

Allison, Archibald: *The Outline of Steel and Iron,* H. F. & G. Witherby, Ltd., London, 1936.

Ashton, Thomas S.: *Iron and Steel in the Industrial Revolution,* The University Press, Manchester, 1924.

Burn, Duncan L.: *The Economic History of Steelmaking, 1867–1939,* The University Press, Cambridge, 1940.

Burnham, Thomas H., and G. O. Hoskins: *Iron and Steel in Britain, 1870–1930,* George Allen & Unwin, Ltd., London, 1943.

Fairbairn, William: *Iron, Its History, Properties, and Processes of Manufacture,* 3d ed., A. and C. Black, Edinburgh, 1869.

Giedion, Sigfried: *Space, Time and Architecture,* 3d ed., Harvard University Press, Cambridge, Mass., 1954.

Jeans, James S.: *Steel: Its History, Manufacture, Properties and Uses,* E. & F. N. Spon, London, 1880.

Steinman, David B.: *The Builders of the Bridge,* Harcourt, Brace and Company, Inc., New York, 1945.

Swank, James M.: *History of the Manufacture of Iron in All Ages, and Particularly in the United States from Colonial Times to 1891,* 2d ed., The American Iron and Steel Association, Philadelphia, 1892.

Woodward, Calvin M.: *A History of the St. Louis Bridge,* G. I. Jones and Co., St. Louis, 1881.

Zucker, Paul: *Die Brücke,* E. Wasmuth A. G., Berlin, 1921.

ELEVEN

Electrical Engineering

Three developments in nineteenth-century engineering have changed the ways of human life and altered the evolution of history. The first was the expansion of the Industrial Revolution described in Chapters 7 to 10. The second was the emergence of civilian engineering as a profession, bringing realization of the importance of scientific and technical education as a prerequisite for engineering practice. The third and most important development, correlated to the second, was the introduction of a new method of approach to the achievement of engineering advances— the method of applied science. The subject of this chapter, the inception and growth of electrical engineering, is an outstanding example of this new and revolutionary method.

Professional engineers, men who earned their living from the practice of engineering, had come into being in France during the seventeenth century. The French also established the earliest schools for instruction in engineering in the eighteenth century. The first civilian engineering school was the celebrated École des Ponts et Chaussées (School of Bridges and Highways) of 1747. However, the French institutions largely employed the apprenticeship method of instruction, and the teaching staff only occasionally gave general theoretical lectures. These schools ceased to function at the start of the French Revolution. The École Polytechnique in 1794, and later other schools, notably the rejuvenated École des Ponts et Chaussées, the École des Mines, and the Écoles d'Arts et Métiers, superseded them. These schools initiated instruction in such basic sciences as mathematics, physics, and chemistry. In the early decades of the nineteenth century, Germany and one or two other continental European countries established engineering schools modeled after the French. However, by 1840 in the United States there were only two schools offering instruction in engineering, the Military Academy at West Point and the Rensselaer School at Troy. In the ten years following the passage of the Morrill Land Grant Act of 1862, the objective of which was to stimulate the establishment of new technological schools by the granting of lands from the public domain, the number of such American schools jumped from 6 to 70. In general, American schools up to 1900 adapted European educational techniques and "had little direct share in the advancement of the art."[1] American schools did not begin to give formal instruction in electrical engineering until toward the end of the nineteenth century.

The rapid rise of engineering science in the nineteenth century extensively altered the practice of engineering and lent considerable impetus to the evolution of technical education. Engineering science may be described broadly as abstract theories such as those of statics and dynamics together with the use of scientific methods to solve engineering problems. Once again it was the French in the eighteenth century who rapidly developed various aspects of engineering science. The rise of engineering science in the eighteenth century was typical of the Age of Reason, when scientific methods began to be used in the study of many questions, notably social problems. When it became obvious in the early nineteenth century that, for instance, a structure or mechanical device scientifically designed to carry maximum contemplated loads or perform a specific function and no more was more economical than one designed on the basis of "experience," engineering science began to develop rapidly. It was also clear that technical schools were far more competent for giving instruction in the new science than was the age-old institution of apprenticeship.

More important than the rise of professionalism and the use of scientific techniques in engineering has been the application of ever-growing scientific knowledge to engineering. As A. N. Whitehead has put it, "The point is that professionalism has now been mated with progress. The world is now faced with a self-evolving system, which it cannot stop."[2] The phenomenal growth of electric power, which is hardly eighty years

[1] William E. Wickenden, "A Comparative Study of Engineering Education in the United States and in Europe," Society for the Promotion of Engineering Education, *Report of the Investigation of Engineering Education, 1923–1929*, 2 vols., Pittsburgh, 1930, vol. 1, p. 822.
[2] Alfred N. Whitehead, *Science and the Modern World*, The Macmillan Company, New York, 1931, pp. 294–295.

old, and of electronics, a more recent development, demonstrates dramatically how rapidly an engineering field evolves when based largely on science. The invention of the method of applied science in the last half of the nineteenth century was one of the most important innovations in the history of engineering.

History of Electricity

Since from its very beginnings electrical engineering has been based on the science of electricity, it will be necessary to review the early development of man's knowledge of electricity. Electricity is one of the youngest branches of physics. The Greeks knew that rubbed amber would attract straw, but subsequently the electrostatic attraction of the rubbed amber was confused for a time with the magnetic attraction of a lodestone. The Italian physician Geronimo Cardan (1501–1576) differentiated these two types of attraction. In 1600 William Gilbert (ca. 1544–1603), a court physician to Queen Elizabeth, published his *De magnete* in which he pointed out the difference between the attraction of amber and that of a lodestone. Amber attracts only small light bodies, not including iron, whereas a lodestone attracts iron only. Gilbert found that other substances than amber, such as glass, various gems, sulfur, hard sealing wax, and hard resin also exhibit a similar attraction when rubbed. The Greek word for amber is *elektron*, and Gilbert used the Latin word *electrum*. For other substances which behaved like amber he coined the word "electrica," which is best translated "electrics." It was in 1646 that Sir Thomas Browne (1605–1682) first used the English word electricity.

In 1629 an Italian Jesuit, Niccolà Cabeo (1585–1650), made an important observation of electrical repulsion. Otto von Guericke of Magdeburg, Germany, who had invented the air pump, built the first electrical machine about 1660, although he seems not to have investigated the phenomena produced in terms of electricity but rather as a manifestation of the *virtus conservativa*, the attractive property of the earth and matter in general. It was Francis Hauksbee (d. 1713) in England who used a static electric machine with a spinning glass globe to conduct the first important electrical experiments. Hauksbee's work stimulated the development of electrical science in the eighteenth century. Although von Guericke had observed electrical conduction, he apparently did not recognize it as such. In 1731 and 1732 Stephen Gray (1696–1736) published papers in the *Philosophical Transactions of the Royal Society* which demonstrated that some substances were conductors of electricity and others were nonconductors, or insulators.

Charles François de Cisternay du Fay (1698–1739) of Paris, superintendent of gardens for Louis XV, constructed the first theory of electrical phenomena. After reading Gray's papers in 1733, du Fay immediately began to experiment with electrical conduction and found that an "electric" on a glass stand and charged by a glass tube would repel other substances repelled by the tube but would attract those also attracted by a charged cylinder of "Gum Lack" (gum-lac). Having investigated this phenomenon further, he wrote, "Chance has thrown in my way another Principle, . . . which casts a new Light on the Subject of Electricity. This Principle is, that there are two distinct Electricities, very different from one another; one of which I call *vitreous Electricity*, and the other *resinous Electricity*. The first is that of Glass, Rock-Crystal, Precious Stones, Hair of Animals, Wool, and many other Bodies: The second is that of Amber, Copal, Gum-Lack, Silk, Thread, Paper, and a vast Number of other Substances. The Characteristick of these two Electricities is, that a Body of the *vitreous Electricity*, for Example, repels all such as are of the same Electricity; and on the contrary, attracts all those of the *resinous Electricity*." [3]

Du Fay assumed that a neutral substance has an equal amount of vitreous and resinous electricity and that it was a characteristic of any one substance that it gives up only one kind of electricity when rubbed and becomes charged with the other electricity. Du Fay's theory explained many electrical phenomena and was a useful concept. Its chief defect was that it did not account for the fact that the charge on any given body also depends on the substances used in rubbing.

Two men working independently arrived at the principle of the electrical condenser. This is a device to hold a charge of static electricity by means of two conducting plates separated by a nonconducting material. Pieter van Musschenbroek (1692–1761) of Leyden discovered what is now called the Leyden jar—a glass bottle or jar coated inside and out with metallic surfaces separated by the nonconducting glass. Musschenbroek reported his discovery in a letter to Réaumur in Paris, who read it to the French Academy of Sciences in January of 1746. In October, 1745, Ewald Georg von Kleist (1700–1748), Dean of the Cathedral of Kammin in Pomerania, had also discovered the principle of the Leyden jar, but his

[3] Charles du Fay, "A Letter . . . concerning Electricity," *Philosophical Transactions of the Royal Society*, vol. 38, pp. 263–264, January–March, 1734.

reports were so inadequate that most of the scientists who read of them were unable to repeat his experiment. The first discharge Musschenbroek received from his "bottle" gave him such a powerful shock that he began his letter to Réaumur with the sentence, "I wish to tell you of a new but terrible experiment which I advise you never to attempt yourself." [4] Volta later gave this device the name condenser, because electricity at that time was regarded as an "imponderable fluid" which could therefore be "condensed." It is now also called a capacitor.

By the time the Leyden jar was only two years old, Benjamin Franklin (1706–1790) had used it in experiments to demonstrate an important new theory of electrical phenomena. Franklin's great contribution to electrical science was his one-fluid theory of electricity, which replaced du Fay's two-fluid theory and, with important modifications, was useful and productive of new experiments for over a century. Franklin expressed his new concept in 1747 when he wrote, "We had for some time been of opinion, that the electrical fire was not created by friction, but collected, being really an element diffused among, and attracted by other matter, particularly by water and metals." [5] When Franklin wrote this sentence he did not know about du Fay's theory. Charles Augustin de Coulomb (1736–1806), French civil engineer and physicist, made the next great advance when in 1785 he verified the inference made in 1766 by Joseph Priestley, English philosopher and chemist, that the force of either attraction or repulsion between two small spheres charged with electricity is inversely proportional to the square of the distance between them. Coulomb's law was the first quantitative law in the history of electricity.

Luigi Galvani (1737–1798), professor at Bologna, found in 1786 that a dissected frog's leg would twitch if touched with a scalpel while an electric machine in the same room was in operation—hence one of the meanings of our word galvanize. After investigating this strange phenomenon further, he eventually found that a frog's leg would convulse when he simultaneously touched a muscle and its connecting nerve with the ends of a metallic conductor, one-half of which was a rod of zinc and the other half a rod of copper. The frog's leg would not convulse when the metallic conductor consisted of only one metal. Galvani was confronted with two possible interpretations of these interesting observa-

[4] *Mémoires pour l'histoire des sciences et des beaux-arts*, Trévoux, p. 2078, October, 1746.
[5] I. Bernard Cohen (ed.), *Benjamin Franklin's Experiments*, Harvard University Press, Cambridge, Mass., 1941, p. 174.

tions; either the two conductors of different metals were producing electricity by contact and the frog's leg was acting as an electrometer detecting the charge, or the nerve and muscle of the frog's leg were producing the electricity with the rods serving as conductors. Unfortunately Galvani erroneously chose the second explanation and coined the term "animal electricity" when he published his paper in 1791.

The professor of physics at Pavia, Italy, Alessandro Volta (1745–1827), at first accepted Galvani's theory of animal electricity. However, in a series of experiments similar to Galvani's, Volta found that he could not explain all of his results on the basis of the animal-electricity theory. Suspecting that the source of the electricity might be the contact between the two metals, he substituted for the dissected frog's leg a sensitive electroscope, a device he had invented in 1782 to indicate the presence of a minute charge of static electricity. With his electroscope he was able to detect an electric charge when two different metals were brought into contact without any frog's legs or other biological material being involved. Pushing his investigations further, Volta discovered the principle of his pile, now called a primary electric battery, which he announced in 1800. For the first time, with Volta's introduction of the electric primary battery, a continuous flow of electric current was available. For electrical science and subsequently for electrical engineering, Volta's discovery of what we shall call current electricity as distinguished from static electricity was of the utmost importance. The discoveries of electrical conduction, the condenser, or capacitor, and current electricity are the three most important eighteenth-century discoveries used in electrical engineering today.

For some time before 1800 various investigators had attempted to discover a relationship between electricity and magnetism. These investigations were in vain because there is no connection between static electricity and magnetism but only between an electric current and magnetism. In 1820 a Copenhagen professor, Hans Christian Oersted (1777–1851), first reported the existence of electromagnetism. Oersted found that an electric current passing through a wire suspended over and parallel to a magnetic compass needle caused the needle to swing out from the parallel position to a position almost at right angles to the direction of the wire. Oersted's finding was a great stimulus to scientific activity. Within two weeks after it was announced to the French Academy, André-Marie Ampère (1775–1836) observed that a coil of wire acts as a magnet when a current flows through it and that two such coils attract and repel each other just as do magnets but without the presence of an iron magnet. He must also be credited with having precisely defined electric potential or electric pressure as distinguished from electric current, in a paper published in 1820. By 1822 Ampère had firmly established by experiment and quantitative analysis the science of electrodynamics, or what may be called electricity in motion.

Meanwhile Michael Faraday (1791–1867), an English chemist and physicist, had shown that a wire suspended with one end free to move and dipping into a bowl of mercury would revolve continuously around a permanent magnet supported in the mercury as long as the wire and the mercury were connected to opposite poles of an electric battery. This discovery of electricity producing motion was the fundamental discovery which eventually led to the invention of the electric motor. In 1826 Georg Simon Ohm (1787–1854), a German school teacher, announced the fundamental law which now bears his name. His discovery was as important as Ampère's distinction between electric potential and current and is equally indispensable in engineering. Ohm made many precise experiments before he was able to state that the amount of current in a circuit is directly proportional to the difference in potential or pressure and inversely proportional to the resistance.

Faraday was convinced that if electricity could produce magnetism, as Oersted had found, then magnetism could produce electricity. He had been searching for electricity which could be produced by magnetism for a half dozen years before he came upon the principle of induction in 1831. It was in this year that he, for the first time, realized that a current was induced in a conductor while the intensity of a magnetic field was either increasing or decreasing or while the conductor was being moved in the magnetic field. Faraday also found that he could produce a current in one of two coils by moving the coils toward or away from each other while a current was flowing in the second coil. He then substituted a magnet for the coil carrying the current and produced the same effect. Using two coils wound on separate sections of a closed iron ring, with one coil connected to a galvanometer and the other to a battery, Faraday observed that when he completed the circuit to the second coil, the galvanometer needle jerked in one direction and then returned to zero. When he broke the battery circuit, the galvanometer jumped about the same distance in the opposite direction and again returned to zero. In

1831, Faraday also demonstrated that by rotating a copper disk between magnetic poles he could generate direct current. It is on these discoveries that the mechanical production of electricity is based. Thus Michael Faraday had found not only the basic principle of the electric motor but also the basic principle of the generator and of the induction coil and transformer.

However, Faraday must share the honors of discovery of induction with the American physicist Joseph Henry (1797–1878). Priority clearly belongs to Faraday who was first to publish his results, in April, 1832. Henry, working independently, had discovered induction before Faraday, but Henry did not publish his paper until later in 1832. The discovery of induction completed the great findings which immediately followed Oersted's detection of electromagnetism. The next important advance occurred in 1865 when Scottish-born James Clerk Maxwell (1831–1879) published a paper entitled "A Dynamical Theory of the Electromagnetic Field." In this work Maxwell presented his equations describing the phenomena which relate the electric conductivity, the dielectric constants, and the magnetic permeability of matter, all with electric and magnetic fields and with mechanical force. One of Maxwell's interesting conclusions drawn from his analyses was that an oscillatory or alternating electromagnetic disturbance would produce electromagnetic waves having the speed of light.

There was little experimental evidence for this remarkable prediction, but in 1887, Heinrich Rudolph Hertz (1857–1894), young German physicist, announced that he had been able to produce and detect waves of "etheric force," vibrating in an "all pervasive ether" and which had electromagnetic properties. Hertz had employed an "electrical oscillator" consisting of two metal balls separated by an air gap and connected with either terminal of an induction coil so that a spark would jump between the balls. His detector was a piece of wire about 7 feet long with a small metal ball at each end. Hertz bent the wire into a ring and varied the distance between the balls. When he achieved the proper distance, a spark would jump the gap in the detector when the oscillator sparked. Hertz was able to show that these new waves (hertzian waves) are subject to reflection and bending or refraction, just as light and radiant heat waves are. Later Hertz calculated that the speed of his electromagnetic waves was the same as that of light and thus confirmed Maxwell's prediction. Hertzian waves are now known as radio waves or wireless waves. Radio and television transmission are directly based on a knowledge of the properties of the waves which Hertz discovered in 1887 and the existence of which Maxwell had predicted nearly two decades earlier.

While these advances were being made in the knowledge of electromagnetic waves, two other avenues of research were opening which were to produce important information for subsequent application in radio or wireless and television. One line of research was on cathode rays, rays projected from the cathode or negative terminal of a vacuum tube through which a current is flowing. Julius Plücker (1801–1868) of the University of Bonn discovered cathode rays in 1859. The other line of research was on the so-called "Edison effect" discovered in 1883 by Thomas Alva Edison (1847–1931). Ten years after Plücker's discovery of cathode rays, his student Johann Wilhelm Hittorf (1824–1914) showed that these rays were propagated in straight lines and that a magnetic field at right angles to the line of discharge would deflect the rays. By 1879 William Crookes (1832–1919), English physicist and chemist, had shown that cathode rays had momentum and also definite energy. In 1897 the English physicist Joseph John Thomson (1856–1940) demonstrated conclusively that cathode rays are negatively charged atomlike particles which he called "corpuscles" and which are now called electrons. Thomson's discovery of the electron was the first physical evidence that such particles exist in nature, and together with the discovery by Antoine Henri Becquerel (1852–1908) of radioactivity in the previous year, 1896, forms the starting point for the remarkable development during the twentieth century of the science of atomic physics.

Thomson's discovery made it possible for him to explain what was known as the Edison effect. Edison had found that a small current would pass from the carbon filament of his then newly invented incandescent light bulb from which air had been pumped to an electrode sealed into the bulb when the electrode was charged positively. No current would flow when the electrode was charged negatively. Two British investigators, William Henry Preece (1834–1913) in 1885 and John Ambrose Fleming (1849–1945) in 1890 and 1896, had studied the Edison effect in great detail. In 1903, Thomson proved that electrons were the carriers of the current observed in the Edison effect. Application of this knowledge has been of great importance in the field of electronics. The radio tube or electric valve is an application of this phenomenon fundamental to communications engineering.

In the following sections of this chapter we shall discuss how knowledge of electricity has been used to establish electrical engineering, including communication and power production and utilization. The first important practical application chronologically was the electromagnetic telegraph.

Telecommunication

During the first thirty to forty centuries after man began to live in villages, roughly seventy-five centuries ago, he began to extend political control over groups of villages and later cities and to carry on trade throughout large areas. Ever since the extension of governmental controls and commerce beyond the limits of the local village, he has had need to communicate messages over considerable distances. Until the nineteenth century men transmitted messages by signal fires, smoke, drums, runners, pigeons, ships, postriders, and by sentinels who relayed shouted messages. Indeed, primitive tribes still use some of these techniques. Our word marathon derives from the Marathon Plain, 26 miles outside the city of Athens. In 490 B.C., the Persians invaded the Greek Peninsula, and the Athenians attacked and defeated them on the Marathon Plain. A young Athenian, Pheidippides, ran the 26 miles to Athens bringing the news that the Athenians were victorious.

In the sixteenth century, Europeans literally explored the world. The colonization which followed their explorations brought about an enormous extension of known distances and greatly increased the volume of commerce. These developments in turn produced more urgent requirements for the rapid communication of messages.

Several men proposed optical telegraphs in the seventeenth century; Robert Hooke's is perhaps the best known. Following the discovery of electrical conduction in 1732 came many recommendations for the use of static electricity to communicate messages. None, however, was practicable. In 1794, Claude Chappe (1763–1805) developed what he called an ocular, or semaphore, telegraph (Figure 11.1), which was the first practical device. His system consisted of a line of towers 6 to 10 miles apart. The operator at each tower, usually using a telescope, could see the tower on either side of him in the line if visibility was good. On the top of each tower was a semaphore consisting of a wooden beam pivoted midway so that it could be rotated in a vertical plane. In addition there were movable arms at each end of the beam. It appears that the speed of transmission could be fairly rapid. For instance, over the 475-mile line with its 120 towers between Paris and Toulon, it is said that the operators were able to transmit a message in ten to twelve minutes. The French government

operated the first Chappe telegraph between Paris and Lille, a distance
of 144 miles. It was a success, and use of the system spread rapidly. The
British subsequently built several Chappe lines connecting London with
various channel ports where it was feared that Napoleon's fleet might land.
In 1800 Jonathan Grout built the first semaphoric telegraph line in the
United States. This 65-mile line connected Martha's Vineyard with Boston and was built to transmit commercial news, especially of the arrival
of ships. Although some of the Chappe telegraphs were still in operation
at the middle of the nineteenth century, the semaphore system was expensive to operate since it required men to be stationed at every tower
and was dependent upon clear weather.

Volta's invention of the electric battery in 1800 stimulated efforts to
develop an electric telegraph. Following Oersted's discovery of electromagnetism, William Sturgeon (1783–1850), English physicist, constructed
an electromagnet about 1825, and the next year Ohm announced his law
concerning the flow of an electric current in a circuit. These discoveries
were essential for the development of practical electromagnetic telegraphy. The demand for rapid communications was so insistent by 1830

that there were literally dozens of men trying to produce an electrical telegraph. In 1820 Ampère had suggested that by using a circuit for each letter and a magnetic needle at the terminal it would be possible to transmit messages. One of the first practical electromagnetic telegraph systems was based on Ampère's idea. Ten years later William Ritchie demonstrated Ampère's proposal on a small scale. In 1832 Paul von Schilling-Cannstadt (1786–1837), a native of Estonia, worked out a system also based on Ampère's suggestion, but he never produced a practical telegraph.

In 1836 a young Englishman, William Fothergill Cooke (1806–1879), saw some electromagnetic experiments in Heidelberg where he was studying anatomy. He was immediately impressed with the possibilities of using an electric current for the operation of a telegraph. Within three weeks he had put together his first instrument; it was the magnetic-needle type and similar to Schilling's device. After Cooke's return to England, Faraday introduced him to Charles Wheatstone (1802–1875) of King's College, London, who had already done important work on an electric telegraph. Later, in 1837, the two formed a partnership. Before they had signed the partnership agreement, Cooke and Wheatstone had constructed in 1837 a telegraph line along the London and Birmingham Railway north from Euston Station, London, about a mile to Camden Town, and successfully transmitted their first messages. The early instruments of Cooke and Wheatstone required five or six wires for transmission and correspondingly five or six magnetic needles, each pivoted between the two sections of a double wire coil. The needles swung to right or left and indicated specific letters, depending on the sender's setting of a dial.

By arrangement with the incompleted Great Western Railway, the partners established a telegraph line along the railway out of Paddington Station and in 1839 extended it 7½ miles to Hanwell. By 1843 this line reached Slough, 18½ miles from Paddington. Their telegraph system received considerable publicity when the news of the birth of Queen Victoria's second son was transmitted over it on August 6, 1844. The device had created so much public interest that there was a charge of a shilling for admission to the office to see it work.

However, the part the telegraph played in the capture of a murderer brought the system its greatest publicity. On New Year's Day, 1845, a Londoner named John Tawell took the train to Slough where he gave a woman a fatal dose of poison. Realizing she had been poisoned, the unfortunate victim began to scream. This screaming aroused the neighbors, and Tawell, alarmed, rushed to the station and immediately boarded a train for London. Someone who had seen him leave the house described him to the telegraph operator who at once wired the news to the Paddington Station, adding that the murderer was on the train which had just left Slough and that he was dressed like a "Kwaker"—there was no "Q" in the Cooke and Wheatstone system at that time. The Paddington operator notified a railroad detective, who spotted Tawell as he left the train and trailed him to his home. Early the next morning the police of Slough arrived and Tawell was arrested. He was subsequently tried and hanged. The Slough murder case proved an excellent advertisement for telegraphy.

While Cooke and Wheatstone were developing their telegraph system which became standard in England, Samuel Finley Breese Morse (1791–1872), an American artist, was perfecting a different type of electromagnetic telegraph. Returning to the United States in 1832 after a three-year stay in Europe, Morse discussed various problems of electromagnetism with a fellow passenger, Dr. C. T. Jackson of Boston, Massachusetts, who apparently told Morse about various experimental electric telegraphs in Europe. Jackson later claimed to be the originator of the idea of the electromagnetic telegraph. This claim is of questionable validity, but there can be little doubt that Jackson was a man with useful ideas. It was he who more than ten years later suggested to Dr. W. T. G. Morton the use of ether as an anesthetic and then claimed a share in that great discovery. He also claimed in 1846 that he had given Christian Friedrich Schönbein the idea he used in developing gun cotton.

After Morse arrived in New York in the autumn of 1832, he immediately set to work on a telegraph, but it was not until 1836 that he completed his first instrument. Shortly thereafter he learned from his colleague Leonard D. Gale of the University of the City of New York about Joseph Henry's improved electromagnets wound with many turns of insulated wire. Morse found that the new magnet would greatly increase the distance over which he could transmit a signal. Morse's receiver at that time was a pen in constant contact with a strip of paper moved by a clock mechanism. When the electromagnet was energized, it moved the pen so that it traced a notch on the paper. Morse formulated messages by varying the time intervals between the notches or by inverting them by reversing the polarity or direction of flow of the sending current of the signal. On September 4, 1837, the same year that Cooke and Wheatstone first operated their Euston Station to Camden Town telegraph, Morse transmitted his first message through 1,700 feet of wire.

Shortly thereafter Morse made two important improvements. First he developed a relay to increase the practicable distance of transmission of messages. In this device the signal activated an electromagnet which closed a contact in a local battery circuit, thereby transmitting the signal to the next relay by means of this second battery circuit. This process could be repeated indefinitely. Curiously, Henry and Wheatstone were developing similar relays at the same time. Morse's second improvement was the widely used dot-and-dash code system and the sounder suggested by his younger colleague Alfred Vail (1807–1859), which greatly facilitated the reading of messages. In this scheme, the operator completes the circuit by closing the sounding key, and the sounding device attached to an electromagnet moves with an audible click. This dot is a very short signal of only a fraction of a second duration, and the dash is about three times as long.

Morse was able to obtain some private financial support, but his efforts to get a grant from the United States Congress were unsuccessful until January, 1843, when Congress appropriated $30,000 for the construction of a 38-mile line along the Baltimore and Ohio Railroad between Baltimore and Washington. On May 24, 1844, the first message—"What hath God wrought"—was transmitted from Washington to Baltimore over this line, and the Morse system of telegraphy was established. After this success, other lines were rapidly built. By 1846, telegraph lines stretched from Washington to Portland, Maine, and westward to Louisville, Kentucky, and Milwaukee, Wisconsin. The development of duplex, quadruplex, and multiplex telegraphy many years later has greatly increased the economy of transmission. The multiplex system enables operators to transmit messages simultaneously over the same set of wires, and perforated tape devices permit far higher speeds than are possible with manual key operation.

Short lengths of cable were laid under rivers as some of the earliest telegraph lines were strung. One of the first cables was that under the harbor at New York by Morse; it was insulated with gutta-percha, a rubberlike substance discovered in 1834. The first submarine cable of considerable length, from Dover to Calais, was put down in 1850. The insulation on this cable proving inadequate, a greatly improved cable was opened for service in November, 1851, and gave satisfactory service for many years. In the next few years a cable connected Denmark and Sweden, and cables were laid in the Mediterranean Sea.

The first successful Atlantic cable began operation in 1866, largely owing to the efforts of Cyrus West Field (1819–1892), after many tragic failures. The Atlantic Telegraph Company had been organized in 1856, and the first attempt to lay a cable between Ireland and Newfoundland was made the following year. This cable broke after 335 miles had been laid, owing to an accident in the paying-out machinery. A second attempt in June, 1858, similarly failed, but later in the summer of that year the company completed, on August 5, the laying of a cable. The "first" commercial message was sent on August 17 after a congratulatory message from Queen Victoria to President Buchanan. This cable remained in operation only a short while for it gradually failed and by October 20 ceased to function. One of the messages transmitted while the cable was still in service was a communication from the British government to Canada, peace having been concluded with China, countermanding an order for the departure of two regiments of soldiers which were about to return to England for service in India. If the cable had not been in operation at that time the regiments would have sailed and considerable unnecessary expense (estimated at £50,000) would have been incurred.

During the next seven years a group of scientists and engineers worked for the Atlantic Telegraph Company to improve the insulation and mechanical strength of the cable, the paying-out apparatus, the receiving mechanism, and the type of signals to be used. Sir William Thomson (Lord Kelvin, 1824–1907), professor of natural science at Glasgow, was largely responsible for the ultimate success of the cable. The company secured the *Great Eastern* for the next attempt, which began at Valencia, Ireland, on July 23, 1865. After 1,200 miles had been payed out, the cable parted in 2,100 fathoms (12,600 feet, 2½ miles) of water, and the project was abandoned for that year. The Atlantic Company failed and was succeeded by the Anglo-American Company, which continued the project. On July 13, 1866, the *Great Eastern* again steamed west from Valencia with a new cable which was landed at Newfoundland on July 27. The *Great Eastern* then conducted a successful search for the cable which had parted the previous year. A new cable was spliced to the recovered end, and the second landing was made at Newfoundland on September 8. Cables have been in continuous service ever since; nineteen cables now cross the Atlantic. In the summer of 1955, HMTS *Monarch* was laying a new type of transatlantic cable, a coaxial cable with a repeater employing three vacuum tubes every 40 miles, for transmitting telephone messages in one direction. After a second cable has gone down in 1956, the system will be able to carry 36 conversations simultaneously. The aggregate length of submarine cables in use in 1955 exceeds 300,000 nautical miles.

The method of sending and receiving messages by submarine cables differs from that of the standard telegraph because the electric capacity of a long cable requires considerable time for charging and discharging it. The current used is therefore relatively weak, because the weaker the current the shorter is the time required to charge and discharge the cable and so to transmit an individual signal. In other words, the weaker the current the more rapidly can a message be transmitted. A mirror galvanometer generally replaced the sounder to receive the signals. The mirror reflects a beam of light from a lamp in such a way that a small deflection of the coil on which the mirror is mounted gives a perceptible motion to the spot of light. The mirror was replaced in 1870 by the siphon recorder developed by Sir William Thomson, consisting of a light coil of wire suspended between the poles of a powerful electromagnet. The motions of the coil are transmitted by silk threads to a small glass tube, one end dipping into an ink reservoir and the other recording on a paper tape.

Some idea of the part played by telegraphy in communications can be gained from an analysis of 650,000 telegrams which the British Post Office transmitted during one week in 1934. Two-thirds of the messages were commercial and one-third were social. There can be little doubt that commercial demand for rapid transmission of messages was the principal incentive for the erection of the early telegraph lines in the 1840s. The telegraph business steadily increased during the nineteenth century, but it has declined during the twentieth. One of the principal causes for this decline has been the telephone.

The incentives for developing the telephone were different from those which produced the telegraph. By the time the first work on telephones was being done, the telegraph had already satisfied much of man's need for speed in the transmission of intelligence. Because of the less imperative incentives for the invention of a practical telephone, there were fewer men who worked on its invention, but they also were men of ability and imagination.

In 1837 an American physician, Charles Grafton Page (1812–1868) of Salem, Massachusetts, discovered that when there are rapid changes in magnetism of iron it gives out a musical note of a pitch depending on the frequency in the changes in magnetization. He called these sounds "galvanic music," and many physicists studied this phenomenon in their laboratories. It remained for Philipp Reis (1834–1874) of Friedrichsdorf, Germany, to construct in 1860 an apparatus by means of which a melody produced in one place could be transmitted electrically and reproduced at a distance. Reis's first apparatus consisted of a frustum of a cone inserted in the bunghole of a beer barrel. The small end was covered by an animal membrane upon which a small platinum wire was fastened by means of sealing wax. The platinum wire formed part of a battery circuit and touched a metal strip in such a manner that it could make and break contact as the diaphragm vibrated and interrupt a battery circuit with a frequency corresponding to the vibration of the diaphragm. The wires of the battery circuit led to a coil which was wound around a knitting needle. The rapid magnetization and demagnetization of this needle caused sound waves of the frequency of the note sounded into the diaphragm. Reis success-fully demonstrated the transmission of musical notes in an improved appa-ratus based on these principles at Frankfurt in 1861. He claimed in a letter to a colleague in 1863 that ". . . if you will come and see me here, I will show you that words also can be made out." [6] This claim has not been fully established, however. Be this as it may, Reis's contribution was im-portant in the history of the transmission of sound.

In 1875, two men in the United States were working independently and unknown to each other on telephonic transmission. They were Elisha Gray (1835–1901), inventor and manufacturer of Chicago, and Alexander Graham Bell (1847–1922) of Boston. Gray developed an instrument very similar to Reis's except that he used a small iron rod attached at one end to the diaphragm with the other end immersed in a fluid of low conduc-tivity that formed part of the battery circuit. The vibrations of the diaphragm, by moving the rod up and down, varied the resistance in the circuit, so that the resulting fluctuating current, depending on the varying resistance of the fluid, corresponded to the original sound vibrations im-pinging on the diaphragm. This fluctuating current was conducted to a receiver consisting of the coil of an electromagnet and a diaphragm on which was mounted a small piece of soft iron. The fluctuating current varied the magnetism and caused vibration of the iron on the diaphragm, which duplicated the vibration of the sending diaphragm and thus the sound. On February 14, 1876, Gray filed in the United States Patent Office a caveat, or formal notice of his claim to the idea of his instrument. He filed this caveat to endeavor to prevent others from patenting his

[6] Alfred R. von Urbanitzky, *Electricity in the Service of Man*, Cassell & Co., London, 1886, p. 662.

idea within the period of a year. Bell applied for a patent for the same type of instrument a few hours earlier on the same day. Bell was finally awarded the patent rights and credit for the invention but only after extended and bitter litigation.

Bell, Scottish-born and university-trained, had specialized in teaching the deaf to speak clearly. He emigrated from London to Canada in 1870 and two years later moved to Boston. Following an attempt to devise an instrument which he hoped would enable the deaf to see spoken words, he began to work on a telephone in 1874. Discouraged by indifferent results with his early telephone, Bell had the good fortune to meet Joseph Henry in Washington. Henry encouraged the young man to continue his work. Bell not only succeeded in producing a liquid-resistance type of transmitter, but he also perfected the magnetic type, the iron diaphragm of which, vibrating in a magnetic field, produced variations in the field, thereby causing variations of current in the coil of wire around the magnet. These current variations, when transmitted to the receiving set, reproduced the vibrations of the sending diaphragm and so duplicated the original sound.

It was on March 10, 1876, that Bell spoke the first words clearly heard over a telephone when he called to his assistant in his bedroom, "Mr. Watson, come here, I want you!" [7] Bell had spilled some acid on his clothes and wanted help, but he forgot the accident when Watson told him how plainly he had heard his summons. Later in 1876, Bell showed his instruments at the Centennial Exhibition in Philadelphia, where they created much interest. A commercial type of telephone was constructed early in 1877, and by September there were 1,300 in use in the United States. The first community central station switchboard for interconnecting subscribers' instruments was placed in operation with 22 subscribers in New Haven, Connecticut, on January 28, 1878 (Figure 11.2). Numerous improvements were made in the years immediately following, and the use of telephones spread rapidly and continuously throughout the world.

The stimuli for the invention of radio, like those for the telephone, were not as intense as the incentives for the invention of the electromagnetic telegraph. By 1887, when Hertz discovered electromagnetic waves, there were in operation extensive wire and cable lines transmitting telegraph and telephone messages. However, the possibility of sending mes-

[7] Thomas A. Watson, *Exploring Life*, Appleton-Century-Crofts, Inc., New York, 1926, p. 78.

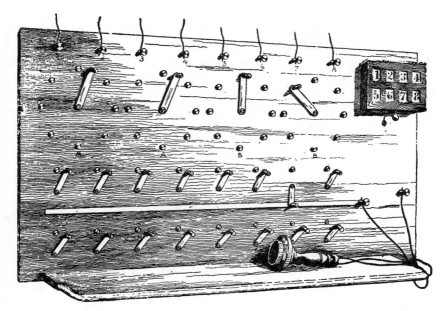

Figure 11.2 First central switchboard, New Haven, Connecticut, 1878 (Courtesy Southern New England Telephone Co.)

sages over lands and across oceans, especially from ships at sea, without wires or cables was an economic incentive. When the radio, or wireless, was first contemplated, its inventors thought of it only in terms of point-to-point transmission, and it was not until after the First World War that radio's many potentialities for broadcasting began to be realized. Today general broadcasting is more important to our society than point-to-point message transmission.

Hertz's discovery was fundamental to the invention of the radio. Hertzian waves carry the radio signal, and without a knowledge of them it would not be possible to have radio. Several physicists, including the English Oliver Joseph Lodge (1851–1940), realized the possibility of using hertzian waves for the transmission and reception of telegraphic messages. In 1894 Lodge described a device which consisted of a receiver tuned in proper resonance to collect waves, a detector, i.e., a relay to amplify the signals and a printer for recording dots and dashes. Lodge's detector was called a coherer, which involved also a trembler, and was based on the phenomena discovered in 1879 by David Edward Hughes (1831–1900). The coherer, which had been used two years earlier for other purposes, was a short glass tube containing loose nickel filings. With its terminals, it formed part of a local battery circuit. Hertzian waves captured in this receiver decreased the resistance of the filings, which "cohered" and allowed the current to flow from the battery through them. After conductivity had been established in the coherer, the filings continued to be a good conductor and therefore indicated no further response to the waves. Lodge restored the resistance of the filings by knocking them loose with the trembler, which tapped the tube. Lodge's receiver worked, but as he was more interested in teaching and scientific research than in practical application, he did not devote himself to improving it for commercial purposes.

In 1894, the same year that Lodge demonstrated his receiver, Guglielmo Marconi (1874–1937), a young Italian inventor, read about Hertz's discovery. He immediately started to devise a practical method for wireless transmission. Marconi made an improved oscillator or transmitter using an aerial carried on a mast and also increased the effectiveness of the coherer. In 1896 he succeeded in transmitting and receiving code messages over a distance of nearly 2 miles. Believing that there were better opportunities to develop his wireless commercially in England, he went there later in 1896. Soon he was sending messages 8 miles. Beginning in 1897, Marconi began experiments with long-distance transmission, which culminated on December 12, 1901, when he heard on his receiver in Newfoundland the three clicks of the Morse code "S" which were sent out from his station at Poldhu in Cornwall, England.

In the meantime, in 1898, Lodge had invented the selective tuner. The coherer as a detector was inadequate, and John Ambrose Fleming, who was then a consultant with Marconi, invented the two-element, or diode, vacuum tube using the principles of the Edison effect described above. The two terminals of this relay were an anode and an incandescent cathode which emitted electrons continuously. Fleming placed the tube, or valve as he called it, in the aerial circuit of a receiver. The alternations of the incoming signals changed the anode from positive to negative, so that during the negative interval it repelled electrons emitted by the cathode and no current flowed. During the positive period the anode attracted electrons from the cathode, and the resulting current operated a printer or telephone receiver. Although Fleming's invention is basic for radio, his diode was not very satisfactory in operation. The American inventor Lee de Forest (1873–) greatly improved the diode in 1906 by inserting a grid in the vacuum tube between the cathode and anode. The grid, or metal screen, was attached electrically to the radio aerial, and the alternating or oscillating charge on it would repel the electrons when negative and reinforce the electron stream when positive, giving a much more effective control of the current than was possible with the diode. Because it has three elements, De Forest's tube is called a triode. Both the diode and triode were operated at partial vacuums.

At the same time that the development of the vacuum tube was progressing, the evolution of continuous-wave transmission as opposed to the make-and-break telegraphic spark was proceeding. Continuous-wave transmission is, of course, necessary for voice transmission. Reginald Aubrey Fessenden (1866–1932) in America, realizing the need for a continuous wave, first produced one with an alternator in 1903. It was Fessenden who invented the heterodyne circuit for receivers utilizing differences in frequency between the sending and receiving circuits. In 1914 Edwin Howard Armstrong (1890–1954) of Columbia University evolved the feed-back circuit, which greatly increased the reception sensitivity of a vacuum-tube receiver, and in 1918 the superheterodyne circuit, which is basic for modern radio and radar reception. By 1919 it was possible to produce efficient transmitters and receivers for voice broadcasts using the inventions of the previous twenty years.

Until 1919 radio had been developed principally for point-to-point communication, like the telegraph and telephone. Since the early 1920s, however, broadcasting has brought receiving sets into individual homes where they provide a varied and popular form of entertainment. Home radios also bring news of all types, including commercial news such as the day's market prices for the farmer. Radio often serves as a means of rapid communication to members of a community during local disasters. It is used to attempt to mold public opinion, and in countries where broadcasting is a monopoly of the government a considerable degree of control can be maintained over the public by propaganda. During the past three decades radio has become one of the most important channels for communications.

The original broadcasting stations used amplitude modulation (AM) of the radio wave, whereby the frequency of the sound waves varies the amplitude of the high-frequency radio waves which transmit the message for receipt and reconversion to sound. Defective electric motors and other apparatus, electrical storms, and electric discharges may add their own amplitude to the wave, causing interference or what is known as static. The most important development in radio since broadcasting became common has been the introduction of frequency modulation (FM), invented in 1933 by E. H. Armstrong of feed-back and superheterodyne fame. FM has a higher fidelity than AM and is relatively free from interference. In the FM system the variation in the radio waves as produced from the sound waves is a variation in frequency rather than amplitude. Since the signals are not dependent on the amplitude of the radio wave, the outside interference referred to is not experienced in this system. It is the FM system which transmits the signals accompanying television broadcasts in the United States.

Television, a word formed by the combination of Greek and Latin words meaning "sight at a distance," signifies the broadcasting of live scenes or moving pictures by radio waves. Television is a direct development from sound broadcasting and consists of converting light into electrical impulses, transmitting these impulses and reconverting them at the receiving station into their original form, namely, light. The foundations on which electronic television rest are three—radio, the cathode tube already described, and what is known as the photoelectric effect, which Heinrich Hertz first discovered in 1887.

Fundamentally, television is a scanning process at both the sending and receiving stations. At the sending station a beam of electrons from an electron gun in a cathode-ray tube called an Iconoscope or Orthicon scans exceedingly rapidly a mosaic of photoelectric cells on a plate on which a camera lens has focused an image. Each minute photoelectric cell has on it a positive electric charge proportional to the amount of light falling on the cell. The electrons in the scanning beam discharge each of the more than 300,000 cells individually, and the resulting electric impulse proportional to the amount of light falling on the cell is collected from a conductive surface on the back of the mosaic and transmitted. The Orthicon thus converts the optical image into corresponding electrical impulses. The electron beam in the present black-and-white television camera in the United States scans each one of 525 horizontal lines of photoelectric cells on the mosaic thirty times each second. Since each of these lines in the devices now standard in the United States has about 600 photoelectric cells, 315,000 separate impulses are sent out in $\frac{1}{30}$ second.

At the receiving station another cathode tube called a Kinescope is so arranged that the electron beam, the size of a pinhead, impinges upon a fluorescent screen, illuminating a spot in proportion to the energy in the beam at a given instant, and thus converting the electrical images back into light. The transmission of the picture consists of synchronously scanning, or sweeping the electronic beams in the sending and receiving tubes across the plate at the sending station and the screen at the receiving station. The scanning is similar to the technique of reading a printed page, line by line and word for word in each line. The electron beam in the receiving set must, of course, be perfectly synchronized in both vertical and horizontal motion at all times with the beam in the sending camera.

An enormous amount of applied research was necessary to bring television to its present status. In 1862 the Italian Abbé Giovanni Caselli (1815–1891) used a crude mechanical scanning device in his partially successful attempt to transmit images over a telegraph wire, and in 1884 Paul Gottlieb Nipkow (1860–1940), a German engineer, further developed mechanical scanning. His device consisted of a disk perforated with a series of holes in the form of a spiral, each hole being just a little nearer the center of the disk than the one preceding it. In 1907 Boris Rosing of the St. Petersburg Technical Institute using an electronic receiver, a cathode-ray tube having an electron gun at one end and a fluorescent screen at the other, endeavored to develop a mechanical scanning transmitter consisting of mirrors on two drums revolving at right angles to one another. The vertical drum revolving at high speed scanned horizontally and the horizontal drum, revolving much more slowly, shifted the scanning line. Although Rosing's system was capable of transmitting and reproducing poorly defined action images, it was far from having any practical value, mainly because it did not have an amplifier. Herbert E. Ives and Charles F. Jenkins in the United States and John L. Baird in England developed mechanical scanning systems in the late 1920s. Ives used 48 lines and Jenkins between 30 and 60. The Jenkins Television Company failed financially before it began broadcasting, and although the British Broadcasting Company began limited television programs in 1929 using the Baird system, mechanical scanning has not been a commercial success.

In a letter to the magazine *Nature*, in 1908, A. A. Campbell-Swinton (1863–1930) of London University suggested a television system using an electronic scanner as well as an electronic receiver. In 1917, Vladimir Kosma Zworykin (1889–   ), one of Rosing's students, began to develop electronic television for the Russian Wireless Telegraph and Telephone Company. After the Russian revolution, Zworykin emigrated to the United States, where he subsequently invented the Iconoscope. Zworykin also developed the Kinescope based on the principle which Rosing used in 1907. The first demonstration of Zworykin's system was at Rochester, New York, in 1929; the number of scanning lines was 120. Following extensive improvements in the details of Zworykin's electronic system, television broadcasting began in the United States on July 1, 1941. The Second World War prevented the extension of television facilities, but beginning in 1946 they expanded rapidly. Commercial television was in black and white until 1954 when color television, which could also be received in black and white, began to be broadcast.

In addition to broadcasting programs for purposes of education and amusement, there are many special uses of radio and television. Radio beams help pilots to find their airports. One of the applications of radio is known as *radar* (*ra*dio *d*etection *a*nd *r*ange). First patented in Germany in 1904 and developed during the 1930s by the armed services in the United States, England, France, and Germany, radar initially found wide use during the Second World War. With this device it is possible to detect objects and to determine how far away and in what direction they are from the observer. Using very short radio waves, a radar set sends out a pulse which is reflected by objects. The set then picks up the reflected rays, or echo, and the distance to the object is determined by the elapsed time between the sending of the pulse and the reception of the echo measured in hundredths of a microsecond, the speed of the waves being a known quantity. The direction of the object is ascertained by a directional antenna.

A radar device on board a ship may pick up any object, such as another vessel, an iceberg, or an exposed rock, even when visibility is zero, as in fog. Many airports are equipped with radar to facilitate the control of air traffic and to assist in guiding planes safely into the airport during periods of low visibility. Meteorologists also employ it to detect and trace storms. A number of other specialized applications of radio waves have been developed, particularly to assist in navigation of ships and aircraft. Some of the devices have been given names made up as acrostics similar to that designating radar, such as *loran* (*lo*ng-*ra*nge *n*avigation) and *shoran* (*sho*rt-*ra*nge *n*avigation). Some of the schemes make use of the known speed of radio waves emitted from the shore or from a ship or aircraft to shore stations. Others employ waves of differing frequencies broadcast from shore stations and analyzed on the ship or plane. Soundings are taken by a Fathometer, a device reflecting from the ocean bottom radio waves sent out by the ship, as sonar does with sound waves. Not the least of radio aids to navigators is the dissemination of time signals, which permit the navigator to check his chronometer accurately.

An example of the interrelation between pure science and engineering lies in radio astronomy. The astronomers of the seventeenth century were able to use the then recently developed technical knowledge of glass lens grinding and metallurgy for the construction of optical telescopes with which they made important new discoveries. Similarly, astronomers in the 1940s, using the knowledge of radio engineering applied to astronomical radio telescopes, have been able to "observe" astronomical phenomena which could not be detected even with the most powerful optical telescopes, by tuning in ultrashort radio waves emitted by various astronomical materials. The electron microscope is another example of the interrelationship between pure science and engineering. The application of electronics and radio engineering was at first based on purely scientific development. Conversely, during the 1940s, microscopists and astronomers have been able to apply the development of electronic and radio engineering to construct the instruments with which they have made dramatic discoveries in pure science.

Electric Power: Generation and Utilization

An electric motor, unless supplied by an electric battery, is not in the strictest sense of the term a prime mover because it must be supplied with energy from an artificial and not a natural source. Electricity, a form of energy, is either converted mechanically from natural energy by means of a prime mover driving a generator or is chemically produced in a battery. The principal prime movers now producing electricity are water wheels, steam turbines, and internal-combustion engines. Engineers are currently developing atomic energy as another source to replace fuel.

The importance of electricity as a form of power is its economy and its flexibility. It is no longer necessary, as it was a century ago, to locate a factory or an industrial city near a stream where water power is available or on a harbor or river to obtain condensing water. High-voltage transmission lines can now carry power from a generating plant to a community hundreds of miles away. Nor is it any longer necessary to allow many remote natural resources to remain undeveloped for want of power. Electric power is transmitted not only into isolated regions but also from far-off water-power sites to centers of population. Moreover, electricity now supplies most homes, at least in the United States, with power for a multitude of purposes—not only for heating, ventilation, refrigeration, washing, water pumping, cleaning, and illumination, but also for recreation and information in the form of moving pictures, radio, television, and telephones.

The battery, originally developed by Volta in 1800, was the principal source of electricity until the early 1870s. It supplied electricity for such projects as the early telegraph, telephone, electroplating, and railroad signals. Michael Faraday discovered the fundamental principle of the electric generator in 1831 when he found that he could produce an electric current by moving a conductor in a magnetic field. Unlike the direct current supplied from a disk, a rotating wire coil would produce only alternating or fluctuating voltage at its terminals as the wire was moved within the magnetic field. As there seemed to be no way to use alternating current advantageously at that time, scientists and inventors evolved a generator which would produce direct current. In 1832, the year Faraday published his discovery of current produced in a copper disk rotating in a magnetic field, Hippolyte Pixii in Paris constructed a machine for generating an alternating current by rotating a permanent horseshoe magnet in the field of an electromagnet. The next year, at the suggestion of Ampère, he added a single commutator to change the current which alternated in each revolution to direct current flowing always in one direction. The commutator was in contact with brushes which collected the current generated by the rotation of the coil in the magnetic field.

In London, from 1833 to 1835, three men, William Ritchie, who had demonstrated in 1830 Ampère's suggestion for a telegraph, Joseph Saxton, and Edward M. Clarke, constructed generators which had commutation devices. A generation later, in 1863, a young Italian professor at Pisa, Antonio Pacinotti (1841-1912), built a generator (Figure 11.3) which possessed marked improvements over the machines of the 1830s. His major contribution consisted in increasing the number of commutator sections from two bars to many separate bars, each connected to a continuous winding which rotated in the magnetic field. Pacinotti's commutator made it possible to increase the amount of current and therefore the power which could be produced by a generator. One feature which Pacinotti's design had in common with earlier generators was the permanent magnet which produced the field. During the year 1866 no fewer than five different men replaced the permanent magnet by an electromagnet consisting of an iron core surrounded by coils through which electric current passed—a remarkable example of simultaneous invention. In England, Henry Wilde, of Manchester, was apparently the first to invent field excitation, using electromagnets excited by the current generated in the

Figure 11.3
Pacinotti's generator, 1863
(From Urbanitsky,
*Electricity in the Service
of Man*, 1886)

rotating armature. Moses G. Farmer of Salem, Massachusetts, and the Siemens brothers of Berlin, also built self-exciting generators in 1866, as did Cromwell F. Varley, of London, and Charles Wheatstone. It was now possible to produce a practical generator for commercial use.

During the same decade that the generator was first being improved, electric motors were also being developed from Faraday's discoveries. In 1821 Faraday had demonstrated that dynamic, or current, electricity in a conductor in a magnetic field could produce continuous motion, but it was not until 1835 that Francis Watkins of London produced in model form the first motor which would do work, albeit with models. The Watkins model motor consisted of stationary coils surrounding a shaft on which a bar magnet was mounted. The shaft had a set of contact makers which sent a battery current through successive coils, causing the magnet and shaft to spin. In 1837 Thomas Davenport (1802–1851), originally a blacksmith, of Brandon, Vermont, made the first electric motors that performed industrial work. Davenport's first motor had a permanent magnet for the field, but he equipped his subsequent motors with electro-magnets. Davenport used his motors for drilling iron and steel and for turning wood in his small shop. His first motors were rotary, but in 1838 he began to evolve a reciprocating motor based on a design used earlier by Joseph Henry. In 1839 a German-born physicist, Moritz-Hermann de Jacobi (1801–1875) of St. Petersburg, Russia, built and tested an electric boat with a 128-cell battery having platinum and zinc electrodes to supply power for the motor. Jacobi's boat had paddle wheels devised by William Robert Grove (1811–1896) of London. In spite of generous aid from the Russian czar, the tests made it apparent that power supplied from a battery was far too costly even for experiments, and Jacobi discontinued them. Most of the men who built these early motors realized that the expense of operation was prohibitive and that the amount of power available from a battery was extremely limited. The great cost of electricity retarded the evolution of the electric motor. It was not generally understood that motors and generators could be made interchangeable until Pacinotti in 1863 produced a machine that was equally effective for either purpose.

Man has always had need for illumination and has made use of many devices since he first tamed fire. In 1800 the principal sources of illumination were candles, generally of tallow, and oil-burning lamps, sometimes using petroleum but usually using animal or vegetable oils. William Murdock produced the first illuminating gas in the 1790s by heating coal and drawing off its volatile components, which were combustible in air. In 1798 he provided gasworks for lighting the Boulton & Watt engine-shop building at Soho in Birmingham, England. In France, Philippe Lebon (1767–1804) obtained a patent in 1799 on gaslighting, using the volatile gases from heated wood, but Lebon died before he had developed his invention beyond lighting his own house and grounds. Pall Mall in London was illuminated by gas in 1807 and Westminster Bridge in 1813. Paris first had a few gaslights on its streets in 1820. In the United States in 1812 David Melville (1773–1856) lit his house in Newport, Rhode Island, with coal gas which he manufactured on the premises, and he installed gas illumination in a cotton mill in Pawtucket, Rhode Island, in 1813. Baltimore was the first American city to have a gaslighting system; a company organized in 1816 began to furnish gas a few years later. The first community to be lit with natural gas was Fredonia, Chautauqua County, New York, where a $1\frac{1}{2}$-inch pipe was driven into the ground near a gas spring in 1821, and the gas obtained supplied thirty street lights for a number of years.

Thomas Drummond (1797–1840) introduced incandescent gaslighting in 1825 by placing a solid stick of lime in a gas flame. The idea had been known for some time but was first applied practically by Drummond. The brilliant light produced by the white-hot lime became known as calcium light or limelight, a word which, because of the use of limelight to illuminate a star performer on the stage, has also come to mean a conspicuous position in public. The widely used incandescent Welsbach mantle, developed in 1885 by Carl Auer von Welsbach, an Austrian, utilized the same principle of bright incandescence at high temperatures in a gas flame. At the beginning of the twentieth century, gas was still an important source of illumination, although the number of electric-lighting installations was rapidly increasing.

In 1801 Humphry Davy (1778–1829), English chemist, had discovered that he could produce a brilliant spark or arc between two slightly separated carbon rods in a battery circuit. Davy's battery, however, was not powerful enough to produce a stable, continuous arc. It was not until 1863, subsequent to a trial installation of an arc light in the Dungeness lighthouse on the south coast of Kent, England, in 1862, that the first practical application was made of arc lights, when they were put into one of the two La Heve lighthouses near Le Havre, France. Inefficient generators of the Pixii type, which had been designed earlier by Florise Nollet (1794–1853), professor of physics at Brussels, supplied the electricity. Arc lighting was not commercially feasible until more than a decade later when electric power far cheaper than that furnished by batteries or by the early inefficient generators had become available.

The Belgian-born Zénobe Théophile Gramme (1826–1901) did more than any other man to develop the generator and the motor commercially. His first hand-driven experimental generator of 1871 was very similar to Pacinotti's, although Gramme seems not to have known of the Italian professor's work. In 1873 Gramme installed his first machines for arc lighting to replace the older generators in the French lighthouses. At the same time he built several generators to replace batteries then being used for electroplating nickel and silver. Gramme's generators were driven by reciprocating steam engines and were superior to the few other commercial generators of his day. In 1876 Gramme further improved his design; he was now able to produce generators that ran at higher speeds, were lighter, and had much greater power capacity. An extensive line of Gramme motors and generators, based largely on the 1876 design, was still being widely sold in the late 1880s.

While working at the Paris factory that turned out Gramme generators, the Russian-born Paul Jablochkoff (1847–1894) invented in 1876 the Jablochkoff candle, a type of arc light. The "candle" consisted of two carbon rods side by side but insulated from each other by the white clay, kaolin, which vaporized as the rods burned down. Since one rod of a carbon arc using direct current burns more rapidly than the other, Gramme designed an alternating-current generator to be used with the

Jablochkoff candle. The first of these new lights appeared in 1878 on Paris streets. Compared with the existing gas lamps they were so brilliant that the system was adopted by many European cities.

Charles Francis Brush (1849–1929) of Cleveland, Ohio, installed the first Brush arc street-lighting system in Cleveland in 1879 and another in New York in 1880. Brush's system was more satisfactory than Jablochkoff's because the Brush lamp would burn twice as long as Jablochkoff's before it was necessary to replace the carbon rods. Brush also designed a generator wherein the voltage was variable and was controlled by the load while the current remained constant. To regulate the arc in the lamp as the carbons burned down, Brush invented an automatic clutch that kept the burning ends of the rods at a constant distance from each other. Many large cities in the United States and Europe installed the Brush system in the early 1880s. The arc light was popular in street lighting and for large indoor areas until well into the twentieth century.

Thomas Alva Edison (1847–1931) began work on an incandescent lamp in 1877. He was by no means the first to try to develop incandescent lamps or even the first to produce one that would function. In fact, such a lamp had been made as early as 1820, and during the ensuing half century scores of men, including, among others, Joseph Wilson Swan (1828–1914) in England, had fashioned incandescent lamps of many designs. However, it was Edison who first developed an incandescent lamp design suitable for quantity manufacture and use. When Edison began his work enough was known about arc lights to make it obvious that they were too brilliant for use in the home. Edison therefore tried to produce a lamp that would give a softer, less intense light. His first lamps devised in 1878 had a platinum-wire filament in an evacuated glass bulb and operated at 10 volts. These series-connected lamps did not prove to be as reliable as Edison had predicted. Moreover, he came to realize that although series operation was satisfactory for street lighting, parallel circuits with each lamp controlled separately (Figure 11.4) would be much more desirable for general use. In an electric series circuit the whole current flows in sequence through each piece of apparatus, whether a lamp, motor, or other device. In a parallel circuit the current divides, so that only part of it flows through each device. To reduce transmission losses he decided to use 110 volts rather than lower voltages—a decision more far-reaching than he could have known. To obtain still further the advantages of higher voltages in distribution without increasing the voltages impressed upon individual lamps, Edison developed a three-wire sys-

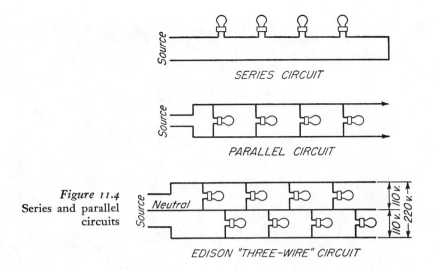

SERIES CIRCUIT

PARALLEL CIRCUIT

Figure 11.4
Series and parallel
circuits

EDISON "THREE-WIRE" CIRCUIT

tem, where, with distribution at 220 volts, each lamp uses only 110, any unbalanced current returning through a "neutral" wire.

The adoption of the higher voltage for the lamps presented new problems in using metal filaments; Edison therefore began to try out various other types of high-resistance filaments. In his first successful high-resistance lamp he used a filament of carbonized thread in a vacuum, but it burned for only two days. After a wide search for a material which, when carbonized in the absence of oxygen, would provide a long-lasting filament, Edison finally chose split bamboo. In 1880 he arranged a public exhibition of 500 of his lamps in and around his laboratories at Menlo Park, New Jersey. This demonstration attracted so much attention that the Pennsylvania Railroad had to run special trains to carry the crowds of visitors to the laboratories. The first application of the Edison-lamp system was in 1879 aboard the DeLong arctic-expedition steamer *Jeannette*, which was equipped with a generator and electric lights that served dependably for two years until the vessel was crushed in the ice. Edison furnished another steamship, the *Columbia*, with electric lights in 1880; each stateroom had a 5-candlepower lamp. Built at Chester, Pennsylvania, for the Oregon Railway and Navigation Company, the *Columbia* operated between San Francisco and Portland, Oregon. Her 1880 power plant was dismantled in 1895. By 1882 Edison had installed over 150 plants in individual residences, hotels, mills, offices, stores, and steamers.

In the meantime the newly formed California Electric Light Company had opened in September, 1879, a little experimental power plant consist-

Figure 11.5 Generator room of Pearl Street Station, New York, first Edison electric lighting central station (From *Scientific American*, Aug. 26, 1882)

ing of three Brush generators, to sell electricity for arc-lamp illumination to customers in San Francisco. Within a year the increasing demand for electricity not only proved the value of this experimental plant but made necessary the construction of a larger plant. These two plants were the first "central stations" for supplying electricity commercially. The development of central stations and subsequently of interconnecting high-voltage transmission networks are among the most significant aspects of electric power for modern society. Although an individual steam-driven generator in a store, office building, or residence was practicable, there were obvious difficulties and limitations in operating a steam boiler and generator in the cellar. The central-station generation of electric power and its distribution by wires was much more satisfactory, as well as more economical, and made electricity widely available.

After the San Francisco installations, the next central station of importance was built on Holborn Viaduct, London. This station began operations on January 12, 1882, and furnished power for 3,000 incandescent lamps from Edison Jumbo direct-current dynamos driven by Armington and Sims steam engines. However, the British Electric Lighting Act of the same year, aimed at preserving the gas-lighting monopoly, forbade the construction of large generator stations and seriously im-

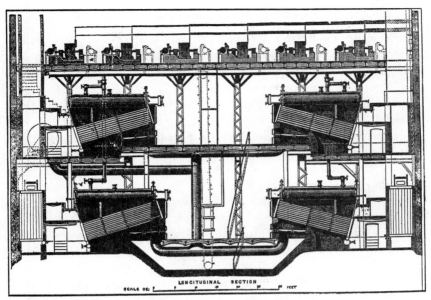

Figure 11.6 Boilers, engines, and generators in Pearl Street Station (From *Electrician*, 1882)

peded the British development of electric illumination. Edison's much-publicized Pearl Street Station in New York went into operation on September 4, 1882. It had six "large" direct-current generators (Figure 11.5) aggregating about 900 horsepower, enough power for 7,200 lamps at 110 volts. Porter and Allen reciprocating engines supplied with steam by Babcock and Wilcox boilers (Figure 11.6) drove the generators. When opened, the Pearl Street Station served some sixty customers with nearly 1,300 lamps through underground conduits distributing power at 110 volts nominal pressure. The conductors were half-round copper bars about 20 feet long and inserted into iron tubes from which they were separated by cardboard washers. The tubes were then filled with asphaltum compound for insulation. The station proved to be a success, and central stations were soon installed in other sections of New York and in other cities. The original Pearl Street Station burned on January 2, 1890.

Although direct-current central stations greatly increased the availability of electric power and although they had the advantage that storage batteries could be used for stand-by emergency service, they had serious limitations because of the necessarily low distribution voltages. These low voltages seriously limit the distance over which direct current can be economically transmitted. A volt is a measure of the electric pressure in a conductor; it is analogous to the hydraulic pressure in a water pipe. The current, measured in amperes, is the quantity of electricity flowing in the conductor and is analogous to the quantity of water flowing in a pipe. Electric power is measured in watts and is a product of volts times amperes. This statement is true in direct-current systems, but in the case of alternating-current systems it must be qualified, since the current and voltage alternations may not be entirely in step with each other, in which case a factor must be employed in calculating power to indicate how far apart they are in the cycle.

Losses in the transmission or distribution of power are in general proportional to the square of the number of amperes flowing. Other things being equal, if the transmission voltage is doubled and the current halved, the power is unchanged, but the transmission losses are reduced to one-quarter. Similarly, all other things being equal, if the voltage is increased tenfold with a corresponding decrease in current, the losses are reduced to one one-hundredth in transmitting a given amount of energy. There are other important factors affecting losses, but it is apparent that the higher the voltage or pressure, the greater the distance over which electricity can be economically sent. In an alternating-current system, transformers step up the voltages at the power plant for long-distance transmission. Alternating current of 220,000 volts is common in long-distance transmission today, and voltages of more than 300,000 are occasionally used. Direct current is not adaptable for the production of widely diverse voltage requirements, but in alternating-current systems the transformer may be designed to produce any desired voltage.

Alternating-current systems were being developed in Europe in the 1880s; one of the most successful was that of Lucien Gaulard and J. D. Gibbs of Paris, who first demonstrated their system in London in 1881. In 1885 George Westinghouse (1846–1914) acquired the American patent rights of the Gaulard-Gibbs system and immediately directed his small electrical engineering staff to develop improvements in the generators and particularly in the transformers. Appointed chief engineer of the Westinghouse Company in 1885, William Stanley (1858–1916) devised an efficient alternating-current distribution system which he installed at his own expense in Great Barrington, Massachusetts, to serve about 150 incandescent lamps for lighting streets and stores. The Westinghouse Company installed a similar plant for commercial operation in Buffalo, New York, in November, 1886. During the next year the Westinghouse alternating-current system became a strong competitor of the Edison direct-current system, which already served many localities.

Recognizing that they had serious competition, the operators of direct-current power systems, who had held a virtual monopoly, became much concerned. In the middle of 1888 they tried to discredit alternating-current power projects by attacks which, among other things, purported to show that alternating current was "a horrible menace to human life." [8] These attacks became very bitter and for a time the two systems developed independently of one another. The direct-current system generally served urban districts, where power demands were concentrated in limited areas, and the alternating-current system supplied outlying territories, where transmission lines were necessarily more extended. Before very long, however, the advantage of combining the two systems to obtain the advantages of each became evident. Direct-current systems then were designed to utilize alternating current for transmission from the power plants to outlying substations which converted it to direct current for local distribution.

The early Westinghouse alternating-current installations produced single-phase 133⅓ cycles per second power at transmission voltages nominally of either 1,000 or 2,000 volts. These early installations provided electricity for illumination only, and it was not until a year and a half after the first Westinghouse installation at Buffalo in 1886 that a reliable alternating-current motor was invented. A number of individuals working independently of each other in the United States and in Europe in the late 1880s developed polyphase-current generators and motors. The Croat-born Nikola Tesla (1856–1943) was one of these inventors, and it was he who was awarded patent rights, after much litigation, for his invention of the polyphase-current system to run an induction motor. Polyphase, as distinguished from single-phase, alternating current comes from coils in the generator, wound to produce two or more separate circuits delivering current to the terminals in such a way that the output of the machine is in two, three, or more circuits in which the alternations are in sequence as the machine rotates. During the next few decades there were innumerable improvements in motors and a tremendous diversification of designs to run almost anything from a toy train to a ship.

However, because the early Westinghouse 133⅓-cycle alternators, as an alternating-current generator is often called, produced power at a

8 "Gibbens Gets Frightened; Experiments on a Dog Prove That There's Death in the Wires," *New York Herald*, July 31, 1888, p. 10.

frequency too high for efficient motor operation, 25- and 60-cycle alternators gradually became standard in the United States. At 60 cycles the human eye does not detect flicker in an incandescent lamp, while at 25 cycles a flicker is quite noticeable, especially in small lamps. So for a time 60 cycles was used when lighting loads predominated, and 25 cycles when power for motors was more important. In countries other than the United States the standard lighting frequency became 50 cycles and power 16⅔ cycles. The large Westinghouse alternators at the Columbian Exposition at Chicago in 1893 were 60-cycle machines. The voltages of alternating- and direct-current were also becoming standardized in the 1890s, and by the end of the century 110 volts was the common American voltage for lighting circuits.

After the development of dependable alternating-current transmission and motors, there remained the problem of converting alternating current to direct current for railroad service and for electroplating and other industries which can use only direct current. About 1892 electrical engineers invented the synchronous rotary converter consisting of a rotating unit receiving at one end alternating current and at the other delivering direct current. The rotary converter was extremely useful in various fields where the alternating-current transmission at high voltages and direct-current utilization were either desirable or necessary. Subsequently, in 1902, the American Peter Cooper Hewitt (1861–1921) originated the static rectifier consisting of an evacuated tank containing mercury vapor, using the principle that current will flow in mercury vapor from metal anodes to a mercury cathode in the bottom of the tank but will not flow in the opposite direction. It thus converts or rectifies alternating current to direct current by suppressing the flow of current in one direction and thus acts as a sort of check valve.

During the period when alternating-current motors and converters were being perfected, the carbon filament incandescent lamp was being steadily improved. This continuing improvement culminated in the introduction of a metallized carbon filament lamp, the GEM (General Electric Metallized) lamp in 1905. The GEM lamp produced about 4.25 lumens per watt compared with the 1.68 lumens per watt of the 1881 and 3.4 lumens per watt of the 1905 carbon filament lamps. A lumen is a unit of light flow, or flux, and is a measure of luminous output of a lamp related to a standard candle source. The GEM lamps were manufactured until 1918 when they were completely superseded by tungsten lamps. The tungsten filament lamp giving 8 lumens per watt, invented by Alexander

Just and Franz Hanaman of Vienna in 1902, was introduced for street lighting in 1907; an improved tungsten street lamp with a drawn-wire filament first appeared in 1911. Both of these early tungsten lamps were vacuum bulbs. Large tungsten lamps filled with nitrogen were first produced in 1913, but it was not until after 1918 that gas-filled lamps became generally available for domestic use. The modern general-service lamp filled with 85 per cent argon and 15 per cent nitrogen produces as many as 22 lumens per watt.

During the twentieth century many new types of lamps, such as the sodium arc and mercury arc, have been introduced for specialized uses. The fluorescent lamp was first available in Germany in the early 1930s and in the United States in 1938. It consists of a glass tube, the inside of which is coated with a fluorescent material, the light being emitted when this material is excited by the ultraviolet rays of an arc passing from the electrode at one end to that at the other. The fluorescent lamp is the most important recent introduction and is another step forward in increased efficiency; it produces 65 lumens per watt, many times the 1.68 lumens per watt of the early carbon filament lamps.

Power Plants

The water wheel and the steam turbine are now the two principal stationary prime movers that generate electric power. In the United States in 1954 hydroelectric production accounted for 23 per cent and steam for 75 per cent of electricity produced for general distribution. Internal-combustion stationary engines generated but 2 per cent; the internal-combustion engine, discussed in the next chapter, is achieving increasing importance for generating electricity in locomotives for transportation.

The Westinghouse Company installed the first hydroelectric power-transmission system in the United States in 1891, six years after the Great Barrington project, to carry power generated at Willamette Falls, Oregon, 13 miles to Portland, where transformers stepped down the 3,300 transmission voltage to 1,100 volts for the city primary distribution circuits. Local transformers converted the current to 50 or 100 volts for lamps; lamp voltages were not standard at that time. The most important early long-distance alternating-current transmission line was also constructed in 1891 from Lauffen to the Frankfurt Electro-Technical Exposition in Germany. The 100-horsepower transmission at 30,000 volts for 109 miles was a convincing demonstration. There had been an earlier experimental transmission line between Miesbach and Munich in 1882, but it had failed because of serious insulation difficulties.

Had it not been for the introduction of an alternating-current system, the power of the Willamette Falls could not have been used to supply Portland with electricity. Direct current could not economically have been sent the 13 miles at either 50 or 100 volts. In fact it was the development of alternating-current systems that made possible the exploitation of sources of water power for the generation of electricity. The first large hydroelectric installation in the United States was in 1895 at Niagara Falls, where three 5,000-horsepower machines with terminal voltages of 2,200 were set up. This plant produced a total of 15,000 horsepower, which compares, for example, with upward of 30,000 mechanical horsepower that the numerous water wheels at Holyoke, Massachusetts, developed. Unlike most generators, the Niagara alternators had fields which rotated outside stationary armatures. The earlier Westinghouse generators had been of the revolving-armature type, but European designers who participated in the planning at Niagara Falls favored the revolving-field design. Subsequent to the Niagara Falls installation, the revolving-field type became standard for alternators, but the field revolved within the stationary armature instead of outside it. All that is necessary is that the magnetic field and the armature coils move relatively to one another; their respective positions are a matter of design and not of principle.

The fields of the Niagara alternators rotated at 250 revolutions per minute. They were mounted on vertical shafts and were directly connected with the water turbines. Since an alternator must rotate at a constant speed to produce a given standard frequency, a governor keeps the water turbine at the correct speed regardless of the head of water impressed on it or the load imposed. Most hydroelectric plants have the generator directly connected with the turbine, and much progress has been made in designing turbines to operate with maximum efficiency at varying loads.

During the second decade of this century, several power companies, striving to increase efficiency and to extend their services, interconnected their isolated power plants by means of transmission lines. These interconnecting lines, now called a network, or grid, permit the operation of the more efficient individual plants to supply the bulk of the load, and thus obtain maximum economy of the system as a whole, and also enable maintaining service in the event of any plant failure. One immediate advantage deriving from these interconnections was the profitable utilization of

Figure 11.7
230,000-volt transmission
lines (Courtesy Public
Service Company of
Indiana, Inc.

many water-power sites. A given water-power plant may have little value
by itself because of low water flow in times of drought. If, however, it
is feeding a system containing steam plants which act as stand-by capacity
in times of plentiful water flow, it may be of great value. With the exploitation of many such marginal water-power sites, systems made up of
hydroelectric and steam plants came into being in the 1920s in various
parts of the world. Modern networks interconnect the grids of individual
systems to obtain these benefits over a wide area.

The improvements in transmission and higher-capacity generators
necessitated larger transformers, and the capacity of transformers has in-

Figure 11.8 Large reciprocating steam engine and generator (Courtesy McGraw-Hill Book Company, Inc.)

creased steadily. Engineers also have improved the design of switches and circuit breakers to handle large amounts of power. From 1906 to 1911 transmission voltages mounted from 13,000 to 150,000. The first 220,000-volt line went into operation in 1921, and there are scores of these high-voltage lines today (Figure 11.7), with some transmissions higher than 300,000 volts. The improvements in transmission made possible the great development of private as well as Federal hydroelectric power in the United States, which began in the 1920s. Not only do government plants generate nearly 50 per cent of the hydroelectric power produced in the United States, but they also are the largest. However, the Kitimat project in British Columbia, with its head of 2,580 feet and capacity of 1,650,000 horsepower, is a private undertaking to produce power for the recovery of aluminum from bauxite. Kitimat, when it is completed, will exceed in size the government-built Grand Coulee plant in the state of Washington.

Until 1900 steam engines generating electricity were of the reciprocating type (Figure 11.8). The largest reciprocating engines driving generators were the 7,500-horsepower engines in service in 1904 in various plants, including New York's elevated railway power plant. These engines operated at 75 revolutions per minute with 175 pounds pressure. Today the steam-turbine-driven generator makes practically all the electricity produced from fuel, although internal-combustion engines and a few gas turbines generate a very small percentage of the total electric power produced.

We have indicated in previous chapters that the principles of the impulse-reaction-type steam turbines have been long known; Hero of Alexandria described a model in which steam, in a metal sphere that could rotate, escaped tangentially from two nozzles at the ends of rotating arms (Figure 3.5). In 1629 Giovanni Branca wrote about a machine in which a jet impinged upon blades projecting from a rotating wheel (Figure 6.10). The first of these turbines was a *reaction* turbine and the second an *impulse* machine. Several inventors attempted to make steam turbines in the eighteenth and nineteenth centuries but did not achieve any more practical success than Hero did with his toy. For instance, in 1784 Wolfgang von Kempelen patented "a reaction machine set in motion by Fire, Air, Water or any other Fluid." Primarily intended to be driven by "boiling waters or rather the vapour proceeding therefrom," this steam turbine disturbed James Watt because it was a possible competitor of his engine. He said little about what he called "Kempelen's engine" on the theory that "lest by talking about it we put him on improvements; for it is capable of them." [9] Among others who worked on the steam turbine was Trevithick who built a "whirling engine" in 1815 with a wheel 15 feet in diameter revolving at 300 revolutions per minute. In the United States, William Avery patented a turbine in 1831, and abroad Binstall received a patent in 1838, Pilbrow in 1843, Wilson in 1845, and Hartham in 1858. All of these patented turbines had one common feature; they would not produce power efficiently. Nevertheless, as Charles A. Parsons once said, most of the fundamental ideas of the steam turbine of today had been suggested or extensively described in the hundred or more patents granted prior to 1880.

In 1889 the Swedish engineer Carl Gustaf Patrik de Laval (1845–1913) built his first functional steam turbine. De Laval's turbines were the single-wheel, single-stage, impulse type with a jet of steam impinging directly on the wheel blades (Figure 11.9); the largest wheel diameter he used was 30 inches and the smallest was 3 inches. Since his turbines

---

[9] Henry W. Dickinson, *A Short History of the Steam Engine,* The University Press, Cambridge, 1939, p. 187.

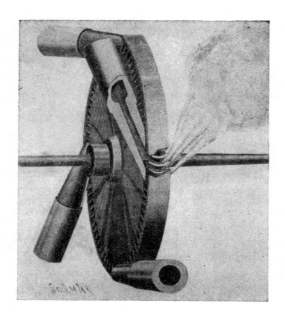

Figure 11.9
De Laval steam turbine
(From E. W. Byrn, *The Progress of Invention in the Nineteenth Century,* 1900; courtesy *Scientific American*)

had only one wheel and one set of nozzles, in which the steam attained high velocity before striking the blades, the speed of the blades at the rim of the wheel was necessarily exceedingly high. Some of his machines ran at speeds as high as 40,000 revolutions per minute, and to eliminate the dangers of "wobbling" at such extreme velocities, De Laval supported the wheels upon flexible shafts to allow the wheel to seek its own center of rotation like a top. He also invented an elaborate helical reducing gear for the operation of an electric generator which could not be run at such high speeds. These gears were often considerably larger than the turbine. De Laval had constructed his first turbine in 1882, and in 1888 he fashioned the flaring nozzles in which steam expands efficiently and attains high velocity before impinging on the blades. By 1897 he was using steam at the high pressure of about 3,000 pounds per square inch, nearly the critical pressure (3,226 pounds) at which a pound of steam has the same volume as a pound of water and the latent heat of evaporation is zero. Although the De Laval turbines were limited in capacity, a considerable number were placed in commercial service.

To obtain efficient operation with reasonably high steam pressures, engineers have modified turbines to permit the steam to expand in several stages, consisting of a number of rows of moving blades mounted on the rotor and interspaced with stationary blades mounted on the stator. If

Figure 11.10 Charles Parsons' first turbogenerator (Courtesy McGraw-Hill Book Company, Inc.)

the expansion of the steam takes place in the nozzles or stationary blades, the action is mainly an impulse one, but if the expansion takes place within the moving blades, the motion is obtained by reaction caused by the steam as it leaves the blades. Modern turbine development has been along these two general lines. Among the many who have been active in the evolution of the multiple-stage steam turbine, two engineers have been most prominent; the first was Charles Algernon Parsons (1854–1931) of England and the second Charles Gordon Curtis (1860–1953) of the United States. Parsons built his first machine (Figure 11.10) in 1884. The unit drove a generator which produced 7.5 kilowatts (about 10 horsepower) at 100 volts. The steam consumption was 130 pounds per kilowatt-hour as compared with perhaps 8 pounds in a modern plant. Development of the Parsons turbine was steady in spite of patent controversies. In 1888 a 32-horsepower unit was installed at the U.S. Naval Proving Grounds at Newport, Rhode Island, and in 1901 a 2,000-horsepower unit was put in operation at Hartford, Connecticut. Curtis developed the multistage impulse turbine. He took out his first patent in England in 1895 and in the United States in 1896. After early discouragements, he built a machine in 1900 with a vertical shaft for the Schenectady, New York, plant of the General Electric Company, and in 1903 a 6,500-horsepower unit also with a vertical shaft, in Chicago. Subsequently the vertical-shaft design was discontinued as capacities, and therefore weights, increased.

Many engineers have participated in the design of impulse-reaction turbines which employ the advantages of both principles. With machines

Figure 11.11 Turbogenerator, Kearny, New Jersey, plant of Public Service and Gas Company. Output 145,000 kilowatts (194,000 horsepower), 20,000 volts. Steam pressure 2,350 pounds per square inch. (Courtesy Public Service Electric and Gas Co.)

being built to use ever-increasing steam pressures and superheat, the horsepower ratings of turbines have skyrocketed. Within twenty-five years after the installation of the last 7,500-horsepower stationary reciprocating engines, there were steam turbines operating at 240,000 horsepower. Turbines are now being installed up to 330,000 horsepower. The development of generators has, of course, kept abreast with that of turbines (Figure 11.11).

The large modern turbine requires a great deal of steam at high pressure and high temperature. The improvements which have led to the boiler of today have been as dramatic as the advances in turbines. Coal, very finely pulverized, natural gas, and residual oil have replaced both raw coal and hand- or mechanical-stoker firing. By 1926 boilers were operating at 650 pounds pressures and at temperatures of 725°F; at mid-century, pressures of 2,000 pounds and temperatures of 1000°F were not uncommon. Modern boilers (Figure 11.12) convert immense amounts of water into steam, and there are in operation single boilers which transform 1,250,000 pounds of water per hour into steam. After doing its work in the turbine, the steam is condensed by passing it through large condensers using cooling water, and the resulting condensate is reused as

Figure 11.12 Modern boiler plant built in the open air. (Courtesy McGraw-Hill Book Company, Inc.)

feedwater in the boiler. To handle such large amounts of steam, many a condenser has more than an acre of cooling surfaces.

Although 70 per cent loss of the heat energy of fuel now common in steam plants is high, it is not nearly as high as the loss in the early plants. Edison's Pearl Street Station in 1882 is said to have used as much as 8 pounds of coal to produce 1 horsepower-hour, or about 10 pounds per kilowatt-hour. By 1900 about 7 pounds were required to generate a kilowatt-hour with the figure dropping to somewhat more than 2 pounds in 1922. During the following thirty years these needs were cut to less than half, and now a kilowatt-hour can be generated for less than the equivalent of a pound of coal (Figure 11.13). Oil and natural gas have become important fuels for steam plants. Of the fuels used in steam plants in 1922, oil and natural gas produced slightly more than 15 per cent of the power generated, but in 1953 this figure was 35 per cent. The total United States electrical production in 1953 was 515 billion kilowatt-hours, or 690 billion horsepower-hours, more than 70 times that produced fifty years earlier (Figure 11.14). In 1953 the United States produced 41 per cent of the total world production and more than four times the amount generated by Russia, the world's second largest producer.[10]

[10] Many of the data in this and other paragraphs have been taken from the *Statistical Bulletins* of the Edison Electrical Institute.

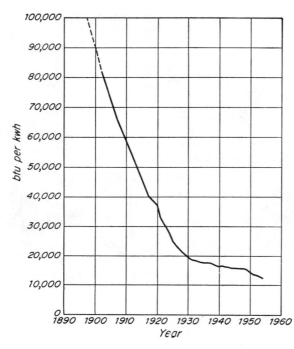

Figure 11.13
British thermal units required to produce 1 kilowatt-hour in United States (Based on data in National Electric Light Association *Bulletin*, 1931, and Edison Electric Institute *Statistical Bulletin*, 1953)

There is room for improvement; maximum efficiencies of 30 or 40 per cent in large steam plants provide opportunities for betterment. This relatively large waste of 50 to 70 per cent of the heat value of fuel is largely due to the roundabout methods now employed for converting the chemical energy stored in the fuel millions of years ago into electrical energy. First the chemical energy is changed into heat, which is converted into mechanical energy, which in turn is converted into electrical energy. It seems inevitable that ultimately a short cut, bypassing this series of conversions, will be found practicable and that electric energy will be produced directly from fuel. Recently, scientists have produced electric power directly from the sun's rays, generating about 100 watts per square yard of surface of their "battery" exposed to the sun. This direct generation of electricity may be the beginning of a revolutionary step in the electric-power field. On the other hand, perhaps the most important new development in stationary prime movers will be the adaptation of atomic energy to the production of electric power.

The history of electrical engineering presented in this chapter has emphasized the developments in the United States. Similar developments have occurred simultaneously in Europe, but with few exceptions it was

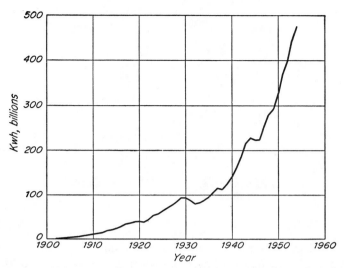

Figure 11.14 Kilowatt-hours of electric energy produced by United States utility companies (Based on data in *Electrical World,* Supplement, Jan. 19, 1946, and Edison Electric Institute *Statistical Bulletin,* 1944, 1953)

Europeans who made the initial, fundamental scientific discoveries and early applications on which electrical engineering is based. This statement is true for the telephone, radio, television, and for the generation and transmission of power. As this chapter demonstrates, American genius has not been as active in fundamental research as in engineering development, engineering production, and engineering service. It has been this type of American genius which has evolved the world's greatest electrical industry.

Bibliography

Cohen, I. Bernard (ed.): *Benjamin Franklin's Experiments . . . With a Critical and Historical Introduction,* Harvard University Press, Cambridge, Mass., 1941.

Howell, John W., and Henry Schroeder: *History of the Incandescent Lamp,* The Maqua Co., Schenectady, N.Y., 1927.

MacLaren, Malcolm: *The Rise of the Electrical Industry during the Nineteenth Century,* Princeton University Press, Princeton, N.J., 1943.

Maclaurin, William R.: *Invention and Innovation in the Radio Industry,* The Macmillan Company, New York, 1949.

Passer, Harold C.: *The Electrical Manufacturers, 1875–1900,* Harvard University Press, Cambridge, Mass., 1953.

Thompson, Robert L.: *Wiring a Continent,* Princeton University Press, Princeton, N.J., 1947.

TWELVE

Modern Transportation

Like the development of electrical communication and power, the emergence of effective transportation during the last half of the nineteenth and first half of the twentieth centuries has been a large factor in the political, economic, and social revolution which has produced our dynamic society. Improved transportation continues to be a factor changing the ways of human life. Cheaper and speedier transportation increases the political unity of a nation, the centralization of political control. It promotes culture by disseminating information, extends great metropolitan areas, facilitates the new organization of industry, effects an ever wider distribution of goods, an increase in wealth, and higher standards of living. Without great transportation and communication systems, the United States would not be the integrated country it is. The National Road described in Chapter 8 was built in recognition of this thesis, to increase the unity and stability of the Union by connecting the East and the West.

The effect of transportation on social conditions is as extensive as its effect on politics. Transportation has exerted a tremendous influence on the spread of populations as well as on the concentration of population in great cities. The shift of population to metropolitan areas since the middle of the nineteenth century is one of the most significant social phenomena; without efficient transportation it would be impossible for huge modern metropolitan areas like London and New York to exist with the present standards of living. Widely available transportation, particularly in the form of the automobile, has changed the whole tempo of our life and has greatly increased the mobility of populations. Transportation has also enormously increased the diffusion of knowledge through the distribution of printed matter, has made remote centers of education accessible, and has reduced the obstacles to many types of scientific investigations.

The economic effects of improved transportation are perhaps more far-reaching than its social effects. Effective transportation brings to the members of a community many benefits it would be impossible to obtain locally, and it tends to reduce the cost of commodities to the consumer because the cost of transportation of raw materials and finished goods is part of the ultimate cost. Transportation also affects rents and land values; improving transportation can increase land values in one area and cause a decrease in another. In the nineteenth century, for instance, the agricultural land values in the United States increased along the railroads as they spread west. Another example of increased land values is in suburban areas, where the increase was brought about by the private automobile and the public trolley car and bus as well as by railroad service.

There are five major types of transportation: railroad, water, highway, airway, and pipeline, disregarding electric-power transmission which in some instances could be considered an important method of transporting fuel in the form of energy. Table 1 gives the distribution of intercity traffic among these five transport agencies in the United States in 1953.

Table 1 Distribution of Traffic among Transport Agencies in 1953 in the United States

Agency	Millions of freight ton-miles	Per cent	Millions of passenger miles	Per cent
Railroads	615,000	52.6	32,450	46.8
Waterways *	185,000	15.8	1,500	2.2
Pipelines	170,000	14.5		
Highways †	200,000	17.1	20,500	29.6
Airways	450	....	14,800	21.4
		100.0		100.0

* Waterways include inland waterways and the Great Lakes.
† Do not include private passenger automobiles.
Source: Data from *Yearbook of Railroad Information*, 1954 ed. Compiled by Eastern Railroad Presidents' Conference.

Within most highly developed countries, railroads transport the largest amount of freight; in 1953 United States railroads carried a total of 2.9 billion tons of freight, whereas the total water-borne commerce, including foreign imports and exports, was 169 million tons. Railroad tonnage was thus more than three times water tonnage.

Railroads

Chapter 9 relates how modern steam railroads began about 1830 in England and the United States and a bit later on the continent of Europe. By 1860 railroads were operating in most of the major territories of the world except Africa, China, and Japan, and by 1883 railroads had appeared even in these areas. In the United States, railroads pushed rapidly westward after the middle of the nineteenth century. The Baltimore and Ohio Railroad reached its western terminus at Wheeling, Virginia, in 1853. The Chicago and Rock Island reached the Mississippi during the year following, but soon aspired to cross into Iowa and even to extend into the rapidly developing prairie territory to the west. Beyond the Great Plains, pioneer engineers were already searching for passes in the almost impenetrable Rockies through which it might be possible to build railroads. It was early realized, and in legislation stipulated by the government in connection with its 1862 land grant to the railroads, that the curves on such a railroad should have radii of at least 400 feet and gradients no steeper than the 2.2 per cent (116 feet to the mile) that had been found feasible on the Baltimore and Ohio. Merely to keep a train moving at a steady or uniform speed on this gradient requires 6½ times the drawbar pull of about 8 pounds per ton that is necessary on straight, level track. It was obvious from the first that locomotives of considerably greater power than any previously built would have to be designed, and that in mountainous country where there were heavy grades, even they would often have to be assisted by added power units either ahead of or behind the train.

The first of the thirteen lines in the United States and Canada to cross the Rocky Mountains was the Union Pacific, which met the Central Pacific coming east at Promontory Point, Utah, in 1869, celebrated by the driving of the Golden Spike. These two roads formed a through route from Omaha, Nebraska, on the Missouri River to Pacific tidewater near Sacramento, California. A dozen other mountain lines crossed the Rockies in the period between 1869 and 1914. In general, world construction increased rapidly from 1860 until the first decade of the twentieth century, but before 1915 construction began to slow down, and much track has since been abandoned, especially in the United States. Competition from other types of transportation has caused this reversal, which has resulted in a net decrease of track mileage after the 1920s. However, there have been important engineering developments since 1860 in all branches of railroad design and construction.

From George Stephenson's 7¼-ton *Rocket* of 1829 to the most recent steam reciprocating-type locomotives of nearly 300 tons there were no revolutionary changes in fundamental design, except for various experimental models which were not commercially successful. The principal development in steam locomotives was a steady increase in efficiency, capacity, and weight on driving wheels. Improvements in efficiency were achieved as in stationary plants by such innovations as superheated steam, mechanical stokers, feedwater heating, etc. The height and width limitations of a locomotive have seriously restricted its efficiency; the maximum efficiencies normally attained in steam locomotives are only about 8 per cent. The efficiency of a thermodynamic cycle is measured by the difference in temperature of the medium employed before and after its use. It is thus desirable to obtain the highest possible initial temperature and the lowest final temperature. Since the condensing of steam is not practicable in a locomotive, the exhaust directly into the atmosphere is at a relatively high temperature, and since boiler pressures and temperatures are necessarily limited, the thermodynamic efficiency is low.

Henry R. Campbell of Philadelphia had invented coupled drivers in 1836. The locomotive *Blackhawk* of that year had a four-wheel leading truck and two pairs of driving wheels, the drivers on each side being coupled to each other by side rods so that they moved together at all times. Locomotives with this wheel arrangement are known as the American type (Figure 12.1). Two years later Joseph Harrison, Jr., connected the two driving-wheel axles on each side with equalizing levers to balance the weight on the axles; he also suspended the weight of the body of the locomotive on three points, the center of the leading truck and the centers of each equalizing lever. This arrangement, which gave stability especially on rough track, has remained fundamental in locomotive design. By 1860 coal was replacing wood as fuel (Figure 12.2), the number of driving wheels had been increased to six in some of the larger locomotives, and the injector for forcing feed water into the boiler had been invented to replace the pumps used up to that time. Henri Giffard, who invented the injector, was also active in building and flying power dirigible balloons.

In designating wheel arrangement of steam locomotives a numerical

Figure 12.1 Early locomotive of American type, New London and Northern Railroad, 1843 (Courtesy New London Historical Society)

formula has been adopted known as the Whyte system. The first digit in the group indicates the number of leading wheels, the second the number of driving wheels, and the third the number of trailing wheels. If there are two sets of driving wheels on articulated trucks, still another digit is added. Beginning in the 1860s there was a rapid development leading to different types of locomotives. The 2-8-0, known as the Consolidation, with eight driving wheels, was introduced during this decade. Engines of the 2-10-0 type were also built at this time, but the 2-8-0 was the most popular heavy American locomotive until the end of the century. The 1860s also saw the introduction of steel into locomotive construction, and in the 1890s the wider, longer, and more efficient firebox burning low-grade coal came into general use. The larger firebox necessitated the introduction of trailing wheels of small diameter which could be placed under the firebox so that it could be built out over the wheels to the full width of the locomotive rather than confined in the space between the wheels.

Campbell's coupled drivers and many other improvements were origin-ally made in the United States, where demands for increased power

Figure 12.2 Coal-burning passenger locomotive, 1864

were greater than in most other parts of the world. In 1876, however, Jules-Theodore-Anatole Mallet (1837–1919), originally of Switzerland, invented the compound locomotive which made the steam work twice, as had already been done in stationary and marine engines. Mallet led the steam first into a high-pressure cylinder, from which it passed after partial expansion into a larger low-pressure cylinder, after which it was exhausted to the air. Although the principle of compounding was first used in the United States in 1889, it was never widely used.

By the beginning of the twentieth century there were three principal types of steam locomotives: the switcher or shunting locomotive, the passenger road locomotive, and the heavy-duty freight road locomotive. In 1900 designers in the United States were concentrating on freight-service engines with high tractive effort rather than high speed, and on high-speed passenger locomotives for relatively light trains. The maximum efficiency obtainable from these locomotives under favorable conditions was on the order of only about 3 per cent, and there was a general realization that efficiency and power output would have to be increased to meet the heavy traffic demands and competition from electric locomotives. It had been an accepted philosophy that the steam locomotive should be as rugged and simple as possible because of the exacting demands of road service and that refinements were of secondary importance.

In 1900, Wilhelm Schmidt of Germany had invented a device for

Figure 12.3 First American electric freight locomotive, Ansonia, Connecticut (Courtesy Charles Rufus Harte)

superheating steam, which was of increasing importance in stationary power production. Schmidt led the "saturated" steam through small tubes inserted inside the boiler flues and thus converted the saturated steam into superheated steam. Maximum steam pressure on locomotives in 1900 was approximately 200 pounds. During the twentieth century these pressures have been increased, but they are even now not generally higher than 250 pounds and seldom as high as 300 pounds. These pressures are much lower than those in stationary power plants and are necessarily fixed by the limitations of the locomotive boiler. The temperature of saturated steam is increased as the pressure rises; increased pressures thus mean higher steam temperatures, and superheating raises temperature still more. The major limitation to the power output of coal-burning locomotives with firing by hand is the amount of coal the fireman can shovel. The constant opening of the firebox door necessary in hand firing means the admitting of streams of cold air and a serious reduction in efficiency. Mechanical stoking and oil firing have removed this limitation and have quadrupled the horsepower production of locomotives. Other detailed improvements in the working parts of steam locomotives have combined to double their efficiency in the twentieth century, but a maximum of 8 per cent even under favorable conditions is still very low compared to that of the steam turbine of the stationary plant or to the internal-combustion engine. One of the first successful oil-fired locomotives was the

Figure 12.4
First electrification of a
steam railroad in the
United States, 1895

Petrolea, built for England's Great Eastern Railway in 1886. By 1900 several American roads were using oil as fuel, but there were never more than 15 per cent of United States steam locomotives fired with oil.

Electric street railways had begun commercial service in the 1880s and developed very rapidly during the next quarter century. The first American electric freight locomotive had operated on the tracks of the Ansonia, Derby and Birmingham (Connecticut) street railway in 1888 in competition with a steam railroad (Figure 12.3). It was not unnatural that electric power should be called upon to help solve the problem of steam railroads. In 1895 the New York, New Haven and Hartford Railroad applied electric power to the operation of its Nantasket Beach branch near Boston (Figure 12.4), the first electric operation of a steam railroad in the United States. Later the same year the Baltimore and Ohio Railroad completed the first main-line railroad electrification in its mile-and-a-half-long tunnel at Baltimore (Figure 12.5) to eliminate offensive smoke and heat conditions. Other electrifications in the United States came during the following decade. The early discussions among engineers as to the relative merits of alternating or direct current for locomotives and railroad motor cars were very bitter, just as some years earlier had been the discussion as to the better method of electric-power distribution for commercial

Figure 12.5 Electric locomotive pulling Baltimore and Ohio train, 1895 (Courtesy General Electric Co.)

and domestic service in cities. The alternating-current distribution adopted in the United States for traction was nominally 11,000 volts, 25 cycles per second, and direct current usually at nominally 600, 1,500, or 3,000 volts. In Europe the alternating-current distribution is usually at 15,000 volts with frequencies either 15 or 16⅔ cycles per second, and direct current usually at 600 or 1,200 volts. If the distribution voltage is more than 600 volts it is necessary to install overhead wires, otherwise third rail is usually employed.

In the United States, only 2 per cent of the total main-line mileage is electrified, but abroad where fuel is scarce, and especially in sections where water power is plentiful, railroad electrification has developed more extensively. While electrification of railroads in the United States has not been as general as abroad, the world's largest steam-railroad electrification, completed in 1935, is that of the Pennsylvania Railroad (Figure 12.6) on its lines connecting New York with Philadelphia, Washington, D.C., and Harrisburg, Pennsylvania. The electrified track totals 2,228 miles and covers 664 miles of road. Very heavy passenger and freight service is handled in this system. Because the power is supplied from the outside, the electric locomotive produces a high output, particularly as compared with the steam locomotive. Furthermore, electric locomotives may be used in sets of two or more units controlled by one driver.

The diesel-electric locomotive began to replace the steam locomotive for main-line operation in the United States during the 1940s, and no main-line steam locomotives have been built by locomotive companies for domestic use since 1949 (Figure 12.7). Railroads and terminal companies

Figure 12.6
Pennsylvania Railroad catenary construction, 1935 (Courtesy Pennsylvania Railroad)

had been using diesel-electric switchers since 1924, but it was not until the late 1930s that orders for diesel electrics began to mount at an increasing rate. The principal components in the diesel-electric locomotive are the diesel engine, the electric generator, and the traction motors. Rudolph Diesel (1858–1913) obtained the basic patents for his engine in Germany in 1892. He first obtained power from one of his engines in 1894, and a later model tested under load early in 1897 showed an efficiency which exceeded that of any other thermal prime mover of the period. Five years later several hundred diesel engines were in operation in stationary power plants. Diesel designed his engine on the general pattern of the gas engine perfected by the German inventor Nikolaus A. Otto (1832–1891) in 1876. The essential features of Diesel's engine are the cylinder equipped with a piston and a fuel-injector nozzle in the cylinder head. When the piston moves downward on the suction stroke, it draws in air. The suction was originally directly from the atmosphere, but nowadays the engine is usually supercharged by air forced into the cylinder under pressure. The return stroke compresses the air imprisoned in the cylinder to a very high pressure and raises its temperature to a point above the flash point of the fuel. When the piston reaches the end of the compression stroke, the fuel injector sprays fuel oil into the cylinder. The hot air ignites the oil, which burns and forces the piston down. In the gasoline engine the explosion of the gasoline-air mixture is obtained

Figure 12.7 Last main-line steam locomotive built by the Baldwin Locomotive Works for use in the United States (Courtesy Chesapeake and Ohio Railway)

by an electric spark, whereas electric ignition is not necessary in the diesel engine. The return stroke expels the products of combustion into the air, and the cycle repeats. This sequence is known as a four-stroke-cycle diesel, but there are many two-stroke-cycle engines where the combustion takes place every other stroke and the exhaust of burned gases and inlet of fresh air takes place simultaneously at the end of the power stroke. Locomotive engines are of either type.

Like the automobile engine, the diesel-locomotive engine requires some type of variable drive mechanism between it and the driving wheels to take care of varying speed and power demands in starting and in climbing hills, while the engine itself works at generally constant speed. Neither the mechanical gear nor the fluid transmission, such as are used in automobiles, is adequate for the high power required in a locomotive. The variable drive in the diesel-electric locomotive (Figure 12.8) is obtained by a direct-current electric generator on the engine shaft which supplies the series-wound, direct-current traction motors with electric power at varying voltages obtained by control of the generator field current. The function of traction motors can be electrically reversed to generate power on downgrades, and this power is absorbed by heating grid resistance. This arrangement acts as a brake on the train going downgrade, since the mechanical energy of the train is changed to heat. This dynamic braking, as it is called, saves wear and tear on the brake shoes and wheels.

American railroads put the first high-speed passenger-service diesel-electric locomotives into operation about 1935. Freight-service diesels appeared in 1938, and the dual-purpose road-switcher type came into use in 1940. The initial cost of a diesel-electric locomotive is relatively high as compared with a steam locomotive of equal nominal horsepower.

However, the diesel locomotive has important advantages. The useful power of a diesel locomotive may be greater than that of a steam locomotive of the same nominal rated capacity because the diesel power is available over a wider range of speeds. The engine terminal expenses are generally less with the diesel locomotive because there is less work to be done by the hostlers at the terminal. There are no boiler feedwater complications, which are troublesome in some parts of the world where steam locomotives are used. The diesel fuel, a relatively high-grade oil, is expensive but easy to handle and store at the terminals. The diesel electric has an efficiency of about 30 per cent, which is four or five times that of the steam locomotive. Moreover, it can be assembled in multiple units under the control of one crew, and these multiple units may produce tractive effort and power concentrations far greater than a steam locomotive. There is no doubt that one of the most important railroad-engineering developments of the twentieth century is the diesel-electric locomotive.

The railroad car has two principal functions: it must carry a load and it must operate as a link in a chain, or train, of cars. Freight-car bodies have become diversified for dozens of types of shipments—livestock, oil, cement, coal, refrigerated products are but a few. The running gear, couplers, air brakes, and other appurtenances have, at least on the North American continent, become standardized to permit interchangeability on all standard-gauge railroads.

George Stephenson, who later built the locomotive *Rocket* described in Chapter 9, constructed the first railway passenger coach in the world. In the early 1830s most of the "carriages" were either converted four-

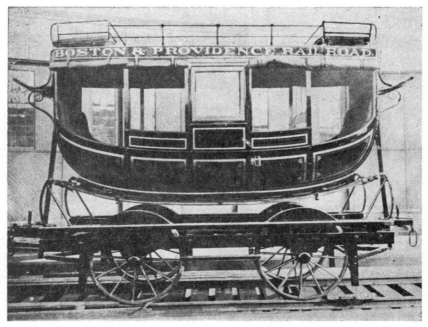

Figure 12.9 Early American railroad passenger coach, 1834 (Courtesy New York, New Haven and Hartford Railroad)

wheel open "waggons" with benches or converted stagecoaches (Figure 12.9). The early freight cars in England were all converted wagons, or lorries, which could carry a load about twice their own weight. These early cars had chains for couplers and were provided with hand-operated brakes only. Since there were no buffers between the cars, there was considerable slack in a train, and whenever the engineer applied the brakes on his locomotive, each car ran in on the car ahead and the whole train ran into the engine. When the engineer started the locomotive and took up the slack in the coupling chains, there was another series of jolts which, as one observer wrote, would "jerk the passengers out from under their hats." In 1831 Ross Winans built horse-drawn, enclosed passenger cars for the Baltimore and Ohio. These cars had seats for 20 passengers and were equipped with two four-wheel bogie trucks and brass bearing boxes similar to those still used on many cars. Although the bogie, or truck, had been patented in England in 1812, British cars did not use them until about 1880. The American railroads began to use Winans's design in the 1830s and soon had cars 40 feet long on two four-wheel trucks.

Both passenger- and freight-car bodies were wood and iron, with wood the predominant material.

Passenger cars in Europe, having developed from the idea of a combination of stagecoaches, were divided into separate compartments. American passenger cars were from the early days much longer than British and Continental cars and because of climate conditions were not divided into compartments. They had from the beginning the typically American center aisle, springs for smooth riding, stoves for heating, candles or oil lamps, and later, water coolers and toilets. They had no vestibules but instead open platforms to permit passengers to pass from one car to another, whereas there was no passageway between European cars for many years.

During the first four decades of railroad development many hundreds of devices for stopping railroad cars were devised and patented. Only a few of them were found to be practically useful. Of these the most common was the hand-operated type which the brakeman set by twisting a horizontal handwheel set on a vertical post at one end of each car. Railways still use hand brakes for the control of detached cars. As train lengths, speeds, and unit loads increased, it became obvious that the engineer should be able directly to control the braking for the entire length of a train. In the late 1860s the successful use of compressed air for drilling the Mont Cenis Tunnel impressed young George Westinghouse (1846–1914) with the possibility of using compressed air to operate brakes. His first patent, covering the straight-air type of brake, was issued in 1869 when he was only twenty-three. In the locomotive cab there was a main compressed-air reservoir from which a hose or pipe, the train line, extended underneath each car, the entire length of the train. This hose connected with a brake cylinder under each car. When the engineer wished to apply the brakes he opened his valve and the compressed air brought each brake cylinder into action almost—but not quite—instantaneously. These air brakes slowed the train or brought it to a sometimes violent stop unless a leak developed in the hose, impairing the brake power, or the train broke in two, in which case the cars had no brakes at all except the old hand brakes.

Westinghouse recorded in his 1910 presidential address to the American Society of Mechanical Engineers, entitled *The Conception, Introduction and Development of the Air Brakes*, the first test of his new brake: "The Superintendent of what was then known as the Panhandle Railroad, Mr. W. W Card, offered to put the Steubenville accommodation train at my disposal to enable me to make a practical demonstration. The apparatus exhibited was removed from the shop and applied to this train, which consisted of a locomotive and four cars. Upon its first run after the apparatus was attached to the train, the engineer, Daniel Tate, on emerging from the tunnel near the Union Station in Pittsburgh, saw a horse and wagon standing upon the tracks. The instantaneous application of the air brakes prevented what might have been a serious accident, and the value of this invention was thus quickly proven and the air brake started upon a most useful and successful career."

In 1872 Westinghouse's second invention, the plain automatic brake, made slowing and stopping simpler, smoother, and more certain. This type was developed from the original straight air brake by adding under each car an auxiliary reservoir with a triple valve which together with the train line is normally kept filled with compressed air at about 75 pounds pressure. When pressure in the train line is reduced or cut off normally by the engineer, the pressure from the auxiliary reservoir under each car, acting through the triple valve on its own brake cylinder, partially or completely sets the brakes. In case the train line is ruptured, all the brakes are automatically set by the pressure of the air in the individual reservoirs. The triple valves are essential features of the automatic air brake. The improvements in air brakes since 1872 have been mainly in details. American manufacturers and railroad men have conducted many series of extensive tests of every feature of brake equipment and of various brake types, beginning with the Burlington tests carried out with freight trains on the Chicago, Burlington and Quincy Railroad in 1886–1887.

In England the first formal series of brake tests was made on the Midland in 1875. The vacuum automatic brake, first used on the Great Western Railway in 1876, was widely adopted after the tests. In England, railway equipment is lighter in weight than in the United States. In 1923 the vacuum brake became standard equipment on English roads; however, the air brake is standard on the Continent. United States railroads have recently developed a brake known as the AB brake that controls freight trains of any length traveling sometimes faster than 70 miles an hour, while new types of brake shoes on passenger cars take care of trains traveling faster than 100 miles an hour. The importance of effective brakes is evident when one realizes that the energy in a moving train is proportional to the square of its speed. For example, a train moving at 80 miles an hour has nearly twice the energy of a train of the same weight moving at 60 miles an hour, and a train at 100 miles an hour has nearly three times the energy. This must be absorbed to stop it. The air brake is more than a mere safety device; it gives the engineer control over the speed of his train at all times and has added to railroad traffic capacity, allowing longer, faster, and more frequent trains. Were Mrs. Carlyle able to ride in a streamliner of the 1950s she would not have to worry as she did in 1836 about "the impossibility of getting the horrid thing stopt."

The lot of a brakeman on early American railroads was not a happy one, especially before the advent of the air brake when he might have to run along freight-car roofs to tighten up many hand brakes whenever a stop was indicated. Another dangerous operation was the coupling of cars in the days when the only device used was the link-and-pin coupler. This procedure required a man to stand between the stationary car and the one approaching it in order to drop the heavy coupling pin into the double link at the proper instant. The first automatic or knuckle coupler which closed and locked on impact was invented in 1873 by Eli Hamilton Janney, a Virginia farmer.

The early American passenger cars were lighted by tiny, dim, and smelly lamps, most of which burned whale oil, later kerosene. Pintsch gas or acetylene gas lamps gradually replaced oil lamps on most roads and were standard equipment for years. Pintsch gas, rich in illuminating power, was made from petroleum; the gas was compressed into portable tanks placed under each car. The first electric lights were installed in a Pullman car in England in 1881, using power supplied by crude storage batteries. In the late 1880s an effective electric-lighting system, the power for which was supplied by a generator, began to replace gas. The first train in the United States to be lit by electric lights was the Atlantic Coast Line's *Florida Special* in 1887. Here the current was supplied from a single generator operated by steam at the head end of the train. This arrangement was not satisfactory for a number of reasons and was therefore supplanted by installing under each car a generator connected by a belt to the car axle and so designed that it could furnish power regardless of which direction the car was moving. This type of lighting system now also supplies power for air conditioning and other demands; in many cases as much as 50 horsepower may be required for each car, or 500 horsepower for a 10-car train. Rugged storage battery units with ample capacity furnish light and power while cars are not in motion.

The Pullman Company in 1907 built the first all-steel car, which was perhaps the greatest single advance ever made in railroad-car construction. Steel cars are obviously much safer than wooden cars. All American passenger cars and most freight cars built in the last thirty years have been of steel or aluminum alloys. The first all-steel cars on the European continent were introduced in 1922, but many European roads still use wood. Air conditioning, or cooling and drying air in summer, was first tried by the Baltimore and Ohio in the form of an ice-cooled car in 1884, but mechanical air conditioning was not generally introduced on American railroads until the 1930s. Roller bearings first replaced Winans's brass friction bearings in the late 1920s; the roller bearing reduces drag, especially in starting. The lightweight aluminum and steel-alloy streamliners and lightweight freight cars appeared in the 1930s. Some of the new freight cars can carry loads four times their own weight.

In the United States the majority of signals are now electrically operated, and most control of trains involves the use of the telephone, which has generally replaced the telegraph. The evolution of electric signaling and control depended directly on the inventions described in Chapter 11. Until the 1920s signals had the one function of preventing accidents; since that time they have also been used to communicate orders to the engineer. Control now means largely the dispatching and routing of trains, and efficient control systems permit a high train density on any line. Over five hundred trains travel each day over the four-track line leading into the Grand Central Terminal in New York, and during morning and evening peak hours there is especially close spacing of trains. Without an efficient method to control this heavy traffic there would be chaos.

One of the first railroad signals was a basket covered with a white cloth which could be raised to the top of a tall pole. The New Castle and Frenchtown Railroad, now part of the Pennsylvania, used this signal in 1832; it was probably adapted from a type of signaling post used during the American Revolution to send coded messages. When the basket of the New Castle and Frenchtown signal was at the top of the pole it meant "clear"; when at the bottom, it meant "stop." A failure of the mechanism would drop the basket to the stop position so that a "false clear" was avoided—a cardinal principle in all signal construction. In later years, even after the telegraph was used for train control, large hollow balls were substituted for the baskets, and the "clear" signal was the origin of the railroad term "highball," meaning go ahead. All early signaling and control worked on a time basis. At perhaps twenty minutes after a train had left a station the signal would be changed from "stop" to "clear"; the assumption was that a following train would not overtake the first if the latter had a twenty-minute start. However, anything from a mechanical failure to livestock on the track could detain the first train, allowing the second to overtake and perhaps collide with it.

The Yarmouth and Norwich Railway in England first used the telegraph for traffic control in 1844, and the Great Western Railway installed a manually operated telegraph signal at its 2-mile Box Hill Tunnel in 1847. American railroads were slow to use the telegraph, and when Charles Minot (1810–1866), superintendent of the New York and Erie Railroad, first proposed about 1849 the construction of a telegraph line to be used to control traffic, he had great difficulty in obtaining approval from the directors. American railroad men had at first little enthusiasm for the telegraph. An English visitor writing in the *London Quarterly Review* of June, 1854, commented on its rarity along railroad rights of way, adding that as a consequence "locomotion in the United States is vastly more dangerous than in England."

Interlocking connects the signal to the track switch so that the signal cannot be set for a train to proceed unless the switch is in the proper position. In the 1850s the British invented interlocking machines which would prevent an operator in a control tower from throwing a switch without properly setting the signal. The first railroad to use interlocking was the London, Chatham and Dover Railway in 1856. Prior to that time a signalman moved a switch and changed a signal by hand independently, a procedure under which human errors were frequent and often disastrous. The Pennsylvania Railroad had installed the first manually operated block signaling system in 1864. The principle of this system was based on the subdivision of the railroad for traffic-control purposes into lengths of tracks called blocks, each of which might be a mile or several miles long. At each end of the block were operator-controlled signals to govern trains approaching from either direction on the single track. When a train entered a clear block the signalman set his signal to stop any following train and telegraphed the operator at the other end of the block who set his signal to prevent any train traveling in the opposite direction from entering the block. The first signalman also notified the operator at the other end of the block from which the train had proceeded, and he changed his signal to "clear." Subsequently a locking device was added to the signals so that the operator could not change the signal from "stop" to "proceed" unless the signalman at the other end of the block unlocked it for him. There are still some manually operated block signals in the United States, but the block signals on heavily used lines are now generally automatic. The Southern Railway in England had tried out an automatic electric signal as early as 1844, but it was not effective. William Robinson developed the first successful track circuit for automatic signals in the United States in 1871. In the automatic signal system an electric current passes through one of the rails of the block from a source of supply at the leaving end of the block and through an electric relay at the opposite end and thence returns through the other track rail. As long as the current flows normally, the relay is energized and keeps the signal at "clear." As soon as a train enters the block the current is short-circuited through the axles of the train and does not pass through the relay which immediately changes the signal to "stop." The relay gives the "stop" signal also in the case of a broken rail which interrupts the rail current. This automatic scheme has made unnecessary manual signal operation at the ends of the blocks.

A train-control system is designed to set the train brakes or give an audible signal in the locomotive cab of a train which passes a signal set against it. Mechanical trip devices to set the brakes had been patented as early as 1880. Two general types of electric train control have been installed on railroads in the United States. These are known as intermittent and continuous devices. In the intermittent arrangement an electromagnet is installed near the running rail at each signal location, and other magnets are mounted on the trucks of the locomotive. If a signal is set at a restrictive position, the track magnet is energized, and if the locomotive engineer does not recognize and manually acknowledge or forestall the indication by operating a switch before the locomotive magnet enters the field of the stationary magnet, the air brakes are automatically applied. Since there is no device connected with this system between signal locations, the control is effective only at the wayside signals and the scheme is therefore intermittent.

What is known as a coded continuous-induction control system, now the most generally used or American method, was first installed in 1927 after much cooperative experimentation. In this system the locomotive truck frame carries an induction coil near the rail ahead of the leading axles. A current is induced in these coils from the magnetic field which surrounds the track rails in which the signal current flows, and vacuum tubes on the moving equipment amplify the induced current and show signals in the cab. The circuit flowing in the track rails is controlled by the coder, consisting of a small motor operating electric switches which interrupt the rail current 80, 120, or 180 times per minute, depending on the position of the signals serving the block or section of track. If the block is occupied, the motor stops and there is no code impressed on the track circuit; the cab signal thus indicates "stop" and the situation must be acknowledged by the engineer. If the block is clear but the next block is occupied, this coder gives 80 interruptions per minute and the cab signals indicate this situation. If two blocks ahead are unoccupied, the code is 120 times per minute, and if three blocks are clear, the 180 interruptions per minute are so indicated in the cab.

Some railroads have installed what is known as centralized traffic control, or C.T.C., on sections of their lines in order to eliminate local signal towers along the route. This system, first employed in the United States in 1925, enables an operator located at a convenient place to control interlocking signals and switches over a large railroad territory. As the trains move over this territory their position is shown automatically on an electrically illuminated track plan so that the operator knows at all times just where each train is. On his desk are small knobs to control signals, and small levers to control switches, perhaps more than 100 miles away. Visible indicators show the position of each switch and signal. The operator may plan just what he wishes each train to do at any time and operate the appropriate track switches and signals to indicate his plan to the locomotive engineer. There are thus no train orders and no operators at any other point. The whole system is handled on a pair or two of wires strung along the right of way without interfering with the automatic block signals which continue to give their own indication of track conditions in the block. It is one of many devices adopted by railroads in the interest of economy that not only save labor costs but also increase traffic volume.

Street Railways

The electric street railway has had a brief but hectic history in the United States. From the late 1880s to 1950 it furnished an enormous amount of urban transportation before being largely displaced by a motorized version of one of its predecessors, the omnibus, which dates back to the seventeenth-century philosopher Blaise Pascal. The first horse-drawn coaches, or omnibuses, began public service on regular schedules in Paris in 1662. They were reasonably satisfactory from an engineering point of view, but once the novelty had worn off their popularity declined and the company failed. The first commercially successful serv-

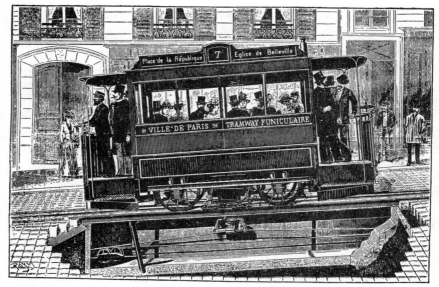

Figure 12.10 One of the first Paris cable cars (From *Street Railway Journal*, January, 1891)

ice began in 1827, also in Paris; two years later the omnibus appeared in London.

Chapter 9 described the little "waggons" pulled on rails by horses or men at coal mines. The 8-mile Surrey Iron Road, the earliest public horse railway, was built in England in 1803; it carried only freight. The first urban street railway, the New York and Harlem, was built in New York City. The lower portion, extending a mile up the Bowery from Prince Street to Union Square, was in operation as early as 1832. Its original cars were converted omnibuses. This horsecar line evolved gradually into a steam railroad as it reached up Fourth Avenue to the Harlem River and beyond. Many other horsecar lines soon appeared on New York City streets where a very few persisted even into the twentieth century. Boston had horsecar lines in 1836; Philadelphia soon followed. Paris street railways date from 1855, London's from a few years later.

In some cities light steam locomotives replaced the horses drawing the passenger cars, but these steam street railways were never developed to any great extent, largely because of public prejudice. In a few cities, especially in San Francisco in 1873 and later in London, Chicago, Philadelphia, Paris (Figure 12.10), and New York, the so-called cable railway was introduced. The cable railway operated on an endless steel cable moving under each pair of tracks at uniform speed. The car operator manipulated a plow inserted through a slot between the rails so that it either released or gripped the cable; when the plow gripped the cable, the car moved at the uniform cable speed. Large grooved pulleys around which the cable wound at the power plant supplied it with motion, and movable carriages controlled by counterweights furnished uniform tension under varying temperature conditions. The cable railways gave good service when they were installed, but in all cases, except on the steep grades in San Francisco, electric traction replaced them.

Ernst Werner von Siemens (1816–1892) built the world's first public-service electric street railway at Lichterfelde, a suburb of Berlin, Germany, in 1881 (Figure 12.11), two years after he had offered visitors to the Berlin Exhibition thrilling rides on his 500-meter road. The Siemens system found its next application two years later on a 6-mile road at the tourist town of Portrush, on the northern coast of Ireland, near the Giant's Causeway. This road climbed 203 feet with a succession of steep grades, and the electric power was furnished by two 52-horsepower water turbines operating under a 26-foot head. This project stimulated a number of other street railways in Germany, France, and England in the next five years.

Somewhat behind Europe, in the United States the electric railway was developed following 1883 by several men, especially by English-born Leo Daft (1843–1922), Belgian-born Charles Joseph Van Depoele (1846–1892), and Frank Julian Sprague (1857–1934), graduate of the U.S. Naval

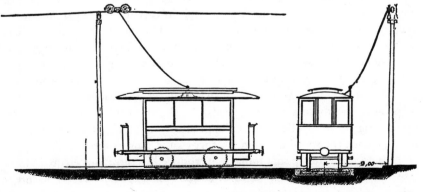

Figure 12.11 Electric street railway in Lichterfelde near Berlin (From *Die Eisenbahn*, 1881)

Academy. In 1887 the magazine *Electrician and Electrical Engineer* listed 21 electric railways in operation on the North American continent. Among these railways perhaps the most prominent was that installed in Richmond, Virginia, by Sprague in 1887 for regular urban public service. With its 40 cars the Richmond line was 12 miles long, and became an immediate success, as did Boston's, which started very shortly thereafter. By 1890 more than one hundred American cities had installed or were in the process of installing electric street railways. After 1900 the system spread rapidly throughout the world, and by 1940 the only remaining horsecars were in Mexico.

The electric street railway usually operated on 500- to 600-volt direct current supplied from an overhead trolley wire. The term trolley was applied to this method of providing electrical contact from the power supply because some of the early cars were operated by means of a trolley or small device on grooved wheels riding on the overhead wire; from this device a flexible conductor led down to the car roof (Figure 12.12). This contraption was soon abandoned in favor of a grooved contact wheel on the end of a pole, which a strong spring held against the wire from below. The name trolley, however, has persisted. The street railway electric circuit is normally made up of (1) an alternator at the power plant; (2) the high-voltage alternating-current power-transmission circuits from the power plant to the substations where conversion from alternating to lower-voltage direct current takes place; (3) the power distribution, or trolley wire, over the track; and (4) the electrical parts of the car, including controller switches and motors. The current returns to the substations through the wheels, rails, and negative feeders, that is, copper conductors to supplement the rails in carrying the current and to keep it out of the earth as much as possible. The motorman starts the car and controls its speed by means of a controller, an ingenious arrangement of switches and resistors. The traction motors are series-wound to provide a large starting torque and connected to the axles by gears so they cannot race. In starting, the current passes in series through the resistance to protect the motors at low speeds. As the motorman increases speed by moving the controller handle, he gradually reduces the resistance, and finally, when the speed is sufficient, throws the motors into parallel.

Charles C. Henry of Indiana introduced interurban electric lines into the United States in 1894 as extensions of street railways. The interurbans, which carried freight as well as passengers, reached their peak operation about 1910, but had almost entirely disappeared in the United

Figure 12.12 "Trolley" collector on street railway car (Courtesy Charles Rufus Harte)

States by 1930. In 1899 the Siemens company, which had built the first electric street railway in 1881, introduced the first trackless trolley, an omnibus driven by electric motors. The trackless trolley has two overhead wires; the second takes the place of the rails in providing the return circuit. Not being confined as the trolley car is to rails, the trackless trolley obstructs traffic less than cars on tracks do.

Sixty years of operation is a relatively brief time. The rapid increase in the number of private automobiles in the United States brought about the decline of the electric street and interurban railway. In the 1890s the trolley car superseded the horsecar and its competitor, the horse-drawn omnibus or stage, in urban transportation. It is significant for the history of engineering, as well as for the history of our metropolitan areas, that sixty years later the omnibus propelled by a gasoline or diesel engine or sometimes by overhead trolley wires has largely replaced the streetcar in the United States.

Rapid transit in large cities dates from the 1860s in London; in America it began with the building of the first elevated railway in New York City. This line, built in 1868, was in Greenwich Street. The original motive power, a cable, was very soon replaced by tiny steam locomotives. New

York was the first city in the world to have an extensive elevated railway system and the proponents of the "El" were convinced that they had found in it the ultimate answer to the city's crying need for quicker, safer, cheaper, and pleasanter transportation than the rough cobblestone streets could provide. Brooklyn followed New York's example, and both cities endured coal smoke from the little locomotives for a whole generation and almost grew to like it. In Chicago the electric locomotives used on the Intramural Railway at the World's Fair of 1893 led to the opening two years later of the Chicago Metropolitan electric system, which, at first 12 miles long, was later considerably extended. Electric motors were used on the Liverpool Overhead Electrical Railway as early as 1893.

When electric motors had been developed with reasonable success for street railways, it was obvious that they would be advantageous to trains of cars as well as to single motorcars. In order to operate a train of motorcars under control of the motorman at the head end, F. J. Sprague introduced the so-called multiple-unit device. In the multiple-unit system a storage-battery circuit is connected to a master controller on the head car by bundles of wires running back through the train. When the controller is operated the arrangement duplicates throughout the train the operation of the master controller, causing all the cars in the train to function as a unit. This device has made possible the ready operation of long electric motorcar trains, each car controlled by the motorman at the head end just as if he were located at the controls of each individual motorcar. Boston's elevated railway, electrically operated, was started in 1901. During this same year New York City's elevated lines were electrified, a total of 117 miles of track. Their progressive demolition began in 1938. On Manhattan Island the avenues are no longer cluttered up with ugly pillars every few feet; instead they are broad open boulevards.

Shortly after the middle of the nineteenth century London began to consider an underground rapid-transit system. In particular there was a demand for shortening the time of travel between the many railway terminals on the edges of the sprawling metropolis and providing quick transportation from these terminals to congested business areas. London, with its narrow and crooked streets, was the first of the world's cities to attempt to solve such a problem by constructing a railway entirely below the surface. The London underground system, later called the Metropolitan District Railway, was begun in 1860. Most of it was constructed, not as a tunnel is usually built, but by making first an open cut, using the cut-and-cover method. This process involved the unprecedented engineering problem of caring for what even then was a maze of subsurface gas, water, and sewer conduits in the more congested areas, the shoring up of buildings, and the replacing of street surfaces. The side walls and most of the arched roof of the tunnels were brick. The original lines of the underground were gradually extended, until in 1884 they formed a 2-by 5-mile oval connecting twelve railway stations with each other. There was close collaboration between the underground and some of the twelve trunk-line railways even to the extent of providing at first for both standard- and broad-gauge trains on certain of the Metropolitan District lines. The motive power used for many years was steam locomotives, which created painful atmospheric conditions. The locomotives were, in general, of the type used on the ordinary British surface railways; they were dependable but noisy and dirty. Up to the 1890s London had in this complicated combination of the Metropolitan District and Inner Circle railway lines the only subway system in the world. It was electrified in 1905.

The first city to follow London's lead was Paris, which began to plan a system in 1871, but for various reasons did not undertake construction until 1898 on the eve of the 1900 exposition. The first section of the Paris *metro* was not unlike the first London underground, a shallow construction roofed over as the work progressed. Where the British had depended largely on brick arches, the French engineers used iron or steel beams and concrete, thus reflecting an advance in construction practice. The motive power in Paris was electric from the first.

A large proportion of the London subways are now of a type quite different from the original underground; they are in fact cylindrical tunnels bored at considerable depths by the shield method. They owe their inspiration to the little Tower Subway of 1869 (Figure 14.10), which led to the electrically operated City and South London Railway opened in 1890 and using 500-volt third-rail power distribution (Figure 14.11). The City and South London is usually regarded as the progenitor of London's "tuppenny tubes" of the Central London Railway in 1900. This type of subway tunnel is peculiar to London. Much of the city is underlaid by a thick stratum of dense blue clay, making shield tunneling and the use of rotary excavators comparatively simple. Rapid progress in construction was possible at depths below water, gas and sewer pipes, telephone conduits, and electric and hydraulic power mains, and the tracks are in general from 50 to 80 feet below the surface. At Piccadilly Circus they are about 100 feet deep; the high ground at Hampstead Heath made a depth there of nearly 200 feet necessary. With the tube railways came rapid improvements in pumping machinery, ventilation, elevators, escalators, lighting, signaling, and train control.

While London's original underground was still using steam power, although the City and South London used electricity, a more modern subway was built in Budapest. This 2½-mile line completed late in 1896 was built mainly to relieve traffic congestion on one of the city's main thoroughfares, Andrassy Avenue. It was built by the cut-and-cover method, and steel beams support the roof. The cars, used only on the subway, have overhead pantagraph collectors and operate at 300 volts. The Budapest subway furnished the pattern for the first subway in America, that at Boston, which was under construction at the same time. Boston's original mile and a half of subway, which provided for trolley streetcars, went into operation in 1898.

The New York subway was begun in 1900, and the first section of 21 miles, including 5 miles of elevated railway, was completed four years later. In this original project there were 5 miles of four-track construction and nearly 5 miles of double-track concrete-lined tunnel. The southern terminus was at City Hall, close to the Brooklyn Bridge, and the line reached up into the east and west sections of the Bronx by two branches, largely elevated. Except that much of the excavation was in solid rock, the engineers of this first line in New York City did not find many problems that had not already been met and solved in London, Budapest, and Boston. The cutting was, in general, rather shallow; at some points the tracks are as little as 17 feet below the street surface. In the more congested southerly section a continuous series of problems arose in connection with providing for the thousands of intersecting sewer, water, and gas mains, steam pipes, pneumatic tubes, and electrical conduits. One particularly difficult piece of construction was the easterly branch crossing under the Harlem River in twin single-track, cast-iron cylinders. Third-rail electric traction at 600 volts direct current has been used in New York's subway as on all other rapid-transit railroads in the United States from the first.

Shipping

Since earliest days, transportation by water has been important; it has become increasingly so throughout the past hundred years. The very existence of the United States results from shipping, for with the exception of some 340,000 Indians and of the few immigrants flown into the country by airplane since 1945, some 160 million Americans either came to the United States in vessels of one kind or another or are descendants of those who came by sea. The commercial importance of marine freight transport is equally great; sail or power craft are the means for carrying on a very large portion of the world's international trade.

The only type of sailing craft to compete successfully with early steamships was the American or Yankee clippers of the 1840s and 1850s. Longer and proportionately narrower than the packets and other sailing vessels had been, the square-rig clippers were built for speed and carrying capacity. Their increased length presented new problems in naval architecture which were to affect the design of steamships. Their ratio of length to beam was 5:1, 6:1, or even 7:1. Such ships have a tendency to "hog" on top of a wave or to "sag" into a trough between two waves in a rough sea. There is also much stress from the towering masts carrying the clouds of sails. All of these stresses must be provided for in the design of the hulls. Clipper ships of 1,000 tons and over were being built and successfully operated in the 1840s; a few reached 3,000 tons. Some were more than 300 feet long, nearly half again as long as the steamship *Great Western* of 1838. Iron frames supporting the wooden planking began to be used as early as 1851. The first steamship hulls were practically of clipper proportions, and it was only with the coming of the early "greyhounds" of the 1880s that steamships exceeded the 7:1 ratio between length and beam.

There are four important general types of power plants in modern marine service: (1) reciprocating steam engines directly connected to the propeller shafts, (2) steam turbines driving directly or through gears, (3) diesel engines directly connected or geared to the shaft, and (4) steam turbines or diesel engines driving electric generators which furnish power to motors turning the shafts. The early steamships described in Chapter 9 had reciprocating engines, as indeed did all commercial steamships in the nineteenth century. The reciprocating steam engine has been very popular; it was not until recently that the combined tonnage of turbine-powered and diesel-powered ships has exceeded the tonnage of ships with reciprocating steam engines.

The innovation in the steamship that sealed the fate of sails was the compound condensing engine. Such engines had been tried on land as early as 1781, and several engineers had improved the design of compounding during the first half of the nineteenth century. With improved design came higher pressures and greater efficiency, but the prejudice against high pressures prevented early wide adoption of compound engines. It was John Elder of Glasgow who first installed a compound engine in a ship, the *Brandon*, launched in 1854. The pressure in this ship's boilers was too low to bring out the full efficiency that can be attained with a compound engine; nevertheless its coal consumption was only two-thirds that of the single-cylinder simple engines of the period. In 1857 Elder's firm launched the 523-ton propeller-driven *Thetis* equipped with a compound engine employing steam at a pressure of 115 pounds. Improved boiler design in the 1850s also increased the steam plants' efficiency. Steamships had been using sea water for boiler feed even though the salt deposits left in the boiler by the evaporation of the water made pressures higher than about 25 pounds dangerous. No ship could carry enough fresh water for a trip of any distance unless the exhaust steam could be condensed and used over again as feedwater. Hence a surface condenser, one of Watt's inventions, was installed in 1860 in the *Mooltan*, which had compound engines. The surface condenser proved to be sufficiently practicable to enable a ship to use over again its fresh boiler feedwater with little addition or "make-up" water even on a long voyage. The *Ajax*, first of the Holt liners, equipped with a two-cylinder compound engine and a surface condenser, steamed 8,500 miles nonstop in 1865 from Liverpool to Mauritius in the Indian Ocean. Her coal consumption was 2.2 pounds per indicated horsepower per hour, whereas that of the *Britannia* in 1840 had been 4.7 pounds. The Holt liners were the first ships to demonstrate efficient and economical long-distance steaming.

However, the two-cylinder compound engine did not retain its supremacy for long. The triple-expansion engine, in which the steam is expanded in three cylinders successively, was patented in 1871 and was first used in the *Propontis* in 1874. The *Ajax's* engine operated at 60 pounds pressure, the *Propontis's* at 150 pounds. The *Aberdeen*, launched in 1881, had triple-expansion engines and improved boilers which enabled her to require only 1.28 pounds of coal per indicated horsepower per hour. In forty years coal consumption in steam navigation had been cut more than two-thirds.

The first completely revolutionary change in the design of marine steam engines came in 1894 when Charles A. Parsons installed one of his steam turbines in his 45-ton experimental *Turbinia*. Operating necessarily at high speed and directly connected to the shaft, the turbine turned the propeller too rapidly to be practical. Three slower turbines, each driving its own screw, produced the record speed of 34.5 knots in 1896. The high speed of the turbine and the fact that it must always rotate in one direction are disadvantages in this type of drive. A separate reverse turbine was later fitted to the shaft to make reversing possible. By 1910, reduction gearing between the turbine and the shaft had been developed to retain the advantages of high-speed turbine operation and the most efficient screw speed. Electric drive was also developed, the constant-speed turbines driving generators which furnished power to the propeller-shaft motors, the speed of which can be effectively controlled regardless of the turbine speed. The *Mauretania* and the ill-fated *Lusitania*, 30,000-ton sister ships built in 1907, had turbines directly connected to their propellers, and the success of the turbines in these two giants paved the way for the general installation of turbines. Because it is smaller and more efficient and requires less servicing than the reciprocating engine, the turbine has also been popular in smaller ships. As late as 1952, however, the tonnage propelled by reciprocating steam engines was still nearly twice that of turbines.

From the increasing rate of motor-ship construction during the years 1945 to 1951, it appears that motor ships using diesel engines are to a considerable extent replacing steamships. In 1903, five years after the diesel engine was first produced commercially, this type was installed in two Russian tankers for service in the Caspian Sea. The first important ocean-going vessel to be diesel-powered was the 7,500-ton *Selandia* built in 1912. The largest motor ships operating in 1954 were 27,000-ton ships. Marine diesels, like locomotive diesels, have a higher thermal efficiency than steam plants, although diesel oil, a highly refined petroleum product, is more expensive than residual or by-product fuel oil or than coal. Oil firing for steamships is now used in 79 per cent of steam tonnage and has greatly reduced costs. For instance, the *Mauretania*, one of the last large hand-fired steamers, required 324 firemen and trimmers to cart and shovel about 1,000 tons of coal a day into her boilers. Even more important than the labor and fuel saving accomplished, oil firing has increased the power output which had definitely reached its limit with hand firing of coal. Oil firing has vastly improved the lot of the fireroom forces by eliminating hardships that had become a scandal in high-speed transatlantic service.

Highway Transportation

Bicycles have never furnished an important form of transportation in the United States—not even during the bicycle craze of the 1890s. In Europe, where distances are shorter and automobiles more expensive, the bicycle still supplies an appreciable proportion of urban transportation. Forty-five per cent of all employed persons in the United States use automobiles in connection with their work, but the percentage in Europe is far less because Europeans depend instead on their bicycles.

Hollow-steel tubing and steel-wire spokes were first used on cycles about 1867, and ball bearings a decade later; both were English inventions. The last invention important for the bicycle, and subsequently for the automobile, was the pneumatic tire. Charles Goodyear, a Connecticut inventor, had succeeded in vulcanizing rubber in 1839, and Robert William Thompson, British civil engineer, had patented a single-tube tire as early as 1847. Thompson's patent lapsed, however, and John Boyd Dunlop (1840–1921), a Scottish veterinary practicing in Belfast, produced in 1888 the first pneumatic tire with an outer casing, or tread, and an inner tube. Bicycles of modern design appeared in large numbers on the streets of England and America soon after 1885 when the English invention of the Rover "safety" wheel with sprocket and chain drive made bicycling a popular and not too hazardous sport. It was the bicycle that in the eighties and nineties stimulated new road construction in the United States and to a lesser extent in Europe. The League of American Wheelmen, founded in 1880 and politically active in the 1890s, constantly fostered public road improvement programs; in England the Cyclists' Touring Club was similarly active.

While the bicycle does not present complicated engineering problems, the history of its development shows an important and curious relationship to the early history of the automobile. The first mass producer of the safety bicycle, James Starley, reinvented the differential gear now so necessary for the automobile. The chain drive used on early automobiles was first used on bicycles. The bicycle thus played an important role in the invention and evolution of the automobile.

The automobile has participated in producing the changes in modern society described at the beginning of this chapter and has produced its own peculiar social innovations, particularly in the United States. More than any other form of transportation it has increased human mobility.

It has greatly improved fire, police, sanitary, and medical protection, but its most important effect has been in the growth of metropolitan areas. Although the railroad was a vital factor in making possible the rapid growth of cities during the last half of the nineteenth century and the first quarter of the twentieth, the automobile has facilitated the rapid growth of outlying residential areas since the 1920s. Cities in the United States are still growing at an increasing rate, but their suburbs are growing 2½ times as rapidly. The number of registered motor vehicles in the United States increased from four in 1895 to more than 55 million in 1953. There is now one automobile for every three Americans—an average of one for each family! The United States has over 70 per cent of the world's motor vehicles, and whereas the railroads haul by far the largest freight ton-miles, it is the private automobile that provides the greatest number of passenger-miles. During the 1930s and 1940s the percentage of motor vehicles owned outside the United States has steadily increased, and the other world areas which have experienced this increase are beginning to undergo the same type of social revolution which has occurred in the United States.

Although the feature most typical of the automobile is its internal-combustion engine, the earliest self-propelled vehicles were steam-powered. The idea of obtaining mechanical work from combustion or explosion in a confined space dates from the seventeenth century, when Jean de Hautefeuille, Christian Huygens, and Denis Papin, working independently, attempted to make gunpowder pumps. Starting in the last decade of the eighteenth century, many men worked on various designs of internal-combustion engines, using coal gas, wood gas, or volatile hydrocarbons as fuel. It was not until the 1860s, when the French inventor Joseph Etienne Lenoir (1822–1900) produced small quiet gas engines, that the internal-combustion engine became a commercial success. Many of Lenoir's engines were installed in factories. In 1862 Beau de Rochas published a remarkable analysis of the cycle of operations for a successful gas engine. There should be four strokes, he said, (1) to draw in a mixture of gas and air, (2) to compress the mixture, (3) to ignite the mixture at the dead point, thereby producing the power stroke, and (4) to exhaust the products of combustion. The Beau de Rochas cycle is that of the familiar four-stroke cycle, or four-cycle engine. Nikolaus August Otto (1832–1891), a German, and his partner Eugene Langen (1833–1895) built their first engine at Deutz. The Otto gas engine of 1867 was noisy, but it consumed only half as much fuel as Lenoir's and ran twice as fast.

Both were essentially similar to the steam-engine arrangement with a re-designed valve and equipped with a flame-ignition system. However, in 1876 Otto produced his famous Otto Silent which used the four-stroke Beau de Rochas cycle; the Otto Silent was the most important invention in the history of the automobile.

Both Lenoir, about 1862, and the Austrian Siegfried Marcus (1831–1898), in 1873, had constructed vehicles powered by internal-combustion engines, but it is not really certain that Marcus's vehicle, which the Aus-trians prize, actually ran. George B. Brayton patented a two-cycle oil engine in 1874 and exhibited it at the Philadelphia Centennial Exhibition in 1876, where George Baldwin Selden, an attorney of Rochester, New York, saw it. Three years later Selden applied for a patent on a horseless carriage powered by an engine of the Brayton type. His patent was not granted until 1895, but Selden, who never built a car, collected substantial royalties from many who did.

It was Karl Friedrich Benz (1844–1929) of Mannheim, Germany, who built the first reliable internal-combustion engine automobile in 1885, after a decade of experience in building and selling small stationary gas engines. Benz's first car was a three-wheeler (Figure 12.13). Equipped with a single-cylinder four-stroke-cycle engine, it had electric instead of flame ignition. There was a differential gear, and the engine was water-cooled. Not having unlimited supplies of water available for cooling his moving engines, Benz invented and patented in 1886 a rudimentary radiator to cool water for reuse. Electric ignition was also original with Benz. These three important features—electric ignition, water cooling, and the differ-ential gear—are on nearly every one of the 70 million automobiles in the world, and it is to Benz's credit that they were all on his first car. Benz's four-stroke-cycle engine had a horizontal cylinder, horizontal flywheel, poppet valves, and a surface carburetor patented by him in 1886, and could produce ¾ horsepower at 250 revolutions per minute. The fuel was benzine (not named for him). The car had one forward speed, and the transmission was a belt which could be moved back and forth on a loose or fixed pulley on a countershaft which had the differential gear and two sprockets at each end with chains to the rear wheels. The two rear driv-ing wheels and the single front wheel had hard-rubber tires. Two years before Benz ran his first car, Gottlieb Daimler (1834–1900) of Cannstatt, Württemberg, who had been a foreman in Otto's factory, built a single-cylinder engine with the high speed of 900 revolutions per minute. Daim-ler fitted a second engine like it to a bicycle in 1885; thus he produced the

Figure 12.13   First automobile—Benz, 1885 (Courtesy Daimler-Benz Aktien-gesellschaft)

first motorcycle. Daimler built his first car in 1886. It had four wheels, the front axle centrally pivoted for steering. Benz and Daimler continued independently to improve and produce cars and to organize manufacturing and sales companies.

The Dunlop pneumatic tire, first perfected in 1888, began to be used on automobiles about 1897. Daimler built a car with four forward speeds in 1888 and added a reverse speed, still using a belt drive, in 1896. Benz adopted a gearbox in 1899. In the meantime he had built his first four-wheel car in 1890, with a fixed front axle having movable stub axles for steering, an important device for which he received a patent in 1893. The French Panhard car of 1894 had its engine in front under a hood, a slanting steering post with steering wheel, and floor pedals; it was thus the first car of the general design that has been almost universally adopted.

The appearance of a Benz car at the Columbian Exposition in Chicago in 1893 greatly stimulated the many American mechanics who had been working on "horseless carriages." Europe, especially France, Germany, and Italy, was at least a decade ahead of the United States in automobile production during the 1890s and even later. The earliest of these American mechanics to turn out a dependable car were the brothers Charles E. and Frank Duryea of Springfield, Massachusetts. They had started with a phaeton to which they had added a 4-horsepower gasoline engine just above the rear axle. Their original car, or "buggyaut," built during 1892 and 1893 is on exhibition at the Smithsonian Institution. During the last few years of the nineteenth century many other American experimenters produced motorcars powered with gasoline, steam, or electricity. While American experimenters were improving internal-combustion-engine autos, a large number of cars powered by the better-understood steam appeared on the streets. During this period electric autos, slow and costly to operate but dependable, were favored by some, especially in cities. Although we must omit even mention of the names of dozens of Americans who, at the turn of the century, had a part in the development of the automobile industry, we cannot neglect Henry Ford, who produced a successful "gasoline buggy" in 1896 and formed the first of his several companies in 1899. Ford began to market this famous Model T auto in 1908; during the next twenty years he produced 15 million of them. Ford's assembly line was his revolutionary contribution to manufacturing methods.

The self-starter dates from 1911. Even at that time American manufacturers were still putting their principal efforts into producing automobiles that were dependable. They had used or were using internal-combustion engines, steam engines, or electric motors for power; chain, bevel-gear, or friction drive; bar, tiller, or wheel steering; and planetary or sliding-gear transmission. These early years have not ineptly been called years of "tinkering and guesswork." During the next decade the major features of design became stabilized. One of the most extensive innovations since 1920 has been the fluid-drive transmission which in various forms has been applied with differing degrees of success to many cars. There has, however, been no revolutionary change in the automobile since the Panhard of 1894, although year by year engineers have made innumerable mechanical improvements resulting in higher fuel economy, increased safety, and greater ease of driving.

Most of the early motor vehicles were designed solely for passengers, but Daimler built a motor truck in 1891 and Benz, the first motorbus in 1895 (Figure 12.14). Each type of vehicle has become an important form of transportation, and there are now great varieties of specialized bodies. Although the automobile bus has generally replaced the electric streetcar in the United States since the first motorbuses were ordered for urban use in 1922, the production of city buses has stabilized since 1925. Except for the period 1942 to 1948, the total production of buses

Figure 12.14 First motorbus—Benz, 1895 (Courtesy Daimler-Benz Aktiengesellschaft)

has been quite constant since 1925. In 1950, 80 per cent of all buses produced were school buses. This significant fact means that the one-room school is rapidly disappearing and that American children are receiving education in larger, better schools—one of the many examples of an important social change made possible by a highly specialized type of automobile.

Perhaps more than any other factor, automobiles have determined the requirements for modern city planning. The needs of the upper middle class had influenced the medieval planning discussed in Chapter 5; throughout subsequent centuries down to the eighteenth, planning was almost exclusively for the great houses and palaces of the wealthy. During the eighteenth century some attention was paid to the planning of imposing squares and great avenues, but it was not until the nineteenth century that city planning began again to take into account the requirements of the comfortably well to do. An example is London's private residential squares in the then suburban Bloomsbury district. These garden squares with their central greenery and residences were completely isolated from traffic. Lower social strata were still neglected.

Later in the century George-Eugene Haussmann (1809–1891) replanned Paris or, more accurately, for the first time planned it in accordance with demands of the new industrial age. He strongly emphasized broad straight avenues not only as arteries of communication but also as a means of controlling the riots which had plagued Paris in the first half of the century. Haussmann knew that broad boulevards would allow the circulation of light and air as well as of troops. The railway age also forced him to plan for the avoidance of traffic congestion in the vicinity of railroad terminals. Haussmann's emphasis on traffic was one of the principal early efforts to solve a problem which has become dominant in the twentieth century. Nevertheless, his plan intermingled traffic, residences, and labor, and neglected housing of lower-income groups.

In 1901 the French Tony Garnier enunciated, or at least exhibited in layout, the principle that was to become fundamental for city and town planning in the twentieth century: the segregation within an organic whole of traffic, residence, work, and leisure requirements. Modern planning has included all social classes and income groups and is based on factors which had been largely disregarded. For instance, preparatory to the designing of a large residential project, the modern planner assembles occupational and vital statistics to determine how many residential units there should be, say, for families of four children, how many for elderly couples, or for single persons, in the several income groups. However, the social fluidity provided by the automobile, electric power, and modern communications has largely dictated the principal requirements of present-day planning, at least in the United States. Automobile traffic is now a major consideration. Of equal importance is the changing character of American cities with their rapidly expanding suburbs. Until the 1940s suburbs were primarily residential, but during the past decade many commercial and industrial enterprises have also moved from the center of cities to suburbs. These changes made possible by the automobile vitally affect metropolitan planning.

Well-surfaced streets and highways are necessary for automobiles. Very few of the world's motor vehicles travel on surfaces which were not built, or at least maintained, for automobiles. Without the modern highway the automobile would be a useless contrivance—little more than a mechanical novelty.

McAdam died in 1836. Broken-stone roads more or less resembling those built by him and Telford in Great Britain, and in the eighteenth century by Trésaguet in France, continued to be built until the early twentieth century. The invention of the stone crusher in 1858 and the steam road roller in 1859 increased the speed and economy with which such roads could be built. In the early 1890s, before the advent of the automobile, it was the bicycle that started the good-roads movement in the United States. New Jersey, in 1891, and Massachusetts, in 1892, were the first to make state funds available for highways; the Federal Bureau of Public Roads was set up in 1893. The efforts of highway engineers of this period were mainly concerned with improving old macadam roads by spraying them with oils, such as asphaltic petroleum and light tar, that would lay the dust for a time and might even act as a binder. The emphasis was on methods for furnishing a protective coating that would prolong the life of the old road. Then came the automobile. In 1895 there were four registered horseless carriages in the United States; only ten years later there were 78,800 automobiles. Many automobilists at first confidently believed and asserted that their vehicles would act simply as road rollers and would compact surfaces. They were soon disillusioned when even the most carefully constructed crushed-stone roads began to disintegrate into loose stones and clouds of dust from the suction of pneumatic tires passing over them at high speeds. By 1910 it began to be clear that the macadam roads of the nineteenth century would be hopelessly inadequate for the twentieth.

The first roads built at the beginning of the twentieth century for the new type of traffic were of the penetration or bituminous macadam variety —a crushed-stone road into the interstices of which a hot bituminous liquid was poured or sprayed. Shortly afterward there was developed bituminous or asphaltic concrete. The bituminous material was mixed with a carefully graded aggregate, which might be crushed stone or gravel and sand, in a stationary mixing plant and afterward spread on the road and compacted while still hot. This type of road is now widely used on streets and highways having a high traffic density. During the same period in which the bituminous roads were being developed, a few roads were built of cement concrete, which had long been used for foundations in city pavements. The early cement-concrete city pavements in Inverness and Edinburgh, Scotland, and in Grenoble, France, were not particularly significant, although they seem to have been reasonably satisfactory. In this same category belongs a cement-concrete pavement laid around the courthouse at Bellefontaine, Ohio, in 1892. The Model T Ford dates from 1908. Within a year Wayne County, Michigan, soon to be a center of the automobile industry, boldly and with prophetic instinct undertook a definite cement-concrete road-construction program which marked the beginning of the modern use of cement concrete for highways. Thereafter, in several states, extensive field and laboratory investigations were undertaken—a vast research program which continued over a long period and covered all phases of cement-concrete road construction.

The great increase in the numbers and speed of automobiles in the 1920s necessitated new types of highways. There was also a need for increased highway funds, which was partially met at first by the Federal Highway Act of 1921, and by 1926 all but four states were imposing the gasoline tax for highway construction; in 1951 all states had taxes which varied from 2 cents a gallon in Missouri to 9 cents a gallon in Louisiana. The Public Roads Administration in 1925 established with the consent of the states the United States system of highways, a system of state roads planned on an interstate basis.

During the past few decades, in addition to the stone crusher and steam roller, an increasing number and variety of power machines have contributed to the effectiveness of highway construction. Early in the twentieth century compressed-air drills, long used in tunnels, simplified rock excavation for highways, and after the First World War, bulldozers and power shovels of increasing capacity supplanted earlier apparatus for handling earth and rock. Concrete mixers in many patterns had already taken the place of the traditional hand mixing. Especially useful have been new rolling and tamping machines and apparatus for careful and rapid finishing of road surfaces. Gasoline and diesel engines have almost entirely replaced horse and human power. Modern road design and construction have more and more depended on physics, chemistry, and soil mechanics; rule-of-thumb and empirical procedures have given place to scientific methods of approach. The steady increase in the numbers, weights, and speeds of motor vehicles during this period has broadened the problem of highway construction. It is no longer sufficient simply to provide a durable and smooth wearing surface; the alignment and profile of the highway must be carefully planned with regard to the safety of automobilists who may be traveling at many times the speed of hesitant early drivers.

The necessary changes in design were a slow development, and United States and European practice advanced at different rates. Automatic traffic lights were installed in American cities in the early 1920s, in British cities a few years later. Floodlighting of some heavily traveled congested areas came in the 1930s, and the bypassing of congested centers by miles of new construction was begun at about the same time. With allowable

Figure 12.15 One of the first cloverleaf highway intersections in the United States, Woodbridge, New Jersey, about 1930 (Courtesy New Jersey State Highway Department)

speeds on some highways approaching that of fast railroad trains, sight lines assumed increasing importance in highway design, and on the more carefully planned roads a driver could now at all times see the highway for a considerable distance in advance, whether he was approaching the crest of a hill or rounding a curve. Most curves were now widened and banked, with spiral easement approaches as on railroads. Despite these precautions, the toll of deaths from motor-vehicle accidents has everywhere continued to increase, although fatalities have remained at approximately a uniform ratio to passenger-miles.

During the past twenty years many multiple-lane highways have been constructed. The increased capacity of such roads, however, has not always been accompanied by increased safety. Superhighways with central grass strips were first built in the early 1930s. New Jersey constructed some of the earliest rotary intersections at grade in the 1920s and the first cloverleaf intersection with grade separation at Woodbridge, New Jersey, in 1930 (Figure 12.15). Express or limited-access highways, designed especially to carry fast-moving traffic, were built after 1925. On an express highway local traffic, even that of abutting residents, does not enter except at designated entrances, and intersections are at different levels with cloverleafs or ramps. The Hutchinson River Parkway in New York with its Connecticut extension, the Merritt Parkway, are examples of early express highways in the United States. Italy and Germany began construction of limited-access highways at a somewhat earlier date, and by 1937 Italy had nearly 300 miles of *autostrade* on which there was no speed limit. The *autostrade* bypassed cities, had no grade crossings, and few gradients steeper than 3 per cent. A considerable portion of the German *Autobahnen*, planned to be part of a 4,375-mile network, had been completed in 1937. Like other express highways they have no grade crossings and few steep gradients. The German government built the *Autobahnen* in particular to facilitate military operations and to provide a four-lane paved highway with central grass strips leading to every important border city.

In general, highway design and construction have lagged behind the capabilities of motor vehicles, but the combination of modern vehicles with modern streets and highways is one of the engineering developments which is rapidly changing the manner of human living.

Aircraft

Airplanes have operated only in the twentieth century. Their evolution during the first half of the century has been amazingly rapid, and at mid-century they are supplying society with its speediest transportation. In fact, the airplane's most important contribution to transportation is speed. Its commercial importance other than its speed is relatively meager, except in certain special instances where it provides access to otherwise inaccessible localities. Nevertheless, fast transportation over vast distances is having its impact on political, social, and economic activities.

Although man's desire to fly is millennia old, he first achieved flight in balloons during the last half of the eighteenth century. The Montgolfier brothers of Annonay, France, constructed balloons of waterproofed linen and used hot air from burning straw as the lifting medium. The first ascent of a human being was made by Jean-François Pilâtre de Rozier in a captive Montgolfier balloon on October 15, 1783. He and a friend made a "free ascent" on November 21, 1783, when they rose to

500 feet and drifted 5 miles in twenty-five minutes. The French physicist Jacques Alexandre César Charles (1746–1823), accompanied by a friend, made an ascent still later in 1783, using a balloon of Charles's design. Charles's law on the effect of heat on gases is named for him, and he was the first to use the new hydrogen gas for a lifting medium. Hydrogen had been discovered by Cavendish in 1766, but its name dates only from Lavoisier in 1783. Ballooning soon became a popular, albeit a hazardous, sport.

Power balloons have had a picturesque history, which dates from 1784 when the Robert brothers attempted to propel a hydrogen-filled balloon by means of light oars covered with silk. Subsequently many others tried to perfect a navigable balloon, but it was the Frenchman Henri Giffard (1825–1882) who built and operated the world's first dependable power dirigible balloon in 1852. After a half century of dirigible development, largely in France, Ferdinand von Zeppelin (1838–1917) built his first rigid ship in 1900. Although balloons and dirigibles have operated extensively in connection with military operations, they have proved impractical commercially. The evolution of balloons and dirigibles had little if any direct effect on the invention and early development of the airplane. It is a strange fact that of all the early contributors to the improvement of the airplane, only the Brazilian Alberto Santos-Dumont (1873–1932) had had any experience with lighter-than-air craft.

The history of the airplane is essentially the history of the engineering solutions of the twin problems of motive power and of aerodynamics. Nikolaus Otto had invented the four-stroke-cycle internal-combustion engine in 1876, and ten years later Otto Lilienthal (1848–1896) began experimenting with gliders to extend his knowledge of aerodonetics or the science of gliding. The two groups of experimenters who followed these Germans worked almost independently of each other, but their combined efforts and those of many other experimenters made possible the invention of the airplane by the Americans Wilbur and Orville Wright.

Lilienthal was a technically trained mechanic who had begun crude glider experiments in his boyhood. Actually, his first gliding flights date from 1886 when he was thirty-eight years old. The published results of Lilienthal's experiments testing the lifting power of curved surfaces had a great influence on the development of heavier-than-air machines. He was killed in a gliding accident caused by difficulties in maintaining longitudinal stability. Octave Chanute (1832–1910), French-born American civil engineer, following Lilienthal, made the first extensive gliding experiments in the United States, beginning in 1896. He used a variety of designs with multiple planes up to five, finally standardizing on a two-decker or biplane. Chanute afterward became the friend and counselor of the Wright brothers and gave them much sage advice.

From 1896 Professor Samuel Pierpont Langley (1834–1906), of the Smithsonian Institution, was experimenting on the Potomac below Washington with model airplanes. Some of these models of 13 or 14 feet spread were powered with small steam engines. Finally in 1903 he built a full-size one-man airplane and provided it with a remarkable gasoline motor designed by Charles Matthews Manly (1876–1927), a Cornell University graduate in mechanical engineering. Manly's engine was a five-cylinder water-cooled radial gasoline engine of 52 horsepower that weighed without accessories only 125 pounds (2½ pounds per horsepower). Unfortunately, Langley's plane twice collapsed as it was being launched from the roof of his houseboat, nearly drowning Manly, the pilot, and Langley became discouraged and refused to experiment further. His plane was recovered and restored and may be seen at the Smithsonian. During the closing years of the nineteenth century and the first of the twentieth, literally hundreds of men followed Lilienthal, Chanute, and Langley in experimenting with kites, gliders, or powered model planes. More than any of the others, however, these three men laid the foundation for the success of the Wright brothers.

Wilbur Wright (1867–1912) and his brother Orville (1871–1948), bicycle mechanics of Dayton, Ohio, were men of extraordinary genius. Developing a decided interest in gliding and mechanical flight, the Wrights read widely on the subject for some years and pored over books and papers by Lilienthal, Chanute, and Langley. The brothers built a wind tunnel, a crude open-ended box 16 inches square and 6 or 8 feet long, with a set of open pigeonholes to guide the air at one end, in which they tested more than 200 types of winged surfaces. They measured lift and drag, using monoplane, biplane, and triplane models. Octave Chanute, who was their confidant, was convinced that they knew more of aerodynamics than anyone living at the time. The approach of the brothers to their problem was from the first that of the engineer—open-minded trial, systematic tabulation of data, the intelligent interpretation of results, together with boldness to follow where the facts led. In 1900 they went down to the windy sand dunes near Kitty Hawk, North Carolina, where at Kill Devil Hill they would have plenty of wind to test their first

Figure 12.16 Orville Wright beside the 1903 airplane (From *Papers of Wilbur and Orville Wright,* courtesy McGraw-Hill Book Company, Inc.)

biplane glider. Here, in December, 1903, they had a powered biplane poised and ready for flight (Figure 12.16). Its framework was of wood; the wings were canvas. Their gasoline motor, a homemade affair, was 4-cylinder, 4-inch bore, and 4-inch stroke, and developed 12 horsepower. The motor weighed 179 pounds (15.9 pounds per horsepower); the whole machine 750 pounds. There were two propellers which turned in contrary directions about 10 feet apart; they were driven by chains that ran over sprockets.

Their first flight on December 17, the first in history for a heavier-than-air device, covered about 100 feet. Two years later, with a 24-horsepower motor, the Wrights made a complete circuit of 24 miles in thirty-eight minutes, and in 1908 Wilbur Wright flew 76 miles in a sustained flight. All of the Wright brothers' airplanes were, like their gliders, biplanes based on Chanute's design. Their most important innovation of warping the wings enabled them to control the plane and to maintain a better stability than any previous experimenters had been able to achieve. By 1903 the Wrights had made substantial contributions to the knowledge of the lifting power of wings, lateral and longitudinal stability, and maneuverability. Their first airplane had elevators and a rudder, as well as their wing-warping device. The Wrights's engine was adequate, but

Figure 12.17 Modern monoplane, a four-engine Douglas DC-7 (Courtesy United Airlines)

much heavier and less successful than Manly's. Their propellers were reasonably efficient and were probably the best that had been made by 1903.

During the quarter century following the Wrights's first flight, the evolution of the airplane was rapid. Henri Fabre made the first take-off from water in 1910 in the south of France, and he was soon followed by Glenn H. Curtiss in California. Igor Sikorsky built in Russia in 1912 the first four-engine airplane that would fly. By the late 1920s the monoplane had become the preferred type of design (Figure 12.17). Speed had increased from 30 miles per hour in 1903 to 280 miles per hour in 1924. Prior to 1940 airplanes used the piston-type gasoline internal-combustion engine almost exclusively. There had been many cylinder arrangements proposed, the number of cylinders varying from 4 to 24, and the engines were either in line or radial (Figure 12.18) about the shaft, air-cooled or liquid-cooled.

It was about 1940 that engineers made the first innovation in airplane motive power. The quest for speed and for operation at high altitudes had become increasingly important, and since it was obvious that the limits of motive power provided by the reciprocating engine and propeller were being approached, tests were made using the gas-turbine jet engine. By 1955 there were three main types of gas-turbine airplane engines: the turbojet, the turboprop jet, and the turbocompound. The history of gas turbines and jet power is extensive, having its faint beginning with Hero of Alexandria, but it was not until the twentieth century that such devices became commercially successful. By mid-century gas turbines had become important new prime movers.

Figure 12.18 A 3,500-horsepower Wasp Major radial engine (Courtesy Pratt & Whitney Aircraft Division, United Aircraft Corporation)

The first commercial installations of the gas turbine were of the stationary type; Brown Boveri and Company of Switzerland set up an explosion-type gas turbine in Hamborn, Germany, in 1933, and three years later installed the first compression-type turbine at Marcus Hook, Pennsylvania. The first gas-turbine electric-power unit was a 2,000-kilowatt Escher Wyss demonstration machine at Zurich in 1940, and by 1954 there were well over one hundred gas-turbine power plants in operation throughout the world. The first gas-turbine locomotive was also a Brown Boveri product, which was put into service on the Swiss Federal Railways in 1941. A decade earlier in 1930 Frank Whittle (1907–   ) in England patented a jet-propulsion engine having a blower-type compressor operated by a gas turbine; Whittle's arrangement is the basic design of the axial turbojet, the most widely used type of jet engine.

The turbojet is a thermal-air-jet engine consisting of a compressor, a combustion chamber, and a turbine (Figure 12.19). The compressor,

Figure 12.19 J-57 turbojet engine. Air intake and compressors at top, combustion chamber, center, turbine and exhaust, bottom. (Courtesy Pratt & Whitney Aircraft Division, United Aircraft Corporation)

driven by the turbine, builds up the pressure of incoming air, which then enters the combustion chamber where part of its oxygen burns continuously with injected fuel. The products of the combustion and the expanded hot air escape through the rear exhaust nozzle at high velocity, but before entering the nozzle they pass through the blades of the turbine that drives the compressor. The reaction to the ejected mass of gases drives the plane forward. The turboprop jet is similar to the turbojet except that it has a larger gas turbine which drives not only the compressor but also a conventional propeller. In the turboprop the propeller uses about 80 per cent of the energy output and the jet the rest. The turbocompound is a conventional propeller piston-type engine, the exhaust gases from which operate a turbine geared to the crankshaft to augment the engine's power.

All jet engines have a higher fuel consumption per unit of output than piston engines, but the turbojet produces much greater speeds especially at high altitudes, is structurally simpler, and weighs about half as much as a piston engine of equal power. However, the fuel consumption of the turboprop compares favorably with that of the large reciprocating engines. The world's largest piston engine, the Pratt & Whitney 28-cylinder radial type Wasp Major (Figure 12.18), producing as high as 3,500 horsepower, consumes fuel at the rate of 0.58 pound per horsepower hour; the 5,600-horsepower Pratt & Whitney Pt2F-1 turboprop has a fuel consumption of 0.64 pound per horsepower hour at take-off and the fuel consumption decreases as altitude increases. Moreover, the Wasp Major produces about 1 horsepower per pound of weight, but the turboprop produces over 2. The increase in horsepower ratings of gas-turbine jet engines has been spectacular. In the fifty years following the Wrights's first flight, the maximum horsepower of piston engines increased from 12 to 3,500, but in fifteen years the horsepower rating of jet engines at operational speeds climbed to 25,000, more than seven times the maximum horsepower of a piston engine. The British Comet, placed in service in May, 1952, was the first commercial airliner to have turbo-jet propulsion, and by 1954 the larger American planes were using the turboprop and the turbocompound.

The helicopter design is the second important innovation in recent aircraft history. The helicopter is a vertically rising device, the main characteristic of which is a large propeller rotating slowly on a vertical shaft. Leonardo da Vinci, who speculated much on aerial flight, suggested the idea in one of his sketches. In 1768 the French mathematician Paucton (1736–1798) produced a design driven by human muscle with two propellers, one to sustain the machine, the other to drive it forward, and many other enthusiasts toyed with the idea. In 1907 a Frenchman, Paul Cornu (1881–   ), constructed a full-size helicopter that lifted him and a passenger clear of the ground for several minutes. Cornu's power plant was a 24-horsepower gasoline engine that drove twin rotors by means of belts turning in contrary directions to counteract torque. The German-American Émile Berliner (1851–1929), his American son Henry A. Berliner, and many others in the United States, France, and Spain experimented with helicopters in the 1920s and 1930s. These men produced a variety of ungainly contraptions that would rise a few feet when conditions were favorable but were "very sensitive to disturbances" and seemed as unpredictable in their behavior as the traditional Missouri mule. Henry Berliner flew a tri-wing, tri-rotor machine for one minute and twenty-five seconds in 1922, and in the same year the Russian George de Bothezat flew a device with four lifting rotors for one minute and forty-two seconds at McCook Field. Years of patient investigations and experiment culminated in the success (Figure 12.20) of Russian-born Igor Ivanovich Sikorsky (1889–   ) in 1939. The development of his helicopter has

Figure 12.20 First successful helicopter with designer Igor I. Sikorsky at the controls, 1939

recently been very rapid. One of the limitations in lifting capacity lay in the vertical shaft on which the lifting propeller was mounted. As power demands increased, the weight of this shaft necessary to transmit the rotating became excessive. The application of jets at the ends of the blades to produce rotation by reaction has removed this limitation, and the field of usefulness of the helicopter is rapidly expanding.

Pipelines

Far less dramatic than railways, ships, automobiles, and airplanes, the pipeline (Figure 12.21) is, nevertheless, one of the most important types of transportation. In 1950 United States oil pipelines carried more ton-miles than intercity motor trucks. Iron or steel pipelines for the transport of oil were an American development which has spread to other oil-producing areas, and steel pipes now carry not only liquid petroleum and its products but also natural gas for thousands of miles.

Edwin L. Drake directed the boring of the first petroleum well in the United States near Titusville, in the Oil Creek valley of northern Pennsylvania, during the summer of 1859. The well was 69½ feet deep and at first produced oil at the rate of about 8 or 10 barrels a day for the Seneca Oil Company of Connecticut. Its yield increased considerably when pumps were installed. During the next few years many such wells were drilled in the Oil Creek region, some of them yielding as much as 1,500 barrels daily. As an illuminant this rock oil seemed from the first

Figure 12.21 A 26-inch pipeline being laid in trench dug by a diesel-powered Caterpillar Buckeye Ditcher (Courtesy Caterpillar Tractor Co.)

to have great possibilities compared with whale oil and candles. Oil Creek was, however, in a wooded region, far removed from population centers. Twenty or thirty miles to the north were several railroad whistle stops. For at least three years the oil was painfully trucked over frightful roads that hardly deserved the name to one of these railroad loading points. This business was expensive; the trucking often cost more than the railroad freight to New York. The other outlet was to the south via Oil Creek and the Allegheny River to Pittsburgh. Here the trucking distance was shorter, and water transportation was cheaper than rail. Flatboats and square-ended barges carried the oil, at first in barrels, then in specially built wooden tanks. Finally in 1862 came the Oil Creek Railroad which reduced the trucking distance materially and provided rail connection to Chicago and New York. In 1865 the first oil-tank car, a flatcar with a wooden tank holding about 45 barrels of oil at each end, made a successful experimental trip to New York, and thus became the ancestor of our modern tank cars, the first one of which dates from 1869.

The next step in oil transportation had been taken several years earlier. A pipeline 6 miles long was laid from Pithole City, Pennsylvania (a short-lived oil-boom town), to the Oil City Railroad at Miller Farm in 1865. Samuel Van Syckel, who built the line, installed three steam pumps that forced 1,900 barrels daily through its 2-inch wrought-iron pipe. In 1874 the first oil trunk line was built from the oil regions to the refineries at Pittsburgh, 60 miles away. This 4-inch line would transport each day 7,500 barrels, more than 300,000 gallons. The first pipeline to bring oil east across the Alleghenies was the Big Benson, a 6-inch line begun in 1878 and completed to the New York area ten years later in the face of determined opposition on the part of railroads and other interests.

Never before had liquid been pumped for a long distance at a constant rate through a pipeline that dipped down into valleys and crossed ridges in its direct overland route. Early experience with water pipelines was of little help; only a few water-supply systems had used pressure pumps; the crude Marly works at Versailles were almost the only ones that did. Cross-country oil pipelines presented entirely different problems in hydraulics. It was soon found that oil could be moved at the fairly satisfactory rate of several miles an hour by the use of single-cylinder reciprocating steam pumps located every few miles. The design and construction of oil pipelines continually presented new problems in the field of petroleum and hydraulic engineering. Until 1874 there were no accurate data on the efficient and economical movement of crude oil or any of its products. Engineers had yet to learn by experiment how to apply the fundamental concepts of hydraulics to the behavior of a liquid as viscous as are some petroleum products.

The most famous of the modern American oil pipelines, known as the Big Inch and Little Inch, were built with government funds during the Second World War. The first, a 24-inch line, extends from Texas to Philadelphia and New York and carried crude oil; the second, 20 inches in diameter, carries petroleum products such as gasoline and fuel oil between Texas and New Jersey. Pipes also carry natural gas to the Atlantic Coast as well as to other sections of the United States.

The first natural-gas pipeline was the line at Fredonia, New York, which was put into service in 1821, but it was not until the twentieth century that natural gas began to be used extensively as a fuel. During the 1930s the natural-gas industry expanded rapidly, and natural gas began to replace gas manufactured from coal. By 1955 gas lines stretched from Texas to New England, and natural gas has replaced not only manufactured gas but also coal in many industries. Gas is not only easier to handle and use, but since it is cleaner than coal, it has greatly reduced the amount of smoke and soot in some industrial areas. Prior to the widespread use of natural gas most of it was allowed to escape into the atmosphere or was burned nonproductively. This appalling waste of a valuable natural resource has been greatly reduced by storage in abandoned wells for subsequent use and the extension of pipeline networks. The development of the productive use of natural gas had been one of engineering's great contributions to conservation of resources and to transportation.

Bibliography

Andrews, Cyril B.: *The Railway Age,* Country Life, Ltd., London, 1937.

Bruce, Alfred W.: *The Steam Locomotive in America,* W. W. Norton & Company, Inc., New York, 1952.

Evans, Arthur F.: *The History of the Oil Engine,* S. Low, Marston & Co., Ltd., London, 1932.

Henry, Robert S.: *This Fascinating Railroad Business,* The Bobbs-Merrill Company, Inc., Indianapolis, 1943.

Magoun, F. Alexander, and Eric Hodgins: *A History of Aircraft,* McGraw-Hill Book Company, Inc., New York, 1931.

Nixon, St. John C.: *The Invention of the Automobile,* Country Life, Ltd., London, 1936.

Wilson, Charles M.: *Oil across the World,* Longmans, Green & Co., Inc., New York, 1946.

Zahm, Albert F.: *Aërial Navigation,* D. Appleton and Co., New York, 1911.

THIRTEEN

Sanitary and Hydraulic Engineering

Man cannot live without water, which is more essential to life than any other nutrient except oxygen. In addition to being a vital necessity for man, water provides society with increased amounts of plant and animal foodstuffs through the irrigation and reclamation of lands. Water is essential for extinguishing fires; indeed, some early modern American water-supply systems furnished water mainly for fire fighting.

Since the beginnings of civilization, bodies of water have served as highways for transportation, and since the Middle Ages, water has been one of the important sources of power. As described in Chapter 5, the water wheel was one of the early devices that relieved men and draft animals of being the principal providers of power, and in the eighteenth century water became of even greater importance in power production with its use in the steam engine. Midway in the twentieth century, water participates in producing electric power as it turns water wheels and steam turbines and cools internal-combustion engines. Industry also uses enormous amounts of water for other purposes than power production; chemical, food, and textile plants, in particular, use water extensively. Indeed, engineers developed some of the first water-purification techniques for industrial uses rather than for human consumption.

Nevertheless, it has been primarily the demand for "pure" water for human consumption in metropolitan areas that has brought about the construction of extensive water-supply systems such as that of New York City. These systems not only store and purify water but also transport it many miles; in a sense, long aqueducts are important forms of transportation, carrying many ton-miles each day. In addition to the provision of pure water, sanitary engineers have devised techniques for the disposal of human wastes to safeguard health. The triumphs of sanitary engineering over epidemic intestinal diseases are among the remarkable achievements in the field of public health.

Sanitary Engineering

Early in their communal life men learned that they could not survive in crowds unless they had systems of water supply and sewage disposal. The discoveries of the archaeologists at Mohenjo-daro in the Indus Valley testify to this fact, as do the ruins of the Minoan civilization in Crete. The records of Rome are filled with evidence of endeavor to provide both of these facilities for the cities of the Empire. Realization of a necessity, however, is not the same as effectiveness in action. It is not praise of Minoan or Roman sanitation to declare that the sanitary facilities of those days were better than the outhouses, cesspools, and primitive wells which persist even in our own time.

The idea that pure drinking water might be better than impure water for human beings appeared in the writings of the Greek physician Hippocrates, who lived in the fifth and fourth centuries B.C. But providing pure water has been an intricate and baffling problem as populations have increased and crowded into relatively small areas. Diseases which experts in public health now trace to defective sanitary systems ravaged towns and cities in the Middle Ages and more recent times; unfortunately, they still do in the world's underdeveloped areas. Municipal authorities of our day, however, enjoy a better understanding of the task than was the lot of their predecessors in London, Paris, Rome, Cnossus, and Mohenjo-daro. And for most of this increase in knowledge the scientists and engineers of the past hundred years are responsible.

Before Louis Pasteur (1822–1895) and Robert Koch (1843–1910) did their revolutionary work in bacteriology, humanitarians and officials concerned with public health had decided that something more had to be done about foul drinking water than to free it from the evil tastes, odors, sediment, and crawling things which anyone could see for himself. The French observer Dr. Alexandre-Jean-Baptiste Parent-Duchâtelet (1790–1836) had remarked in 1836 that he did not know the amount of foreign matter which water must contain to be dangerous. He was certain that the daily use of filthy and disgusting water had no effect on animals; the same might be true of the men in a woolen mill he had visited. Yet "some principles of infection which defy analysis" could nevertheless exist. He was right in both respects. Mere putrescence in water, however repulsive, has not been proved fatal if there are no disease-producing organisms present. But in this endeavor to eliminate putrescence from drinking water, health officers and engineers empirically came upon the principles of infection that hitherto had defied analysis.

Dr. John Snow (1813–1858), a pioneer in anesthesia, sensed the truth and published in 1849 an essay on the communication of cholera by contaminated water. His experience in the famous Broad Street well epidemic of 1854 in London led Dr. Snow to suggest to local authorities that they remove the handle from the Broad Street pump. This was done, and as soon as the use of the contaminated water was discontinued, the epidemic subsided. Revising his essay, Dr. Snow anticipated remarkably the theory of the germ origin of infectious disease, but other leading men in public health did not agree. The dominant idea continued to be that putrescence itself in water caused disease. Roused by the persistent inroads of cholera, the British Parliament passed the Metropolis Water Act in 1852, requiring that after three years all water supplied for household use within the area of London must be filtered, except that pumped directly from deep wells into covered reservoirs. Slow filtering through sand beds, in imitation of nature, had been practiced for some years to make water pleasanter to sight, taste, and smell. Soon after 1852 the practice began to spread throughout Great Britain, since it seemed to check disease as well, and by about 1900 the public supplies for many major cities were being filtered.

The work with purification in Britain attracted two Americans to the study of European accomplishments for use at home. They were Ellis Sylvester Chesbrough (1813–1886), city engineer first for Boston and then for Chicago, and James Pugh Kirkwood (1807–1877), a native of Edinburgh, Scotland, who had come to America as a young man in 1832 to build railroads. Chesbrough's investigations and report of 1858 led to a municipal system which in the end properly related the sewage disposal of Chicago to its water supply. Meanwhile the city continued to empty its liquid wastes into the Chicago River and Lake Michigan, in hopes that submerging and diluting in the lake water would remove all noxious elements. In 1867 Chesbrough ran a 5-foot cast-iron brick-lined tunnel 2 miles out through blue clay 30 feet under the bed of the lake. At the end of the tunnel was a crib, where water was taken in. A pumping station on shore raised the water from the level of the lake into a 154-foot tower, whence it flowed by gravity through the city's mains.

The population of Chicago, however, had leaped from some thirty thousand inhabitants in 1850 to more than three hundred thousand in 1870. Pollution of the lake outran tunneling from the shore. Even at 4 miles out the water was unsafe. It was obvious that the flow of sewage into Lake Michigan had to be checked. First the engineers tried pumping some of it over the watershed behind the city into the Illinois and Michigan Canal and the Illinois River. Then dredging, begun in 1892 and completed in 1900, permanently reversed the flow of the Chicago River into the Des Plaines River and gave the people of the back country the benefit of Chicago's waste. This was no treat to neighbors on the south although it was effective in safeguarding the drinking water of the city. The sanitary experts could not be sure that the diluting and oxidizing which occurred on the journey downstream removed the harmfulness of Chicago's sewage before it reached the next urban center dependent upon the watershed. The state of Missouri sued the state of Illinois and asked for injunctions to prohibit Chicago from dumping sewage into the St. Louis water supply. The court dismissed the case when the evidence showed that the pollution had been so removed or diluted by the time it reached St. Louis that it was not a health menace. Nevertheless, the condition of the drainage canal and the upper Des Plaines River became so intolerable that Chicago was forced to install modern sewage-treatment plants to eliminate the nuisance.

Years earlier, in 1865, it had been in the search of means for providing the city of St. Louis with clear and palatable water from the Mississippi River that Kirkwood was sent to Europe. The result was his report of 1869 on the filters which he studied in England, Scotland, Ireland, France, Germany, and Italy. But the authorities of St. Louis decided not to undertake to separate the water of the Mississippi from its mud by means of the sand filters which Kirkwood advocated. Poughkeepsie, New York, engaged him to apply his principle in 1872 to the Hudson River, where he constructed the first sand-filtration plant in the United States. The water of the Hudson was less turbid, though quite as harmful, since a state hospital discharged raw sewage into the river only 2,000 feet above the water intake and continued to do so for sixty years. Typhoid deaths continued in Poughkeepsie at an erratic rate, but owing to several improvements in the purification plant and the treatment of the hospital sewage beginning in 1933, the deaths rapidly dropped to zero during the 1931 to 1935 period.

A fundamentally important scientific advance was the theory of Louis

Pasteur (1822–1895) that infectious disease is caused by germs or bacteria. Pasteur announced his formative theory in 1857, and by the 1880s Robert Koch (1843–1910) had firmly established the science of bacteriology. The application of Pasteur's theory and a knowledge of bacteriology to methods of supplying pure water have brought about some of the greatest successes of sanitary engineering. Aware of the discoveries of Pasteur and Koch, biologists, chemists, and engineers on both sides of the Atlantic experimented with the filtration of polluted waters to see if the microorganisms which caused diseases were water-borne and could be removed. Karl Joseph Eberth (1835–1926) and Edwin Klebs (1834–1913) had identified the typhoid bacillus in 1880. In Britain on April 16, 1886, Percy Faraday Frankland (1858–1946) told the Institution of Civil Engineers that he had removed most of the bacteria from samples of London's water by filtering slowly through sand. In Berlin, Germany, during the same year, Karl Piefke found that completely sterilized sand alone did not retain microbes; certain organisms with gelatinous coverings had to develop a "living slimy layer" which caught the bacteria and accomplished most of the purification.

In America, Hiram Francis Mills (1836–1921) and his younger associates at the Lawrence Experiment Station in Massachusetts made proof of this observation in dramatic fashion beginning in 1887. George W. Fuller, one of Mills's associates, returned from study under Piefke in Berlin in the early 1890s. Their joint investigations revealed that filtering intermittently through sand lowered the count of the typhoid bacilli in the polluted water of the Merrimac River. The intake of the water supply for Lawrence was only 8 miles downstream from the outlets of the sewers of Lowell. The work of the Lawrence Experiment Station attracted wide attention at home and abroad. Its young men, Edmund B. Weston (1850–1916), George W. Fuller (1868–1934), Allen Hazen (1869–1930), and George Chandler Whipple (1866–1924) went elsewhere in the United States as engineers and consultants to carry on work in sanitary engineering. Their efforts brought prompt results. Between the years 1886 and 1925, the death rate from typhoid in the urban centers of the United States fell from 50 per 100,000 of population to less than 5.

Slow filtering through sand became general in the United States soon after 1900. But rapid filters, first devised to purify muddy water for paper mills, presently displaced more than half of the older type, after independent studies in the 1890s by Weston and Fuller had shown the possibilities of using rapid filtration for treating municipal supplies. Rapid filters are more economical in their use of land and materials. As the standards of purity became increasingly exacting, preliminary sedimentation with the addition of a coagulent like aluminum sulfate clarified the water before filtering and thus lightened the burden upon the filters. It has become common procedure in recent years to add a disinfectant, chlorine, to filtered water as a final precaution.

As nineteenth-century cities grew in population, engineers were called on to provide increased water supplies. Since Watt's day some cities had depended more and more on water pumped by steam from local rivers which were often, like the Thames, grossly polluted. Many important cities, however, followed ancient Rome's example and reached back into distant highlands in order to tap sources that were relatively pure. At mid-century engineers began to realize, as perhaps the Romans had not, the importance from a sanitary point of view of providing extensive storage reservoirs close to the sources. Moreover, in the last half of the nineteenth century, as knowledge of sanitation began to catch up with engineering ability, the techniques of modern water supply began to evolve on a healthful basis.

The Marseilles Aqueduct, built 1839 to 1847, draws its water from the Durance, an Alpine tributary of the Rhone, at a point 28 air miles north of Marseilles. Its course is a devious one; the 51-mile aqueduct, built in general on a uniform slope, passes over many small bridges and through no fewer than forty tunnels. At Roquefavour, 5 miles west of Aix, it crosses the steep little Arc River Valley on a three-tiered bridge (Figure 13.1) similar to but even more striking than the Roman Pont du Gard Aqueduct at Nîmes, 60 miles to the west. De Mont Richer's imposing Roquefavour aqueduct bridge of cut stone is 1,300 feet long and towers nearly 300 feet above the river. This canal aqueduct is generally 30 feet wide on top and 7 feet deep. It not only supplies Marseilles, but it also brings water by a series of branches for irrigation to the Crau, a vast arid basin to the west, toward Arles, the area which has been served by the Crapponne Canal since the sixteenth century. By 1855, Glasgow, Scotland, had begun its aqueduct from Loch Katrine, 26 miles away, and in 1869, Vienna, Austria, tapped the mountain springs 59 miles to the west in the Schneeberge. With these enlarged supplies of water came dams of greater height, forerunners of the huge structures of our day. The Furens Dam, built of granite in 1861 to 1866 on a tributary of the Loire River near the mining city of St. Etienne, France, was then the tallest in the world, 184 feet. It curved slightly upstream as an arch

Figure 13.1 Bridge-aqueduct of Roquefavour (compare with Figure 4.2) (From *Minutes of Proceedings, Institution of Civil Engineers, 1854–1855*)

against the pressure of the water. The largest dam in Britain was built in 1881 to 1890 on the eastern slope of the Berwyn Mountains in Wales to supply Liverpool more than 50 miles away; its aqueduct reaches 68 miles.

New York City Water Supply

The history of New York's water system is an excellent illustration of the engineering advances which made possible the many extensive metropolitan water supplies of the twentieth century. For nearly two hundred years after its founding New York City had no public water-supply system. The inhabitants depended on private or public wells and pumps. The Tea Water Pump, on the north side of the present Park Row, was famous through the years for the apparent purity of the spring water it furnished for tea making and cooking; it was even distributed through the town in carts by the "tea-water men," somewhat as a luxury. The first successful pipeline system was a private enterprise. The Manhattan Company, incorporated in 1799 by Aaron Burr and his friends, proceeded to sink wells, build tanks and reservoirs, and lay mains of bored logs. The city's population then was about sixty thousand. Within a generation the company laid 25 miles of wooden mains which sometimes supplied daily as much as 700,000 gallons of not too pure or even too palatable water to about 2,000 homes. The water was pumped by two 18-horsepower steam engines from a number of large wells, some

432

Figure 13.2 New York City's first municipal supply reservoir, built 1829 (From *The Family Magazine*, 1839)

of them not far from the Collect Pond, into elevated tanks or reservoirs from which it flowed by gravity. The Collect Pond was an irregular-shaped body of fresh water, several city blocks in area, lying east of Great George Street (now lower Broadway) and a short distance north of the present City Hall.

During this time the city itself proceeded to build what may properly be called its first public municipal system. In 1829 it established an elevated cylindrical cast-iron tank reservoir with a diameter of 43 feet and height of 20½ feet, enclosed in an octagonal stone building (Figure 13.2). This reservoir, holding 230,000 gallons, was "far uptown" on 13th Street west of Broadway, then called Bloomingdale Road, and just below Union Square. The water surface was 104 feet above sea level. The well and a 12-horsepower steam engine which pumped 21,000 gallons daily were some five blocks away at Jefferson Market, Sixth and Greenwich Avenues. This project was for fire protection, and the mains were 12-inch cast iron. Up to this time the city had not been solely dependent on the Manhattan Company, for there were forty or more public cisterns from which the many fire companies pumped their own water. Even as the Manhattan Company started, about the turn of the century, to lay its mains, the Common Council was considering the recommendations of the English engineer William Weston that the city get its water from the Bronx River at "Lorillard's snuff factory" some miles to the north in Westchester County. The report was not acted on, but strangely enough the Rye Ponds, sources of the Bronx River, were incorporated into the Kensico Reservoir on the city's Catskill Aqueduct system, some 118 years later, in 1917.

About 1830 there was renewed agitation as the city's population continued to grow (220,000 in 1832) for a source of water supply outside of rocky Manhattan Island. Many sources were suggested, among them the Housatonic River in Connecticut, the Passaic in New Jersey, and even the Hudson. Opinion gradually crystallized in favor of the Croton River, a tributary of the Hudson some 40 miles north of the City Hall. The city voted a $2,500,000 issue of "water stock," and in 1834 the New York State legislature authorized the Croton Aqueduct project. Construction began in 1837. Appointed chief engineer in 1836, John B. Jervis, later an outstanding railroad engineer, directed and completed the project.

In 1842, with New York's population at 360,000, the Croton supply of 220 gallons daily per capita seemed adequate. The Old Croton Dam, built in 1842, of granite ashlar masonry, was 50 feet high with a very wide spillway 166 feet above sea level. Croton water had been brought 45 miles to the city in a horseshoe-shaped brick-and-stone aqueduct about 7 feet wide and 8 feet high. In the main the devious course of this aqueduct had followed the contour of the ground nearly parallel to the Hudson River, so that, as in the old Roman aqueducts, the water took a generally uniform slope, which was about 13 inches to the mile. At the crossing of the Harlem River, the lofty monumental High Bridge of 15 semicircular arches was in appearance not unlike certain late Roman structures, such as Segovia and the Pont du Gard. It was, however, built a few feet below the hydraulic grade line, and so carried the water under some pressure in two 36-inch cast-iron pipes to which a 90-inch wrought-iron pipe was later added. High Bridge has been replaced by a steel-arch bridge. After reaching Manhattan Island the aqueduct continued down Tenth Avenue to a receiving reservoir south of West 86th Street in what later became Central Park, and from there down Fifth Avenue to the final or distributing reservoir "at Murray Hill, a short drive from the city." This reservoir was on the west side of Fifth Avenue between 40th and 42d Streets, where the New York Public Library now stands (Figure 13.3). The water surface was 115 feet above sea level, about 8 feet above the roofs of the houses at the bottom of the hill. The Croton

Figure 13.3 New York's Murray Hill Reservoir, Fifth Avenue and 42d Street, 1842 (Courtesy New York Public Library)

Aqueduct was completed late in 1842. New Yorkers were properly thrilled when they realized the generous supply of "pure and wholesome water" that had thus been brought to their very doors. Their confidence, it was said, in "the abundance of the source relieves from all solicitude as to adequate supplies for the multitudinous population of hereafter." Boston at that time had only a most unsatisfactory supply, and Philadelphia's water, pumped from the nearby Schuylkill, was not too pure or inviting. In October of 1842 the Croton Aqueduct was finally put into use—not, however, without appropriate pomp and circumstance. John Tyler, President of the United States, headed the list of notables who, unable to be present, wrote congratulatory messages. The October 14th parade was a gala affair, "the most numerous and imposing procession ever seen in an American city." Seven miles long, it took more than two hours to pass the City Hall. Four thousand fire fighters from ninety-two companies participated, firm in their belief that hand-pumping engines were now outmoded. Temperance and total-abstinence societies to the number of thirty also marched, convinced as they were that Croton water was to be the beverage of the future for all proper New Yorkers. Those who attended the official City Hall collation which followed found that they could drink either Croton water or lemonade, but no wine or spirituous liquors. "It was a well arranged republican repast." The ceremonies closed with "nine hearty cheers for the City of New York and perpetuity to the Croton water." "Magnificent fountains in the [City Hall] Park and Union Square . . . formed the most novel . . . feature of the day." [1]

Within a few years, following a succession of dry seasons, the Croton Lake storage supply was substantially augmented by more and more reservoirs, almost all of them on the Croton River or its many tributaries. Meantime the city's population increased steadily. Finally, during the years 1885 to 1893 a new Croton Aqueduct, its horseshoe cross section of 160 square feet more than three times that of the old one, was built connecting Croton Lake with the Central Park reservoir. Its form was the same as the old one, but instead of following a meandering route and being laid close to the surface, it was generally straight and bored practically altogether as a tunnel on the hydraulic grade through rock. There were 30 shafts, averaging 127 feet deep, the deepest nearly 400 feet below the ground surface. The 30-mile tunnel was the longest in the world at the time and for long afterward. The Harlem River Valley was crossed this time by a pressure tunnel or inverted siphon, 300 feet deep and nearly 7 miles long, carrying water at a depth of 420 feet below hydraulic grade. This river crossing was an unprecedented procedure, easily the boldest undertaking on the entire aqueduct; nothing like it in magnitude had ever been attempted anywhere. It was made possible and sure by an extensive system of diamond-drill borings down through the various underlying strata and well into hard limestone and gneiss. The entire tunnel excavation brought out a number of new features. Incandescent lamps were used as early as 1886; ventilation was provided mainly by air discharged from the compressed-air drills. The average progress of excavation in the tunnel headings was 25 to 40 feet per week. At the time of the completion of the new Croton Aqueduct in 1892 the population of the city had increased to nearly two millions and the daily per capita consumption to about 100 gallons. The capacity of the new aqueduct was in excess of 340 million gallons daily, considerably greater than the anticipated yield of the entire Croton watershed.

Just as the New Croton Aqueduct was being finished, work began under the direction of Chief Engineer Alphonse Fteley on the New Croton, or Cornell, Dam several miles down the Croton River from the

[1] Charles King, *A Memoir of the Construction, Cost, and Capacity of the Croton Aqueduct,* printed by the author, New York, 1843, *passim.*

much smaller dam completed in 1842. This dam, with its total height of nearly 300 feet, flooded out the old dam, raised the water level 36 feet, and transformed Croton Lake into a reservoir 19 miles long, holding about a third of the more than 100 billion gallons stored in the entire Croton system. The new dam is a graceful structure of stone masonry, and when completed in 1905, was the tallest dam in the world. Although Fteley had prepared the original plans and directed the construction during the first seven years, he did not complete the dam; William R. Hill succeeded him in 1900 and was in turn succeeded by J. Waldo Smith, who completed the major construction.

For some years what to many seemed an audacious, even foolhardy, plan had been under consideration for securing a greatly enlarged supply from several streams in the Catskill Mountain area, 70 to 90 miles farther north and a few miles west of the Hudson. After extensive studies by a board of distinguished engineers, including a comparison with several other schemes, the undertaking gradually took shape; construction started in 1907 and continued until 1937. During part of this time as many as 17,000 workmen were daily engaged on the project. The cost has already exceeded 250 million dollars. This Catskill water-supply system began as a project to bring to New York City the water from the relatively pure Esopus Creek, a tributary of the Hudson. Here, about 14 miles west of Kingston and 80 air-miles from New York, was erected the Olive Bridge Dam, an epoch-making structure 252 feet tall, built largely of concrete with large boulders embedded, known as cyclopean masonry, and faced with concrete blocks. From the Ashokan Reservoir thus formed the Catskill Aqueduct extends southerly in a somewhat winding course through Ulster and Orange Counties to a point midway between Newburgh and West Point. Then it swings east and at Storm King Mountain drops vertically to a point 1,500 feet below the hydraulic grade or natural flow line as it passes 1,100 feet below the Hudson in a 14-foot cylindrical tunnel through bedrock. On the east bank it rises vertically, and the natural gradient is resumed, except for a few inverted siphons, until it reaches the Kensico Reservoir 3 miles north of White Plains. This reservoir will hold about a month's supply for the city.

Fifteen miles beyond the Kensico Reservoir is the Hill View Reservoir, east of Yonkers and just outside the New York City limits. In its 98 miles the average fall of the Catskill Aqueduct is about 2.2 feet to the mile. It brings water into the Hill View Reservoir at an elevation of 295 feet above sea level as compared with the 115 feet elevation of the old

Murray Hill Reservoir. The Hill View Reservoir's function is to equalize the difference between the steady flow in the aqueduct and the hour-to-hour varying use of water in the city. Two deep distributing tunnels, aggregating 38 miles in length, deliver water from the Hill View Reservoir to the several boroughs of the city. These tunnels were bored through solid rock to depths down to 750 feet below the ground surface or 500 feet below sea level and are lined with concrete. One of the tunnels extends the length of Manhattan Island; the other, farther east, is 500 feet below the sea throughout and supplies sections of Queens and Brooklyn Boroughs. The tunnels meet in Brooklyn and eventually, by cast-iron and steel conduits laid under the Narrows, deliver this water by gravity to the terminal Silver Lake Reservoir on Staten Island in the Borough of Richmond at an elevation of 228 feet above sea level.

Two features of the Catskill Aqueduct are of especial engineering interest. One is deep underground, while the other is in the open air and daily visited by sightseers. The underground feature is the 3,000-foot-long pressure tunnel under the Hudson, mentioned on page 437. At the point of crossing it was necessary to be sure of the character of the bed of the river for hundreds of feet down since a satisfactory pressure tunnel could not be bored through glacial drift or seamy or unsound bedrock. Twenty-four vertical holes, bored from scows anchored in the river, furnished only inconclusive evidence as to subterranean conditions. It was not until four diamond-core-drill holes, two from each side, at two different inclinations, had been bored entirely through granite and had reached midstream 900 and 1,450 feet below sea level, respectively, that the engineers obtained adequate information to enable them to plan a horizontal tunnel at 1,100 feet below the river. This tunnel is circular in cross section, 14 feet in diameter, and is lined with plain, not reinforced, concrete averaging 17 inches thick. The water exerts a pressure of more than 600 pounds per square inch on the walls.

The second feature concerns the purity of the water. The Catskill Aqueduct flows by gravity, without pumping, to most parts of the city. Coming as it does from a sparsely settled area free from limestone, the water is soft and so exceptionally pure that filtering is unnecessary. It is, however, screened, chlorinated, and aerated as it leaves the Ashokan Reservoir and again below the Kensico Reservoir. At each of these points 1,599 nozzles in the aerator break up the water and throw it high in the air in a fine spray—a spectacular sight. This aeration not only liberates any odor-producing gases in the water but also charges it with oxygen.

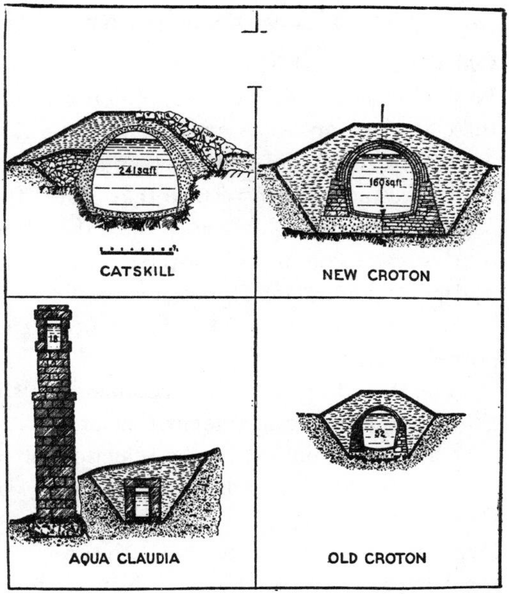

Figure 13.4
Cross sections of Roman and New York Aqueducts (From Johns Hopkins University, School of Engineering, *Lectures on Engineering Practice*, 1921–1922; courtesy Johns Hopkins University Press)

The ancient Roman aqueducts described in Chapter 4 were all of simple rectangular cross section. The Catskill Aqueduct is in large part horseshoe-shaped, with a very slightly concave invert, or floor, and an arched roof, all of concrete, mostly without steel reinforcement. This economical cross section was used where the aqueduct followed a hydraulic (natural-flow) grade and lay either close to the surface of the ground or in tunnel, for some 69 miles in all. None of this aqueduct is raised on lofty arches as was the Old Croton at the Harlem River Valley crossing; the Catskill Aqueduct used inverted siphons at almost all river and stream crossings. Speaking broadly, where this aqueduct follows just below the natural surface of the ground for perhaps 55 miles, it is a much glorified version of both Croton Aqueducts. Where it drops vertically hundreds of feet below the surface, carrying the water under pressure in deep pressure tunnels, or inverted siphons, the Catskill Aqueduct is an enlarged version of the Harlem siphon, the boldest undertaking on the New Croton Aqueduct. In cross section the Catskill Aqueduct (Figure 13.4) is roughly half again as large as the New Croton and nearly five times as large as the Old Croton. A tall man could walk erect through the Old Croton, a small auto could be driven through the New Croton, while a railroad train could fit into the Catskill. In carrying capacity the

Old Croton could in its prime deliver some 90 million gallons daily; the New Croton can deliver 340 million gallons, the Catskill somewhat more than 600 million.

Above the Ashokan Reservoir the Catskill Aqueduct was in 1926 extended 34 miles northwest, to take water from a reservoir built on Schoharie Creek which flows north into the Mohawk River. The main feature of this extension is the 18-mile Shandaken Tunnel, most of it on the hydraulic grade. Here is perhaps the first instance in history where water is conveyed through a watershed between two streams flowing in opposite directions.

The first Catskill water reached the city in 1917. The city's rapid growth soon made it evident that an additional supply would be necessary. Investigation led to the selection of several extensive and sparsely settled watersheds to the west of the two which were supplying the Catskill Aqueduct, all of them on tributaries of the Delaware River. The Delaware Aqueduct extends southeasterly from the Rondout Reservoir, some 10 miles east of Liberty, New York, and, like the Ashokan, about 85 air miles from New York's City Hall. From this reservoir the aqueduct is straight, crossing first under the Catskill Aqueduct and then under the Hudson River 5 miles above Newburgh to the West Branch Reservoir on the upper reaches of the Croton watershed in Putnam County. From here it extends southerly by way of the Kensico Reservoir to the Hill View Reservoir. Throughout its entire length it is in deep tunnel, roughly from 300 to 1,000 feet below the ground surface. At its lowest point, nearly under the Kensico Reservoir, it is 660 feet below sea level, or 1,500 feet below the surface of the Rondout Reservoir. In form it is circular, varying in diameter from 13½ to 19½ feet inside diameter, or slightly larger than the pressure tunnels on the Catskill Aqueduct. The shafts are 1.8 to 5.2 miles apart; the deepest is 1,551 feet deep. The rate of excavation of the tunnel was 135 to 270 feet per week as compared with 55 to 70 feet on the Catskill and 25 to 40 feet on the New Croton. This increased rate reflects sixty years' progress in excavating machinery and methods. The combined length of the Delaware Aqueduct and one of the two city tunnels is 105 miles (Figure 13.5); it is the longest continuous tunnel ever built for any purpose anywhere.

Not all of New York City's water is supplied by the Croton, Catskill, and Delaware Aqueducts. A small fraction still comes from the sandy Long Island area east of Brooklyn. Previous to its inclusion as a community in Greater New York in 1898, Brooklyn's complicated water

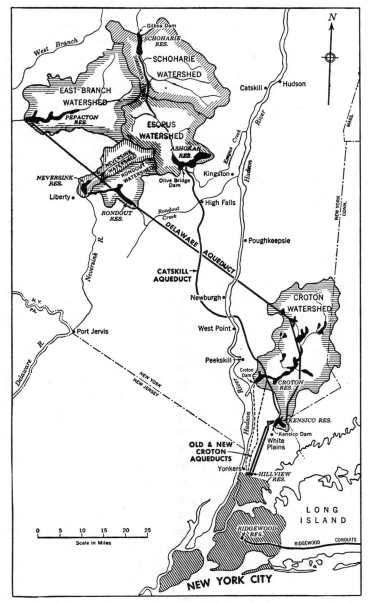

Figure 13.5 New York City's several aqueducts, showing sources (Based on Board of Water Supply maps)

supply came mainly from a great number of ponds and deep-driven wells in this area. Some of it was furnished by private companies, and all of it required at least one pumping and considerable purification. The largest system was the Ridgewood, which supplied nearly 100 million gallons daily, about a third of the total. Most of these systems still form part of New York's municipal supply.

Increasing demands have tended to deplete the city's reservoirs at times of long-continued drought. Provision has therefore been made for obtaining additional water during extreme emergencies from the Hudson River in the vicinity of Chelsea, a few miles south of Poughkeepsie and near a shaft of the Delaware Aqueduct. This emergency supply may amount to about 100 million gallons per day. Since the river water, while not brackish at this point, is essentially impure, it must be suitably screened, superchlorinated, and otherwise treated before it is admitted into the aqueduct.

Sewage Disposal

Sewers had been originally designed to carry away storm waters; the famous Cloaca Maxima in ancient Rome was mainly a storm-water sewer. Human wastes and other offensive matter had been either buried or removed by public authority or private contractor. It was only with the general installation of water closets, some years after Joseph Bramah's modern invention in 1778, that noisome cesspools or the dilution and discharge of sewage into nearby bodies of water became general practice and compelled municipal engineers to construct sanitary sewer systems. Following a conflagration that destroyed much of the old city, Hamburg, Germany, built in 1843 a complete sewer system which had for one of its features a thorough flushing each week with river water. Chesbrough began a similar system for Chicago in 1855. Two years later Julius Walker Adams (1812–1899) developed for Brooklyn, New York, a system that discharged into tidewater.

Edwin Chadwick (1800–1890) in England had proposed that two sewer lines be laid in each street, one for storm water, the other for sewage. John Phillips, surveyor for London, advocated such an improvement in 1847, and others stressed the need. But it was not until the 1880s that there was real progress. The model town of Pullman, Illinois, near Chicago, and the city of Memphis, Tennessee, in 1880, appear to have been the first American communities to have this double system. At present in the United States, four times as many communities use separate sewers for storm water and for sewage as have single systems for both, but experience has shown that complete separation is practically hopeless. London had admitted sewage into its storm sewers in 1815; Boston did so in 1833, and Paris in 1880. But by that time people were beginning to suspect that the evil which they thought they were banishing from their homes in sewer pipes was creeping back through wells and water faucets.

The custom of diverting sewage in open ditches and furrows upon arable land may be older than the records of civilization. Certainly it persists today in many parts of the world, and its spread is strenuously urged by some conservationists. There are sewage farms in England and America as well as in China and Japan, but sewage farming works well only in arid regions on cheap, large areas of sandy soil. The practice was first attempted in the United States for the State Insane Asylum at Augusta, Maine, in 1872. But the idea is as dangerous as it is distasteful to the American public; no truck gardener who boasts of his cow manure would care to advertise that he had raised his vegetables from soil equally enriched with human waste. And there is always the chance that the microorganisms which cause disease have not been completely destroyed by nature.

Following the discoveries of the Lawrence Experiment Station, a chemical precipitation plant was built in 1889 for the city of Worcester, Massachusetts. It applied principles which for some years had been developing in Britain. Iron compounds such as ferric chloride formed a precipitation, leaving a fairly clear and sterile liquid. The drying of sludge in open beds came in 1891, and by 1894 trickling filters had been introduced. These, with nozzles to spray liquid sewage into the atmosphere, had proved successful in speeding up the bacteriological reaction and the oxidization which are essential to rendering the wastes innocuous. The whole process of purifying sewage has become increasingly complicated in recent years. In most communities the sewage is a mixture of household and factory wastes. Standards of purity vary according to whether the effluent discharges into a lake, a stream, or tidewater. Methods of purification differ but the end is the same—to remove any foreign matter which is or may become noisome and to destroy microorganisms which cause disease. By using extensive sewage-treatment processes it is possible to convert sewage effluent into water of high purity. About 1918 New York City built for Mount Kisco, New York, then a village of three thousand, a sewage-treatment plant designed to purify its sewage for discharge into a tributary of New York's Croton River water supply. The plant was an unusual one and no expense was spared by the city in its construction and operation. The sewage was put through a septic tank, then filtered and disinfected with chlorine. C.-E. A. Winslow, dean of American public health authorities, has estimated that "the final effluent probably was one of the purest streams of water to be found in New York State." [2]

Rivers and Canals

The renaissance of water power, beginning with the world's first major hydroelectric installation at Niagara Falls in 1895, has led to the construction of the world's largest dams. Some of these huge dams not only supply hydroelectric power, but like the Hoover Dam described below, they also provide municipal water supplies, irrigation, navigation, and flood control. Engineers increase the amount of arable land by supplying arid areas with additional water, by preventing other areas from being flooded, and by draining marshy or even submerged areas. Engineers thus make important contributions to increased food supplies.

What may be called the boat canals described in Chapter 8 were important transportation routes which accompanied the Industrial Revolution. After the middle of the nineteenth century railroads largely supplanted boat canals in the United States, although some boat canals are still important avenues of commerce in Europe. While very few were constructed after 1850, a number of rivers have been improved for navigation since that time. The most important canals built since 1850 have been ship canals, which provide an entirely different service from that of the earlier boat, or barge, canals. Nevertheless, engineering experience gained in building boat canals was of value in the construction of ship canals.

Engineers have been improving rivers for navigation for many years, but it has only been since the inception of the Industrial Revolution that river improvement has converted streams into effective inland waterways for boats. The history of the improvement of the Ohio and Mississippi Rivers, for instance, illustrates how engineers have developed and maintained navigable rivers.

Chapter 9 relates the history of the *New Orleans,* the first steamboat

---

[2] Charles-Edward A. Winslow, *Man and Epidemics,* Princeton University Press, Princeton, N.J., 1952, p. 104.

to navigate any river west of the Alleghenies. When in 1811, four years after Fulton's historic trip on the Hudson, Nicholas G. Roosevelt left Pittsburgh in his Fulton-designed craft and headed for New Orleans, nearly 2,000 miles away, he took many chances. The voyage of several weeks would test the power plant and machinery of a boat such as those that had made the relatively short trip on the placid Hudson. Roosevelt's greatest problem was to find and keep within the meandering channel, negotiating the several troublesome rocky rapids on the Ohio and swinging safely around the innumerable hairpin or oxbow bends of the Mississippi, each with its unpredictable bar, or crossing. Even if he succeeded in locating the channel and slipping over the bars, he had to dodge the terrifying snags, sawyers, or planters. These roots or limbs of trees that the current had undermined might send his vessel to the bottom with little warning. Until he came within sight of the crude levees that extended perhaps 100 miles above New Orleans, he saw no evidence of any attempts to improve or control the stream. Many of the levees, in fact, had been built by individual planters merely to furnish some protection against too frequent flooding of rich bottom lands. Some of them dated back nearly a century to French days when the authorities required abutting landowners to build and maintain them. Such levees could not, however, be considered as aids to navigation. A map of the new town, Nouvelle Orléans, dated 1728, shows along the water front La Levée, short for *la levée de terre*, raised land. Henry M. Shreve made a voyage in 1814, with his 80-foot *Enterprise*, between Pittsburgh and New Orleans. In 1816 came his *Washington*, destined to be the prototype in luxury of Mississippi steamboats for generations.

Like other grèat rivers, the Mississippi discharges through an extensive delta which begins 100 miles below New Orleans and extends 12 to 15 miles out into the Gulf of Mexico. Here the river, through silt brought down and deposited in the delta, empties by six branching mouths, or passes, into the Gulf. At the Head of Passes, where the wide and deep river begins to fan out and divide, a shoal has formed. A more shallow bar has marked the outlet of each pass. As far back as the early years of the settlement of Louisiana, the problem of shoal water in the passes had been recognized. In the year 1723, when the French town Nouvelle Orléans was barely five years old, Adrien de Pauger, an engineer, using a canoe, made for the French colonial government an examination of the lower reaches of the river and reported on a plan for fortifying it and improving the South Pass. In his report he speaks of a bar over which there is only 9 or 10 feet of water. Then he suggests closing some of the other passes with sunken vessels and with trees brought down by the current and building rough jetties or cofferdams which will not only serve as quays but would fix the current of the river. "It is indubitable," he adds, "that by this means the pass will gradually enlarge itself." "This undertaking," he concludes, "will not involve great expenditure," [3] for good cyprus wood was easily accessible. De Pauger, an optimist, knew his river hydraulics but was a century and a half ahead of his time.

There was no extensive improvement of the delta until about 1874, at the time James Buchanan Eads had finished his St. Louis Bridge, 1,250 miles upstream. The lower river, from Baton Rouge down, was at least 40 feet deep, except at the passes where the depth at some points was barely 16 or 18 feet. For the clipper ships of the 1850s this depth had usually been sufficient. Eads contracted with the United States government to deepen the South Pass to 30 feet and maintain this depth for a period of twenty years. The deepening was finished and the twenty-year maintenance period began in 1879; the Eads heirs received their final payment in 1900. Intermittent dredging is still necessary, however.

The Eads plan was quite simple and not too unlike a glorified version of de Pauger's of 1723. It consisted in dredging shoals at the ends of South Pass and building jetties along each bank, thereby coaxing the narrowed river gradually to scour out a deeper channel for itself. The jetties, separated by 1,000 feet, were begun by sinking successive layers of woven-willow brush mattresses framed with yellow pine timbers. When a sufficient number of these mattresses had been weighted down into the soft bottom by riprap or loose stone until they offered a stable foundation, the concrete jetties were started atop them and completed at some feet above sea level. The plan was in large part modeled after that used earlier at the Sulina Pass of the Danube. The combined length of the east and west jetties was a bit more than 4 miles. The Southwest Pass, somewhat wider than the South Pass, has been improved since 1900 in much the same manner and now constitutes the most widely used channel through the delta.

Congress, in 1820, authorized the first complete survey of the Mississippi and Ohio Rivers. As a result of this survey, army engineers recommended that both rivers be cleared of snags and other obstructions

---

[3] Charles E. A. Gayarre, *Histoire de la Louisiane*, 2 vols., Magne & Weisse, New Orleans, 1846, vol. 1, pp. 195-196.

to navigation; Congress shortly appropriated some money for it. This difficult work proceeded slowly for several years until 1826, shortly after the formation of the U.S. Army Engineer Corps, when Henry M. Shreve was appointed Superintendent of Western River Improvement. Shreve had, a few years earlier, designed and built a unique twin-hulled steamboat for pulling giant trees out of the channel by winches and disposing of the almost inextricable heaps of gnarled and twisted tree trunks. This snagboat, the *Heliopolis*, the first of its kind, seems to have represented the earliest application of steam engineering to the improvement of the Mississippi and Ohio Rivers.

The Ohio River, between its source at Pittsburgh, Pennsylvania, and Cairo, Illinois, where it joins the Mississippi, very early presented serious navigational problems, not only from snags and similar obstructions, but also from rapids and bars. The problem presented by bars was the subject of analysis in 1821 by a board of engineers which was appointed by Congress to make an examination of Western rivers. Based on the report of this board of engineers a congressional act in 1824 provided for experiments on certain bars to determine the practicability of a scheme to construct dikes to concentrate the flow of water within a limited width where it would scour a channel. The first experimental wing dam was constructed at Henderson Island on the lower Ohio. The effect of this dam was to increase the minimum low-water depth of the channel from less than 20 inches to nearly 3 feet. A second experiment was conducted by Henry M. Shreve at Grand Chain, also on the lower Ohio, in 1830, with indifferent success.

Not until many years later was a plan evolved for making the entire river navigable under all conditions. In the 1870s a party of army engineers was sent abroad especially to study control methods used on French rivers. As a result, a series of movable dams on the French pattern was projected for the temperamental Ohio, that is, dams whose navigable pass could be opened up in time of high water. The first of these dams at Davis Island, about 5 miles below Pittsburgh, was completed in 1885. The Ohio has now been made navigable for the entire distance, 980 miles, between Pittsburgh and Cairo, for draft of 9 feet. There are 53 dams; each with its lock, 110 by 600 feet, and slack-water navigation of at least 9 feet is provided for throughout. The total fall of the river is 430 feet, and the pools vary in length from about 5 miles to about 55, averaging 18. The mean difference in level between successive pools, the lift at a lock, is less than 8 feet.

The first use of the Mississippi for power purposes was in 1823. In that year, at the Falls of St. Anthony above St. Paul, the United States government built at what later was called Fort Snelling a sawmill and a mill for grinding wheat. This flour mill was the beginning of the industry that later would make Minneapolis widely known. Steamboat navigation on the Mississippi had begun with the *New Orleans*, which left Pittsburgh in October, 1811, and reached New Orleans in January, 1812. The upper-river navigation began with the twenty-day voyage of the *Virginia* in 1823 from St. Louis to St. Paul. At that time, below the St. Anthony Falls, there were a number of rapids, including those at Rock Island and Keokuk; otherwise the slope of the river was relatively slight and quite uniform. At Rock Island the river divides into two channels which flow on either side of the island on which the United States Arsenal is located. In the natural condition of the river a fall of about 15 feet at low water occurred here in a distance of 2 miles. The first Rock Island power development was undertaken in 1843, a low dam and water wheels near the head of the island in the west channel. In 1846 a second power site was developed in the other channel. After a number of changes, three dams at this point now supply power of a few thousand kilowatts for the Moline Water Power Company, the government arsenal, and the operation of the locks.

The Keokuk, or Des Moines, Rapids, 120 miles farther downstream, were examined by young Lieutenant Robert E. Lee, in 1837. As a result the channel was deepened by blasting and rock excavation several times during the next few years. After the Civil War, navigation around the Des Moines Rapids at Keokuk was substantially improved by means of a lateral canal 7 miles long with three locks, each 80 by 350 feet. This canal was opened to navigation in 1877. It was a noteworthy engineering project of the times and provided minimum navigation depth of 5 feet. In 1913 the Mississippi River Dam Company constructed the Keokuk hydroelectric plant, at that time the largest in the world. The concrete dam then installed there is 4,649 feet long and the normal head is $34\frac{1}{2}$ feet. The powerhouse section of the dam provides ultimate housing for thirty 10,000-horsepower turbines, 15 of which were installed initially. A lock, 110 by 400 feet, was included for navigation, replacing the old canal. Between St. Louis and St. Paul-Minneapolis there are now 26 low dams. Hydroelectric power, as has been indicated, is developed only at Keokuk and on a much smaller scale at Rock Island. Each of the 26 dams has its lock or locks, which provide for 9-foot navigation throughout.

The dams have a variety of controllable spillways, including roller gates, a European development comparatively new in this country. The submergible-type spillways have been installed at many of the dams.

Dredging came in the last quarter of the nineteenth century. A crude steam dredge had been used in England before the end of the eighteenth century, but it appears not to have been too successful in deepening the harbor of Sunderland and was remodeled in 1804, using a more powerful 6-horsepower Boulton and Watt engine. In this same year, Oliver Evans built for the Board of Health of Philadelphia his Orukter Amphibolos, a double-acting "machine for cleaning docks," the first American dredge. Early attempts at deepening the channel of the Mississippi were crude, local, and generally ineffectual. Most of them were actually harrowing, scraping, or dragging operations. Following the establishment in 1879 of the Mississippi River Commission, dredging operations at several points were undertaken and have continued intermittently ever since. The chain, or bucket, type of dredge, which was the earliest type and proved so successful on the Suez Canal and on the Clyde and other European rivers, was little used in the United States. Some dredging was done on the Mississippi, beginning late in the century, with the dipper type of dredge. The dredge most widely used on the Mississippi, however, has been the hydraulic, or suction, type. In 1877 Eads had the first one, the *G. W. R. Bayley*, built at Pittsburgh from his designs. Such dredges seem to have been first suggested by the French hydraulician Henri Emile Bazin in 1867. General Q. A. Gillmore made the first American use of such a dredge—a centrifugal drainage pump he called it—in 1871 for deepening the channel of the St. Johns River in Florida. Following its use on the South Pass maintenance dredging, many such dredges were built and used on the Mississippi River.

Another phase of navigation improvement has been the straightening and shortening of the river channel. The earliest attempt to shorten the channel, made in Shreve's day at the mouth of the Red River and named for him, was not permanently successful. More recently, in the face of some opposition from the older river men, portions of the lower Mississippi have been straightened by cutoffs. Since 1929 the low-water channel between Baton Rouge and Memphis has been straightened by 16 cutoffs, reducing the channel's length by some 170 miles. The famous Greenville Bends, where the river made five sharp turns, have been replaced by two sweeping curves. The advisability of removing these and other bends had been a much debated question—there were even many engineers who opposed. Since 1928 the extensive United States Waterways Experiment Station at Vicksburg has been studying cutoffs and other related river-control problems in much detail, using a series of model studies on a scale never before attempted in engineering. Here, in a laboratory area of many acres, elaborate models of any particular section of a river that is being studied are carefully constructed to scale. From these studies, supplemented by a mass of statistical material that has accumulated through the years, engineers can predict with some assurance the effect that a proposed improvement will have on the river for miles above and below it. Twentieth-century engineering has provided restraining reservoirs on many of the Mississippi tributaries, and floodways on its lower reaches, to serve as additional channels into which, in time of flood, the river may disgorge its swollen waters. The time-honored system of levees has, however, been aptly called the backbone of flood control in the lower Mississippi area. Along the 1,000 miles of main stream from Cairo to the Gulf there are now 1,600 miles of levees, almost all of them built since 1850. In addition, since 1880 there have been constructed several hundred miles of substantial stone and concrete revetments, located at 97 strategic points and serving to prevent scour and the resultant meandering, or change of channel location.

In 1953 the U.S. Army Corps of Engineers eliminated the last great obstacle to navigation on the Mississippi. The chain of rocks immediately north of St. Louis has always seriously impeded river traffic. The 8.3-mile Chain of Rocks Canal has an average depth of 32 feet, is 300 feet wide at bottom, and 550 feet wide on top. Its main lock is the largest in the Western Hemisphere. This lock, 110 feet wide, can lift or lower a 1,200-foot-long train of barges.

Ship canals serve two purposes. Some, such as the Sault Ste. Marie Canals and the Manchester Ship Canal, provide ships with access to areas which they could not otherwise reach, and others, like the Suez and Panama Canals, shorten the distance which ships have to travel between two points. The Suez Canal reduced the voyage between Liverpool and Bombay from 10,680 to 6,223 miles, an important saving in time and fuel in transit of ocean-going ships. We have already referred to the earliest attempts to connect the lower Nile with the Red Sea. Later there were several other attempts, some of them measurably successful, such as those of Darius in the fifth century B.C., of Ptolemy Philadelphus two centuries later, and of Trajan in the second century A.D. In Roman days there was considerable traffic, by way of the lower Nile, and this canal, between the Mediterranean and the Red Sea regions, even to far-off India and China. Napoleon's engineer Lepère, who reported in 1798 that the Red Sea was higher than the Mediterranean, only postponed the inevitable attempt to pierce the isthmus of Suez and thus provide a more direct water route to the Orient. A generation later the advent of steamship navigation made this project increasingly imperative. The British in the 1840s urged a railroad across the Isthmus; the French insisted on a canal.

The Suez Canal owes its inception and successful completion in largest part to a French diplomat and promoter, Ferdinand Marie de Lesseps (1805–1894). It was envisioned by him in 1852; construction began in 1859, and the canal was completed in ten years. De Lesseps' role was largely in the fields of international diplomacy and finance and his enthusiasm was contagious. The work itself involved no particular engineering difficulties; it has been called "a magnificent ditch digger's dream." In its entire length of 101 miles there are no locks. The highest ground traversed is only 50 feet above sea level and one-fifth of the canal passes through shallow natural lakes. The alignment is quite direct, the northerly half nearly straight. The work was begun with very crude and inadequate machinery and equipment. Earth was carried in many cases on the shoulders of poorly paid laborers, some of them women requisitioned from Egypt, but after a few years the corvée, or forced labor, was abolished. Then modern machines including chain or bucket dredges (Figure 13.6) were introduced and proved profitable. In fact, the experience gained on the Suez contributed not a little to the development of dredging and materials-handling equipment the world over. The initial cost of the canal was nearly 150 million dollars. Verdi's "Aida" was composed to celebrate its opening in 1869.

The original depth was 27 feet. The canal of 1869 would not have accommodated the *Great Eastern* comfortably for her draft was 26 feet. The canal has been enlarged and deepened more than once and it now accommodates vessels drawing as much as 34 feet. Drifting sand has made almost continual dredging necessary. Precise and extended leveling observations have now established the fact that the mean level of the Mediterranean is higher than that of the Red Sea by only some 10 inches. Quite recently the capacity and usefulness of the canal have been increased by the simple expedient of adding an extensive bypass, so that for some miles there are now two independent canals, one each for northbound and southbound traffic. The present surface width of the canal varies between 400 and 500 feet.

Figure 13.6   Suez Canal dredging (From J. E. Nourse, *The Maritime Canal of Suez*, 1884)

Fantastic dreams of a canal cutting through the 40-mile-wide isthmus of Panama date from the sixteenth century. Concrete plans in great variety began to take shape in the American mind about 1825, nearly a century before the canal became an actuality. The Mexican War and the discovery of gold in California turned the attention of the people of the United States to the importance of improvement of communications between the coasts, something better than the long trek "the plains across" or the equally tedious sea voyage "the Horn around." A company was formed in 1848, the year of the California gold rush, to build a railroad across the isthmus at its narrowest part. Two young contractors, George M. Totten and John C. Trautwine, both afterward to become well-known engineers, made the necessary surveys and supervised the construction of the Panama Railroad. Construction began in 1850, and the line was finished five years later at a cost of 8 million dollars and the lives of 1,200 employees, mostly Chinese, Negro, and Irish. The line crossed the Culebra Divide at an elevation of 287 feet above sea level. The 48-mile road was a profitable venture from the start, especially during the years preceding the opening of the transcontinental line in 1869.

The French organized a company in 1876 to build a canal across the isthmus, following much the same route as the railroad, which they bought in 1881 for more than 20 million dollars. The moving spirit in the venture was the aging promoter Ferdinand de Lesseps, the congenital optimist who had successfully completed the Suez Canal a few years earlier. De Lesseps, then "the most decorated man in Europe," was not an engineer—he "didn't like engineers"—and he had a genius for over-looking or minimizing difficulties. However, with Gallic enthusiasm, the French began their canal under the direction of their ablest contractors and engineers in 1882 and kept at it desultorily and spasmodically in the face of countless discouragements for twenty years. They spent more than 200 million dollars on it. Only about a third of this amount, some said, had gone into actual construction; another third had been practically wasted, and the rest had been stolen by minor officials. Still the canal was only about 40 per cent complete when the United States paid the French company 40 million dollars in 1903 and took over canal and railroad. The French plan was modified radically in some of its details, and the canal was first put into use in 1914, although it was not com-pleted until 1920. It cost the United States more than 500 million dollars up to 1920. James Bryce once described the Panama Canal as "the greatest liberty man has ever taken with Nature."

The French failed in their efforts to complete the canal partly be-cause they had attempted an extravagantly expensive project that they were not able to finance properly. A second factor, equally important, was that their workmen died off by the thousands from yellow fever and malaria, much as the Egyptians had in the days of Neccho. The success of the American effort was due not only to adequate, even gen-erous, financing, but also to two monumental medical discoveries, made in 1898 and 1900. The first was that the Anopheles mosquito transmits malaria from one person to another; the second was that the *Stegomyia fasciata* mosquito carries yellow fever. It was Colonel, later General, William C. Gorgas (1854–1920) of the United States Army Medical Corps who directed the effective eradication of these mosquitoes through-out the area traversed by the canal and eliminated these two diseases of which the yellow fever was the more deadly.

The physical difficulties at Panama were enormous. The climate is deadly to most races, as the French had learned. However, President Theodore Roosevelt gave great impetus to the project. Operations began in 1904, and John F. Stevens (1853–1943), a railroad civil engineer ap-pointed chief engineer in 1905, directed the excavation. Stevens resigned in 1907, and Colonel, later General, George W. Goethals (1858–1928)

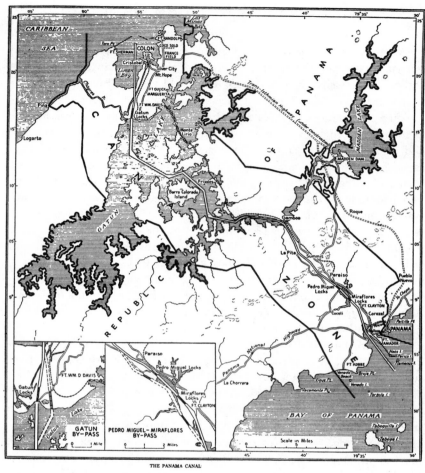

Figure 13.7 The Panama Canal (From N. J. Padelford, *The Panama Canal in Peace and War*, 1942; courtesy N. J. Padelford and Harvard University)

succeeded him. Goethals supervised the construction of the locks and saw the canal through to completion. On the Atlantic side a temperamental, often tempestuous river, the Chagres (Figure 13.7), had to be controlled. Next came the Continental Divide, a ridge 312 feet high which must be sliced through before the canal could be dropped down to the Pacific. The engineers blocked off the Chagres River by a huge earth dam, forming the inland Gatun Lake with a shore line of 1,100 miles at an elevation of 85 feet above the sea. A devious channel through

Figure 13.8 Dredge picking up 20-ton boulder from bottom of Culebra Cut on the Panama Canal (Courtesy Panama Canal Company)

this lake constitutes the middle half of the entire canal. Beyond the lake the Culebra or Gaillard Cut was sliced through the ridge. Nothing like this excavation, carried to a total depth of 272 feet, had ever been attempted anywhere. The material composing the ridge was of uncertain and unpredictable composition, and countless slides taxed the patient ingenuity of engineers for years, even after the rest of the canal was complete and in operation. Potential slides are still a source of worry in 1955. The excavation and transportation of this material presented problems of a magnitude never before encountered. The construction of the canal furthered the development of large-scale machinery for excavating and the handling of materials, much as the Suez project had done on a much smaller scale two generations earlier (Figure 13.8).

To attain the level of 85 feet above the sea, three steps of twin concrete locks with chambers 110 by 1,000 feet were built on the Atlantic side and three more on the Pacific side, giving a total of twelve lock chambers in all, six in each direction. Gatun Lake furnished most of the water for lockage, but since it proved inadequate during prolonged drought periods, an additional supply was later provided by damming the Chagres River near its source and bringing the water of the upper Chagres into Gatun Lake. The lock gates presented delicate and interesting problems in the fabrication and machining of steel, which were a challenge to the designers of gate mechanisms. The first ship passed through the canal in August, 1914, but shortly thereafter slides in the Gaillard Cut closed it; regular traffic did not begin until 1915.

About 65 per cent of the traffic through the Panama Canal is between the Atlantic and Pacific coasts of Central and North America. The distance from New York to San Francisco through the canal is only 40 per cent of the voyage via the Magellan Straits. Thirty per cent of the canal traffic travels between Europe and the Pacific coasts, and the voyage from Liverpool to San Francisco is only fifty-eight per cent of that via Magellan.

Irrigation and Reclamation

Like man, domesticated plants cannot live without water nor can they thrive with too much of it. Irrigation of land to make it arable probably began about 6000 to 4000 B.C. in the Near East during the stages of the great food-producing revolution. For perhaps six or seven thousand years, engineers in some countries have increased food production by providing water to land normally too dry to support cultivated plants. Restraining floodwaters so that they do not drown out cultivated lands is a more recent undertaking. The most dramatic reclamation project is in Holland. This long-term Dutch project is not only one of flood control—it is reclamation in the most literal sense of the word.

Perhaps the earliest North American irrigation projects were those of the Spaniards in Mexico during the latter half of the sixteenth century. Several Augustinian and Franciscan friar-engineers planned and constructed not only irrigation and land-reclamation projects but also more than one monumental and well-designed aqueduct, not unlike those of the Roman type with which they had been familiar in Spain. One of the most important of these Mexican aqueducts, built just after the middle of the sixteenth century, was the Zempoala Aqueduct of Padre Francisco de Tembleque. It brought water, mainly for household use, but doubtless also for irrigation, 28 or more miles to the little convent town of Otumba between Mexico City and Vera Cruz. The open conduit followed the contour of the ground, except for the crossing of three valleys, where it was carried at a height in some places exceeding 100 feet by three series of narrow and bold masonry arches, respectively 13, 47, and 67 in number. The water channel was very small—less than a square foot in area. The aqueduct is said to have served for 123 years. The supply was temporarily interrupted late in the seventeenth century, and no one knows when the venture was abandoned permanently. The series of arches were photographed in 1888 when some of the arch centers of adobe were still standing. It is of interest that the structure is in an earthquake region.

It was not until three centuries after the construction of the Zempoala Aqueduct that the first irrigation projects in the United States were undertaken, 1,500 miles to the north of Mexico City. Between 1860 and 1870 the Mormons in Utah constructed 277 canals, totaling 1,043 miles and bringing water to 115,000 acres. Utah did not become a state until 1896, and cooperation with the United States dates from 1902. Where and how the Mormons acquired their engineering skills is not certainly known, perhaps on New York State canals or in Great Britain. It is not unlikely that, beginning very gradually with the crudest of tools and equipment, the Mormons proceeded by trial and error, under gradually increasing legislative grants provided by what was then the state of Deseret. During the first decades of the twentieth century a number of high dams were built in the Western United States to furnish water for irrigation. One of the most notable of these is the Arrowrock Dam, completed 1915, on the Boise River in Idaho, and the Owyhee of 1932, on the Owyhee River in Oregon. The latter was the first to exceed 400 feet in height.

The U.S. Bureau of Reclamation completed in 1936 what was then the world's most extensive river-control project, the damming of the Colorado River where it forms the boundary between southeastern Nevada and northwestern Arizona. The great Hoover Dam, the highest in the world, is its central feature, holding back the waters of Lake Mead, an artificial lake of unprecedented size. The project is one of flood and silt control, irrigation, municipal water supply, and hydroelectric power; it includes also a national recreational park. The irrigation areas lie more than 200 miles below the dam, both west and east of the river. The municipalities using the Colorado River water are a group of 14 cities in the Los Angeles area of California, more than 200 miles to the southwest across a mountain range. Hydroelectric power is furnished to these same cities and to many other California, Nevada, and Arizona communities.

The Hoover Dam was projected as early as 1919, at first as a river-control project. The Colorado, in its lower reaches, had for years been difficult, indeed almost impossible, to control, even with expensive levees; extremes of flood and drought followed each other all too frequently. At the beginning of the twentieth century the river had been tapped for irrigation by private interests at points close to the Mexican border, but the farmers of California's rich Imperial Valley to the west of the river clamored for a less temperamental stream. They were especially vociferous after the river's 1905 to 1907 rampage. In 1904 engineers had cut a narrow channel through the western levee 4 miles to the south of the border, confidently expecting that enough river water would be diverted through it to scour the silt out of the almost dry parallel Imperial irrigation canal. An untimely and unexpected series of floods enlarged the opening into a wide crevasse through which silt-laden Colorado water poured down to drown out the Imperial Valley, the lowest point of which at Salton Sea is nearly 250 feet below sea level. In a short time the sea was enlarged until it covered an area of 400 square miles, at some points 50 feet deep. Nearly two years of costly, patient, and often discouraging efforts to restore the river to its natural bed were successful only after a pair of railroad trestles of 90-foot piles were built across the crevasse and 3,000 carloads of rock, followed by small stones and gravel, had been dropped into the current below, all within 15 days, to form a dam. For its boldness and magnitude this procedure was unique in the annals of engineering. From that time until the completion of the Hoover Dam in 1936 the complete control of the river was increasingly certain.

The engineering studies made of the lower Colorado River basin during the decade following 1919 were epoch-making in their comprehensive thoroughness and were carried on in the face of almost insurmountable physical difficulties. Finally a dam site was chosen about 100 miles downstream from the lower end of the celebrated Grand Canyon and 200 miles from the Mexican border. Here, the engineers decided, could be built a dam high enough to provide a lake of capacity sufficient to impound the entire normal flow of the river for two years. The dam would fit neatly between the canyon walls but would have to be more than 700 feet high. At that time the few high dams in the world were only slightly more than 400 feet tall; one was in France, two or three in the western United States. Finally, in 1931, the labor contract for the construction of the Hoover Dam was let to a combination of six large Western contracting firms, but the United States government supplied the bulk of the materials for the project throughout. For the Bureau of Reclamation, the chief engineer was Raymond F. Walter and the chief designing engineer, John L. Savage. Situated as it was, 25 miles from a town or railroad, the first problem of the contractors, after a branch railroad had been provided, was to create a community for the thousands of men who would be engaged. Never in history, not even in Khufu's day, had so many skilled men been required on a single project—the number here reached a peak of 5,250. All were provided for in a model city built entirely for the purpose, Boulder City, Nevada. Electric power was brought 222 miles across the desert, from San Bernardino, California, to serve the new city and the contractors for every power purpose.

Like a number of earlier and much smaller dams, the Hoover Dam is curved in plan, convex upstream, a concrete arch wedged between canyon walls. Before it could be constructed the entire flow of the river had to be drawn off by means of four diversion tunnels, two on each side of the canyon, which in themselves were noteworthy projects, each 50 feet in diameter and ¾ mile long. The artificial cooling to facilitate the proper setting of the enormous mass of concrete containing 5 million barrels of cement was an unusual and almost unprecedented operation. The refrigerating system, with more than 570 miles of 1-inch tubes embedded in the entire concrete mass when it was poured, accomplished the curing of the more than 3 million cubic yards in about two years. This huge mass, with its carefully grouted contraction joints, has been called "truly monolithic."

Lake Mead, formed by the Hoover Dam, is the world's largest artificial lake, 115 miles long. Below the Hoover Dam several others have been built, most of them to provide water for irrigation. The largest is the Parker Dam, 120 miles below the Hoover. Here a reservoir 450 feet above sea level furnishes water through the Colorado River Aqueduct to the Metropolitan Water District of Southern California. The Colorado River Aqueduct is not only the longest in the world; it was at the time it was built the only important aqueduct that pierced a mountain range. The water it brings from the Colorado River flows by gravity for nearly all of its length of 242 miles. However, at five points on the eastern half of the aqueduct, electrically driven pumps, powered from the Hoover Dam, raise the water a total of 1,617 feet.

In addition to its great length and its equally unprecedented profile, several features make the Colorado River Aqueduct unique. For its first 200 miles it crosses an arid, almost rainless and practically barren, desert where more than 150 miles of modern surfaced roads had to be built before construction could proceed. At points along the aqueduct 10 wells of 16 inches diameter were drilled and connected with pipelines that practically paralleled the aqueduct and supplied water necessary to the construction. There are 144 inverted siphons, all of monolithic concrete

Figure 13.9 An inverted siphon on the Colorado River Aqueduct (Courtesy Metropolitan Water District of Southern California)

and totaling 29 miles in length (Figure 13.9). Three of these cross what are known as active earthquake faults where the monolithic construction was interrupted each 20 feet or so by special joints that could resist shear but could not resist tension. One of these siphons at a valley crossing is about 5 miles long. Except for the 63 miles of open, lined concrete canals, the cross section of the Colorado River Aqueduct is practically the same as that of each of the New York City aqueducts, roughly 15 or 16 feet wide and high. It will eventually deliver to Lake Matthews, an artificial reservoir some 60 miles east of Los Angeles, about 1 billion gallons of water daily, which will double the present supply of southern California. New York City's two principal aqueducts, whose combined length is nearly equal to that of the Colorado River Aqueduct, are both gravity supplies and in general traverse the Hudson River Valley. The total yield of New York's sources is expected to be about 1½ billion gallons daily.

In the course of many centuries Holland has wrested about a quarter of its land from the North Sea; some 3,000 square miles are therefore below sea level. The greatest reclamation project of all times has required the incessant labor of countless generations, at first using only crude implements. The Frisian Dutch who inhabited the coastal marshes a few centuries ago lived in almost constant peril from the sea and from the three rivers of which Holland forms the deltas. In December of 1287, fifty thousand Frisians are said to have been drowned in a single night. The idea of using windmills for pumping seems to have come from the Near East, but they were not used on a large scale until about 1600. Perhaps the outstanding leader in the use of windmills for reclaiming land in Holland was Jan Adriaensz Leeghwater (1575–1650), who called himself a "mill constructor and engineer." The professional work of Leeghwater took him to the North Sea coastal areas of France, Germany, Denmark, and Poland. One of his better-known followers was Cornelius Vermuyden (ca. 1590–1656), who spent much of his life reclaiming the fen lands of eastern and southeastern England in the days of Charles I and Cromwell.

About 1640 Leeghwater wrote the unique little book *Het Haarlem-mer-Meer-Boek*, by which he is best known. In it he set forth his revolutionary plan for pumping out the Haarlemmermeer, a shallow freshwater lake or "sea" some 72 square miles in area which lay a few miles southwest of Amsterdam. He proposed installing 160 windmills, each with its pump. The plan was doubtless too ambitious even for the persistent Dutch; it was not undertaken in Leeghwater's day, nor in fact until nearly two hundred years after his death. In his book he had cautioned against undue haste. Later he assures his countrymen, "The draining of lakes is one of the most necessary, most profitable and the most holy works in Holland." [4]

Nearly two centuries later than Leeghwater's day, about 1840, after the Haarlemmermeer had enlarged itself perilously close to Leiden and Amsterdam as the result of hurricanes, the Dutch decided that the time had come to turn the Meer into a fertile polder. By this time steam-pumping engines of the type used in the Cornwall tin and copper mines were available. Three were bought and set up around the edges of the lake. They were of the compound vertical type, with one high-pressure cylinder inside the other. The three engines seem to have developed a total horsepower somewhat in excess of 1,000, and each engine operated a number of pumps cleverly arranged in a circle around it. The lake was pumped dry in less than four years, between 1848 and 1852. The Haarlemmermeer water surface was about 14 feet below high water in the

[4] John van Veen, *Dredge, Drain, Reclaim*, Martinus Nijhoff, The Hague, 1948, p. 45.

North Sea; the polder is about 1½ feet lower still. The water was raised into a canal and then up to the level of the North Sea, a few miles away. Pumping continued intermittently after 1852 merely to keep the land arable, one of the pumps being in continuous service for 84 years. Modern pumping engines have replaced the original engines, but the tower and one original engine, the Croquius, have been preserved as a museum.

Until the completion in 1932 of a 17-mile dam which shut out most of the salt water, the Zuider Zee was thought of simply as a tidal gulf projecting southeasterly from the North Sea into the heart of Holland. Geologists say, however, that it was not always salt, but that North Sea storms about the year 1300 gradually forced an opening into what had been a large fresh-water lake. Plans for shutting out the North Sea began to take shape as early as the 1840s at the time the Haarlemmermeer was reclaimed, but the project was much too vast for the mechanical appliances available. It had to await the development of hydraulic dredging, diesel engines, electric power, and bulldozers. Following the First World War a plan finally crystallized and shortly secured enthusiastic popular support. The dam is 600 feet wide at the base with its crest 20 to 22 feet above sea level and carries double lanes of highway and an electric railway track. Secondary dikes which surround the four polders are slightly lower. The area of the polders is more than two-thirds that of Rhode Island.

The Zuider Zee reclamation project has been the most ambitious that Holland, or indeed any nation, has undertaken. When completed, half of the area will be dried up and restored to cultivation, adding 900 square miles to Holland's arable land. The two dams that shut out the sea water have already been completed. Both are of earth fill, generous in their width, and faced with masonry. One is over 17 miles long, the other 1½ miles. The project involves surrounding four areas, or polders, with dikes and pumping them dry. The northwest polder was pumped dry by pumping units located near two of its corners. One station comprised 6-cylinder diesel engines of 400 horsepower each, direct-connected to centrifugal pumps with horizontal axles; the other contained electric motors operating on 3,000-volt current and direct-connected to centrifugal pumps with vertical axles. This 77-square-mile polder, the smallest, was producing crops by 1936. A considerably larger second polder had been unwatered as the Second World War began. The cost of the entire project may exceed 200 million dollars, almost half as much as that of the Panama Canal.

Bibliography

Baker, Moses N.: *The Quest for Pure Water*, American Water Works Assn., New York, 1948.

International Engineering Congress, San Francisco, 1915, *Transactions*, 12 vols. in 13, vol. 1, *The Panama Canal*, San Francisco, 1916.

Kirkwood, James Pugh: *Report on the Filtration of River Waters, for the Supply of Cities, as Practised in Europe*, D. Van Nostrand, New York, 1869.

Kunz, George F.: *Catskill Aqueduct Celebration Publications*, The Mayor's Catskill Aqueduct Celebration Committee, New York, 1917.

Mack, Gerstle: *The Land Divided*, Alfred A. Knopf, Inc., New York, 1944.

Metropolitan Water District of Southern California, *The Great Aqueduct*, Los Angeles, 1941.

New York City, Board of Water Supply, *The Water Supply of the City of New York*, New York, 1950.

Siegfried, André: *Suez and Panama*, Harcourt, Brace and Company, Inc., New York, 1940.

Veen, John van: *Dredge, Drain, Reclaim*, Martinus Nijhoff, The Hague, 1948.

FOURTEEN

Construction

Modern skyscrapers, bridges, and tunnels are truly daring accomplishments of the construction engineer. Skyscrapers rise over 1,000 feet, bridges span 3,000 and 4,000 feet between piers, and tunnels bore through miles of rock. Like most other engineers the construction engineer must be able to draw together many types of engineering knowledge and many different engineering techniques. He must also use numerous materials and above all he has to employ power in countless ways during construction. Engineering began to be diversified into its various specialties in the nineteenth century, just as did natural science and other areas of knowledge. By the middle of the twentieth century, however, the engineer was integrating the many specialties on large projects. The construction engineer is not the only engineer who effects such integration. For instance, the petroleum engineer, with whom we have not dealt in this book, has to bring together many types of engineering to plan and to build a large oil refinery. The integration of the engineering specialties to produce specific works is one of the important engineering developments of the twentieth century, and construction engineering illustrates this integration particularly well.

Skyscrapers

The skyscraper in America has become of considerable economic importance. With the great population migration from rural to municipal areas, there has come a significant change in living habits. Today the majority of Americans spend more of their lives inside buildings than out of doors. Architects and engineers have therefore striven to make buildings more attractive and livable because of the increasingly high

464

Figure 14.1
Jenney's Home Insurance
Building, Chicago, 1884
(Courtesy Yale School of
Fine Arts)

percentage of time which people spend within them. In fact, when the
Home Insurance Company in Chicago gave the commission for the design
of its new office building to William Le Baron Jenney in 1883, the company specified that the building should be fireproof and should have a
maximum amount of natural light in each room (Figure 14.1). It was
this demand for a safe and comfortable building that led to the construction of the first skyscraper. We have already noted in Chapter 10 the
three main factors that made modern skyscrapers possible—the elevator,
the skeleton framework, and relatively cheap steel. It is not too much to
say that the limitation on the height of buildings is a matter of economics
rather than of engineering.

The hydraulic elevators described in Chapter 10 had served a number
of the older type of notable tall structures like the Pulitzer Building in
New York City. They continued to be used in many tall buildings well
into the twentieth century, long after electric motors became common.
The first successful electric elevators date from about 1889. In the earliest
of these, worm gears were used to revolve the drums, and it was necessary
for the motor to rotate these gears if the elevators were to move. However, if power should fail and the motor did not rotate, the car would

not fall because the drum could not work backward through the worm gears. A later safety development was the electromagnetic brake on which the brake shoe, normally held away from the drum, sets automatically in the event of a power failure or other abnormal condition. There have been many changes in elevator mechanism within the last two generations, but in the main they have all been improvements along established principles rather than radical departures in construction. Escalators, or moving stairways, date from about 1900; they have much larger capacities than elevators for buildings of few stories.

The first nine stories of the Home Insurance Building of Chicago were completed in 1885. This building was the first in which an appreciable quantity of steel was used in the framework. More important, however, a large portion of its exterior was only a curtain wall of masonry that did not even support itself, but was entirely supported by the framework. At least this was true for all except the lower stories of the LaSalle and Adams Street fronts which together with the rear walls were of solid masonry. The columns were of two types: round cast-iron and built-up box sections of wrought iron, but on several of its upper stories some Bessemer steel replaced wrought iron. This building, in spite of its moderate height of 10 stories, was the world's first skyscraper. While it was being demolished in 1931 a technical examination showed that, although it did "not fulfill all the requirements of a skeleton type," it was "a notable example." It was the first office building ever constructed that used, even in part, the principle of skeleton construction.

Some four years later, 1889, two blocks north on LaSalle Street, a taller structure of much the same type, the Tacoma Building, was completed. Chicago residents were properly proud of this building; a contemporary guide book, issued for visitors to the Columbian Exposition of 1893, boasts of its "towering above its surroundings to the dizzy height of twelve clear stories" (it actually had 14). The entire outer walls, with large window areas, were curtains of brick and terra cotta which carried no load but served merely to keep out the elements and to let in light. The walls of the Tacoma Building could be, and were, commenced independently at various floor levels, rather than starting from the ground; here was the first consistent use of such a method. The columns and lintels in the building were all of cast iron. Both wrought iron and Bessemer steel were used in the skeleton, which was riveted rather than bolted. Some have called the Tacoma Building the first skyscraper. It had five hydraulic passenger elevators. Following the Tacoma Building came, also in Chicago, a rapid succession of skeleton buildings. The second Rand-McNally building, completed in 1890, was the first to be supported on an all-steel frame. The Great Northern Hotel, 14 stories tall, "a magnificent architectural achievement," stood from 1891 to 1940. Finished in 1892 just before the Columbian Exposition, the 21-story Masonic Building was the tallest building of the period; Chicagoans called it the finest building in the world—"a gorgeous edifice."

Spread footings to distribute loads go back many centuries. In the earliest Chicago tall buildings raft footings gradually replaced the traditional spread foundations. These footings often consisted of several layers of railroad rails crossed at right angles and used to reinforce a substantial bed of concrete. In many of these buildings a number of inches of settlement was anticipated and the building was therefore started higher than necessary. Wooden, later concrete, pile foundations going down in many cases 50 or 75 feet gradually superseded raft footings in Chicago.

The first skeleton construction in New York City was the narrow Tower Building in lower Broadway, only 129 feet tall and finished in 1889. A tablet on its lobby wall called this building the "earliest example of skeleton construction." Careful historians have vigorously denied this claim. New York City's first building of skyscraper proportions was the Manhattan Life Insurance Building on Broadway opposite Trinity Church; finished in 1894 and 347 feet tall, it is still standing. It was the first building for which engineers used compressed air to sink pneumatic caissons in the foundation work. Fifteen caissons of various sizes and shapes were used, enabling excavations to be carried down through more than 50 feet of mud and quicksand to bedrock. The caissons were then filled with concrete, topped with iron beams supporting brick piers. The Manhattan Life Building was also the first to have extensive windbracing, added metal members in the framework to guard against the collapse of tall structures in violent gales.

In 1902, following the epoch-making Manhattan Life Building, came the more famous Fuller, or Flatiron, Building, farther up Broadway. It had six vertical hydraulic elevators which are still in use. Among dozens of other structures like the Singer Building and the Metropolitan Life Insurance Tower in New York, the Equitable and Woolworth Buildings of 1913 were the best known. For 15 years the 760-foot Woolworth Building was the world's loftiest structure except for the Eiffel Tower which tops it by more than 200 feet. New York now has more than thirty buildings exceeding 500 feet in height, Chicago has at least nine, other

United States cities have 10 in all. These buildings presented in varying degrees certain engineering problems that had barely begun to arise late in the nineteenth century, problems of foundations, of stability under wind pressures, of elevators, of watersupply and sewerage, of heating and ventilation, and of distribution of electric power. The earliest skyscrapers contained little if any concrete. At the turn of the century, as more and more American portland cement became available, and in part influenced by European builders who used less steel, American architects and engineers began to design many-storied buildings of reinforced concrete. Reinforced-concrete skyscrapers are now competing with steel-frame structures in buildings of moderate height.

Since 1931 the Empire State Building (Figure 14.2) in New York City has been the tallest structure in the world—1,250 feet. It tops New York's Chrysler Building by more than 200 feet; these two are the only buildings taller than the Eiffel Tower. The Empire State Building is distinctly a milestone in tall-building construction, the culmination of a development that began nearly half a century earlier in the Home Insurance, Tacoma, and Tower Buildings. Its unprecedented height posed a number of problems concerning the stability of the structure itself. It was founded on bedrock at a depth of about 50 feet below the street level. The area of the base is about 2 acres, roughly half that of the Eiffel Tower. Its total weight of 302,500 tons, including 67,000 tons of steel, is divided among more than one hundred rolled fabricated steel columns, the largest of which each carries 5,000 tons down to the base. These columns rest on broad and heavily reinforced bases of concrete supported in turn on bedrock. The entire frame is, of course, elastic and has ample windbracing throughout. The top has been observed to "give" about 1½ inches during gales approaching 100 miles an hour. The steel columns are subjected to a constant compressive stress by the sheer weight of the structure itself; the eighty-fifth story is thus actually 6½ inches lower than its elevation would have been had there been no compressive stress.

The water-distribution system in the building consists of a series of water tanks at different levels; if the water system were not thus divided, a faucet on a lower floor would be subjected to a pressure of about 600 pounds a square inch, too great for even modern plumbing. The building has its own water pumps; the Catskill gravity water would reach to only a fraction of the height. Steam pipes posed the problem of expansion from heat. Radiators on the top floor, if rigidly attached to vertical feed pipes without provision for expansion, would rise more than a foot off the

Figure 14.2
The Empire State Building
(Courtesy *Engineering
News-Record*)

floor when the pipes were hot. The building has 69 electric elevators of which 63 carry passengers; their maximum speed is 1,000 feet per minute. The Empire State Building was ready for occupancy May 1, 1931, 14 months after the first steel was placed on the footings—a record speed for construction. At the peak of operations there were 3,400 workmen of 50 different trades employed on the 104 floors.

Suspension Bridges

In most bridges, as in skyscrapers, there is much structural steel. The first steel bridges, described in Chapter 10, preceded the first skyscrapers. Like the skyscrapers, the large bridges of the twentieth century are notable engineering accomplishments, the design and construction of which brought into play many types of engineering, scientific, and administrative skills. The Brooklyn Bridge dates from 1883. After seventy years' service it is carrying traffic of a type of which the Roeblings never dreamed. Its span of 1,595½ feet was exceeded by the cantilever spans of the Forth Bridge in Scotland in 1889, but not by any suspension bridge until 1903 when the Williamsburg Bridge was built across the East River, more than a mile northeast of the Brooklyn Bridge, with a span only a few feet longer than that of the Brooklyn. Then in 1924 came the somewhat longer but much lighter Bear Mountain Bridge, which crosses the Hudson 40 miles to the north of New York City.

During the next few years two suspension bridges, both in the United States, of still longer spans were built. The Philadelphia-Camden Bridge span is 150 feet longer than the Brooklyn, half again as wide, and considerably heavier. This suspension bridge was the first to exceed the spans of the Forth cantilever structure. The Ambassador Bridge at Detroit is still longer, 1,850 feet span; when built in 1929 it had the longest span of any bridge. At this point suspension-bridge spans seemed to have reached their economic limit. The cost of the Brooklyn Bridge had been about 15 million dollars, that of the slightly longer Philadelphia-Camden Bridge more than twice as much—36 million. By the early 1920s the increase in motor-vehicle traffic was taxing the capacity of the many Hudson River ferries. The year 1921 marked the incorporation of the Port of New York Authority, established by the States of New York and New Jersey to finance, construct, and maintain tunnels, bridges, and various facilities in the New York City area. The problem of financing such construction, which would eventually be paid for by tolls, now proved easy to solve. The Authority soon planned the George Washington Bridge at 179th Street, extending to the Palisades at Fort Lee, New Jersey, with a span of 3,500 feet. Designed by Othmar H. Ammann, construction of the bridge began in 1927, and it was opened in 1931. It does not yet carry all of the traffic for which it was designed; a lower deck, for either rapid-transit railroad tracks or motor vehicles, can be added

Figure 14.3 Golden Gate Bridge, San Francisco (Courtesy Redwood Empire Association)

later. The bridge has cost to date more than 55 million dollars and is rapidly being paid for by the motorists who use it.

A bridge across the Golden Gate at San Francisco was first suggested in the early 1920s. The success of the George Washington Bridge project indicated how, by the formation of an Authority, such bridge construction could be publicly financed, particularly on a much-traveled arterial highway like that extending north from San Francisco. Joseph B. Strauss was the chief engineer. The spot chosen for the Golden Gate Bridge (Figure 14.3) is at the narrowest part of the entrance to San Francisco Bay, where it is a mile wide. The configuration of the bottom determined that the span had to be slightly longer than 4,000 feet. The main part of the channel is more than 300 feet deep; however, for some distance from the south (San Francisco) shore the depth increases only gradually, reaching 65 to 80 feet at 1,100 feet out. The south pier was accordingly located here, and the north pier on the Marin County shore, 4,200 feet distant.

The location of this bridge presented a combination of difficulties that had not been encountered in previous structures. Earthquakes are not uncommon in this entire region. The site is practically in the open sea, a storm-swept area subject to heavy ground swells and cross winds, with tidal currents reaching a maximum of nearly 7 or 8 miles an hour. The bottom is hard smooth rock, basalt on one side, serpentine on the other. One of the first problems was how, under such circumstances, to anchor the south pier at such a depth. The north pier, on dry land, presented no unusual difficulties. For the south pier, the engineers built a steel trestle 22 feet wide and 1,100 feet long out from the shore to the pier location, each vertical tubular steel pile anchored into the solid rock floor. Around the pier site a boat-shaped wall of concrete about 170 by 311 feet was built, having been started from bedrock at a depth of 100 feet below sea level. This wall was originally intended simply as a protective fender inside of which a pneumatic caisson was to be set up. However, since heavy seas made the use of a caisson impossible, the concrete fender wall, 30 feet thick at the base, was finally built up to serve also as a cofferdam. The concrete in this wall was deposited under water through a tremie (elongated hopper). The water inside the wall was pumped out, and the pier was built in the open air. The concrete in the fender wall and pier was made from a newly developed high-silica cement, highly resistant to sea water. Tests showed that it had the high compressive strength of 4,000 pounds per square inch 28 days after it had been poured.

The two towers, or pylons, are 700 feet tall, their tops 746 feet above sea level. Each is made up of twin cellular shafts built up of riveted angle irons and plates. The shafts are spaced 90 feet apart on centers and are connected by transverse framing at four points above the bridge floor. The towers were designed to resist stresses due not only to dead loads, live loads, and wind, but also to earthquakes. They are considered especially notable not only on account of their size, but also because they were designed to satisfy architectural as well as structural requirements. Silicon steel with a tensile strength of 80,000 to 95,000 pounds per square inch and carbon steel about two-thirds as strong were used in the towers.

The main supporting cables are 36⅜ inches in diameter as compared with 16 inches on the Brooklyn Bridge and 36 inches on the George Washington Bridge. Each cable is composed of 61 strands of 452 wires each. The cables do not slide or roll over the tops of the towers; instead they are rigidly fastened to them and therefore pull them back and forth longitudinally, with variations in temperature and loading. A tower may be deflected from the vertical as much as 18 inches toward the channel or 22 inches toward the shore from a combination of these two causes. The cables hang like gigantic hammocks between the tops of the towers.

At 50-foot intervals double suspenders of wire rope support the 25-foot-deep stiffening trusses which in turn carry the floor system. There is enough flexibility in the central floor system to allow for a lateral deflection of about 28 feet from winds of 60 miles an hour or for upward and downward deflections totaling some 16 feet. A unique protective device saved the lives of 19 workers during construction. It was a safety net of manila rope, woven with a 6-inch mesh, which extended under the entire structure and 10 feet out on either side. The Golden Gate Bridge, with approaches, was built in four years; the Brooklyn Bridge required fourteen, the George Washington Bridge, four. The actual cost of construction was slightly more than 27 million dollars. Its estimated capacity is 5,000 motor vehicles per hour.

Steel Arch Bridges

The Eads bridge at St. Louis, as has been noted, was erected between 1869 and 1874. During the following quarter century some eleven iron or steel arches longer than its central span of 520 feet were built. Three of these were completed during 1898, the Niagara-Clifton Bridge across the Niagara River between the United States and Canada, and two across the Rhine, one at Düsseldorf, the other at Bonn, in Germany. The arch of the Niagara-Clifton Bridge was for years the longest in the world—840 feet; the Bonn Bridge had the longest arch span in Europe—721 feet. At the end of still another decade Gustav Lindenthal (1850–1935) was completing the Hell Gate Arch Bridge across the East River at New York (Figure 14.4). It is an imposing and beautifully proportioned structure, carrying the four tracks of the New York Connecting Railroad across the East River, thus forming a rail connection between the New York, New Haven and Hartford Railroad and the Pennsylvania Railroad. Its span is 977½ feet. Lindenthal designed it for moving loads heavier than those carried by any steel arch in the world—3 tons per lineal foot on each of the four tracks. It was completed in 1917.

Fifteen years later Australian and British engineers had completed across Sydney Harbor, New South Wales, Australia, an arch with a span of 1,650 feet (Figure 14.5) of the same type as the Hell Gate. It carries, however, besides four electric railway tracks, a six-lane highway and sidewalks, making its total deck width 160 feet. No wider bridge has ever been built. The engineers designed it to carry a total live or moving load of more than 6 tons per lineal foot. The arch span of the Bayonne Kill van

Figure 14.4 Hell Gate Bridge, New York (Courtesy New York, New Haven and Hartford Railroad)

Kull Bridge at New York, which O. H. Ammann had designed and which was built during the same period, was purposely made about 2 feet longer, but it carries only highway traffic. There were no particularly puzzling foundation problems in the building of the Sydney Harbor Bridge, for bedrock is only 30 feet down. In this respect the Eads bridge and the Hell Gate Bridge taxed the ingenuity of their engineers considerably more. The larger part of the structural steel at Sydney Harbor was rolled in Britain, more than 12,000 miles away. London engineers prepared plans and calculations for the structure as a whole. The steel was fabricated in extensive and well-equipped temporary shops built on the north shore of the harbor, opposite Sydney. Each half of the arch was erected as a cantilever and anchored back by 128 wire cables, each 2¾ inches in diameter, over a concrete pier into curved tunnels cut into bedrock. As the structure progressed, each steel member was loaded on a lighter and towed out until it was directly under the spot in the structure where it belonged; electrically operated traveler or creeper cranes of 120 tons capacity, one on each cantilever, then hoisted the member into position. At the Eads bridge the material was floated on barges, and then men on each cantilever laboriously hauled it up by hand.

The arch trusses of the Sydney Harbor Bridge are of silicon steel, the lateral and deck members of carbon steel. All truss members are rectangular in cross section, made up of plates and angles riveted together. The heaviest portion of the truss, the lower chord, which in this type of

474

Figure 14.5 Sydney Harbor Bridge (Courtesy *Engineering News-Record*)

bridge is the principal supporting member, is divided internally into three sections. The upper and lower chords of the Eads bridge are, it will be remembered, 30-inch cylinders of chrome steel. The curve of the Sydney Harbor arch, that is, the lower chord, is parabolic; the upper chord reverses its curvature as it approaches the piers. The arch, as stated above, was built out from each pier as a cantilever, anchored by the 128 wire cables, until the ends came to within about 40 inches of meeting. Hydraulic jacks then slackened off the cables until the ends met and the keys connected. The temperature range allowed for at Sydney was only 60°; at the Eads bridge it was assumed as about 160°. The dead weight of the Sydney Harbor span is about 28½ tons per lineal foot, that of the Hell Gate Bridge is just over 26 tons.

Reinforced-concrete Bridges

Stone-masonry structures approached perfection late in the eighteenth century with the bridges of Perronet. Until early in the twentieth century the world's longest stone arch was that of the Cabin John Bridge spanning Rock Creek on the Washington, D.C., Aqueduct completed in 1864. Its span is 220 feet. Comparatively few stone arches have been built since. The longest stone arch ever built was the Syra River Bridge, at

Plauen (Saxony), Germany. Its span is 295 feet, and it was completed in 1903.

The use of concrete in modern engineering construction dates back a little more than a century. For many years engineers were cautious about using it except in large masses. In the nineteenth century the French were the chief pioneers in concrete construction. Poirel, in 1833, used massive precast concrete blocks. Some 16,000 concrete blocks, averaging 20 tons each, were used a generation later for the jetties at the Port Said entrance to the Suez Canal. Among the earliest large concrete bridges were those of the Vanne Aqueduct which brought water to Paris through the Fontainebleau Forest in 1870. There were nearly 3 miles of these aqueduct bridges, some of 125-foot span. The aqueduct pipes, also of concrete, were 6½ feet in diameter.

The idea of reinforcing concrete with iron or steel should be credited chiefly to Joseph Monier of Paris whose patent of 1867 covered the construction of basins or tubs of cement with embedded iron netting. Monier, the proprietor of a commercial gardening establishment, came gradually to realize that his idea could be extended to such structures as railway sleepers or ties, even to floors, footbridges, arches, and pipes. The several Monier patents led to the Monier system of reinforced structures, with iron bars embedded in the concrete and crossing each other at right angles. A few years later Monier built, in France, a reservoir, or tank, more than 50 feet in diameter. Monier was not the first in this field; another Frenchman, Joseph L. Lambot, in 1849, and at least two Britishers, William B. Wilkinson, in 1855, and William Fairbairn, in 1864, had preceded him. Monier, however, with no technical training or experience but only a keen native intuition, seems to have been the first to combine iron and concrete scientifically, so that they acted together as a unit, the metal taking nearly all of the tension and the concrete most of the compression. Other patented systems that followed Monier's were those of Hennebique (French), Melan and von Emperger (Austrian), Ransome, Thacher, and Turner (American). All of these techniques were developed in the closing years of the nineteenth century and found wide use in Europe and the United States. The underlying theory of reinforced-concrete design developed only slowly, and the art advanced for some years in part by the expensive method of trial and error.

The earliest reinforced-concrete bridges in the United States were light footbridges of modest spans in parks. American engineers at first looked askance at combining iron or steel with a relatively untried and unpredictable material such as concrete—at least in thin slabs. A 20-foot-span reinforced-concrete arch built in a San Francisco park in 1889 seems to have been the first in the United States, and in 1894 the first American reinforced-concrete highway bridge for heavy traffic was built in Iowa; its span was 30 feet. European engineers had by this time built many reinforced-concrete bridges; in Switzerland there were three of the Monier type with spans of 128 feet. Rolled-steel rods gradually supplanted wire netting.

After 1894 reinforced concrete came gradually into favor as architects and engineers relied more and more confidently on the results of experimental research and experience. By 1911 the Tiber at Rome for the first time in history had been crossed by a single arch, the Risorgimento Bridge, a reinforced-concrete structure of 328 feet (100 meters) span and slightly less than 33 feet (10 meters) rise. On completion, the bridge was subjected to unique strength and stiffness tests. Eleven hundred soldiers marched over it at double-quick, and three 15-ton road rollers abreast crossed it forward and back; the greatest deflections were of the order of $\frac{1}{10}$ inch, and the vibrations were negligible. The Risorgimento Bridge was the longest masonry arch in the world for some years. In the twenty years following the completion of this bridge engineers learned much of the possibilities of reinforced concrete. Continental European engineers, especially, learned how to produce, when necessary, a dense concrete whose tensile strength was considerably greater than had previously been considered possible. Subjecting the steel reinforcement to preliminary stretching also has the effect, when the stretching force is released, of artificially precompressing the concrete and thus counteracting tensile stresses which may be due to shrinkage in setting and to temperature change as well as to loading.

The Plougastel, or Albert-Louppe, Bridge over the river Elorn, France, is one of the boldest concrete bridges thus far built (Figure 14.6). E. Freyssinet was the chief engineer. This structure consists of three 612-foot arches spanning the river where it flows into the harbor of Brest. At the time it was completed in 1930 its arch spans were the longest ever built in concrete; only two exceed it today, one each in Spain and Sweden. It carries a single-track railway and above it a roadway 26.2 feet wide across a river in which the tidal range is 26 feet and the current is sometimes very strong. The river piers and the abutments rest on rock. The

Figure 14.6 Plougastel or Albert-Louppe Bridge, Brest, France (Courtesy Société Technique pour l'Utilisation de la Précontrainte)

arches are cellular in construction. Between the lower and upper slabs (intrados and extrados) are four spandrels, or vertical walls, which bind the two slabs together. The outermost of these spandrels constitute the arch faces. The arch is built of concrete with relatively little reinforcing steel to take care of secondary stresses such as those due to shrinkage of the concrete in setting.

The piers were built in two sections, and the lower section or base shaft presented no great difficulties. The pier tops, or umbrella cantilever sections, about 22 feet high, 52 feet wide, and very heavily reinforced, were precast on shore and floated into position, where they were attached to the piers at low tide to form the bases of the individual arches. These pier tops are almost entirely of a curved V shape and are designed to support the two arches on either side, forming indeed the first few feet of each arch. Only one arch center or supporting frame for the concrete form, constructed almost entirely of wood, the ends being tied together by cables to take the horizontal thrust, was used in turn for all three arches. It was built close to shore and floated out on two large barges, one at each end, and was secured to the umbrella sections at the pier tops. The concrete was poured in four successive stages, or layers. The intrados slab formed the first stage, two of the four vertical walls the second, the other two the third, and the extrados slab the fourth. A very few days after the concrete in the extrados slab was poured, the center was detached and moved to the next span. Construction materials were handled by twin cableways nearly half a mile long running across the river from bank to bank, some 250 feet above low tide.

There are striking differences between the problem presented to the engineer of this bridge and that which Eads had to meet sixty years earlier. The Elorn, like the Mississippi, could not be entirely obstructed, for the Brest harbor is often busy. Arch centers were provided at Plougastel, one arch at a time. Eads could not have used centers; he was obliged instead to build his arches out from each pier by the cantilever method, hoisting the separate parts by derricks from barges, using only hand power. The Plougastel arches were slightly longer than those of the Mississippi bridge. The rise at the center of the arches is about 110 feet for Plougastel, about 43 feet in the Eads bridge. The Plougastel traffic is the lighter, both on the bridge and under it. The French engineers had many types of apparatus that Eads lacked—cableways, diesel engines, electric power, concrete mixers, and telephones, to name only a few. The Plougastel foundation problems were much simpler.

The Swiss Robert Maillart (1872–1940) was the first to use reinforced-concrete slabs as active-bearing structural elements. Maillart dispensed with supporting beams and developed in 1900 a new structural principle. Active-bearing concrete-slab floors are an appropriate type for buildings with nonbearing walls; the slabs can be cantilevered out to support curtain walls, floor by floor. Although C. A. P. Turner (1869–   ) used slab construction in the Johnson-Bovey Building in Minneapolis in 1906, and there are many applications of this type of construction especially in roof design, American engineers were slow to adapt the principles in bridgebuilding, largely perhaps because in the United States labor costs are higher and material is more plentiful than in Europe.

Railroad Mountain Tunnels

Modern tunnels may be classed in groups according to the purpose they serve and the methods employed in their construction. This section will discuss in detail two of these groups: one mainly through solid rock, the other through pressure-exerting subaqueous material. France had taken the initiative in canal-tunnel construction in the seventeenth century with a 500-foot tunnel bored through solid rock on the line of the Languedoc Canal. Most canal tunnels, however, date from the latter half of the eighteenth century and the first few years of the nineteenth, and the greater number are in England where the aggregate length of 45 exceeds 40 miles. The use of hand drills and blasting powder together with improved surveying instruments and methods gave contractors of this period a decided advantage over the ancient Romans in the construction of water tunnels. In a French tunnel of about 1800 were developed the first systematic schemes for supporting by timbering any loose rock and earth inside a tunnel more than a few feet wide. One of the two American canal tunnels, completed in 1828 on the line of the Union Canal near Lebanon, Pennsylvania, is now preserved as a local monument to a bygone transportation era.

The first railroad tunnel in the United States, like the Lebanon Canal tunnel, is also in Pennsylvania, four miles east of Johnstown. It was bored in 1831 to 1833 and was abandoned after 1857 when the Pennsylvania Railroad Company bought the line and straightened it, making the tunnel unnecessary. Two early epoch-making railroad tunnels were the Hoosac Tunnel, on what is now the Boston and Maine Railroad in northwestern Massachusetts, and the Mont Cenis Tunnel piercing the Alps between France and Italy. The Hoosac had been suggested in 1825 for a canal to connect Boston with the Hudson Valley. It was begun in 1854 and required twenty-two years to complete (1876), while the Mont Cenis, nearly twice as long, was begun three years later (1857) but was finished five years earlier (1871) than the Hoosac. Both of these tunnels served as laboratories in which many an experiment in excavation methods was tried out. While rock excavation in the Hoosac began with hand drills (Figure 14.7), these were shortly replaced by steam drills and a few years later by some drills operated by compressed air. Black powder served for the first few years in blasting at the Hoosac until nitroglycerine replaced it. The Hoosac was probably the first tunnel in which blasts were set off by electricity, a method that had been used in only a few localities earlier. The following is a detailed description of the Mont Cenis Tunnel.

Some of the sixteen or more passes through the Alps, like the Brenner, may have been used by sturdy-wheeled conveyances since Roman days. Augustus Caesar's engineers may have built a short-lived carriage road across the Little St. Bernard pass. Napoleon's engineers, Polonceau and others, built several transalpine roads, of which the best known was the Simplon. The road over the Mont Cenis Pass, some 40 miles east of Grenoble and an equal distance south of Mt. Blanc, was finished in 1810

Figure 14.7 Work at a Hoosac Tunnel heading (From *Science Record*, 1872)

and is still much traveled. The pass is 6,772 feet above sea level and lies on a quite direct line between Lyon and Turin. The Mont Cenis road thus serves to link Savoy in France with Piedmont in Italy, both, until a century ago, part of the Kingdom of Sardinia. Railroads across the Alps seem to have owed much of their inspiration to the success of Benjamin Henry Latrobe in carrying the Baltimore and Ohio Railroad over the Alleghenies from Cumberland, Maryland, to Wheeling, Virginia, in 1842 to 1853. Carl von Ghega (1802–1860), who had examined and written about the Baltimore and Ohio, built the first transalpine railroad. This line was the bold Semmeringbahn across the ruggedly picturesque Styrian Alps, the vital link in the Vienna-Trieste line; the tunnel was built in 1848 to 1854. The summit level on this 34-mile road at the Semmering Pass, some 50 miles from Vienna, is 2,892 feet above sea level. The summit tunnel was less than a mile long, but there were many short tunnels and viaducts. The maximum gradients were 2½ per cent and the sharpest curve had a radius slightly over 600 feet. The second railroad across the Alps went through the Brenner Pass, the lowest of the passes (4,497 feet) through the principal Alpine range, some 200 miles to the west of the Semmering. This road furnishes the connection between Innsbruck in the Austrian Tirol and Verona and Venice, Italy. It was built by Karl von Etzel (1812–1865) and was completed in 1867. There is no summit tunnel; the maximum gradients were the same as those on the Semmering, but the curves were easier.

The third railroad across the Alps was called the Mont Cenis, because it follows in large part the line of Napoleon's Mont Cenis Pass road. It was realized at least as early as 1840 that there should be a connection between the railroad systems of Savoy and Piedmont in order to expedite travel from Lyons, Paris, and London down through central Italy. By 1863 the Italians had ceded Savoy to the French, and the railroad on the French side had reached St. Michel in the valley of the River Arc and that on the Italian side had progressed to Susa in the Dora River Valley. Between the two railheads there was a gap of 48 miles along the old Mont Cenis Pass road. However, some years earlier it had been pointed out that the Arc River near Modane, Savoy, just above St. Michel, was only about 8 miles in a direct line through the mountain from Bardonecchia in the Dora River Valley, Piedmont. It was even then predicted that a tunnel driven through the mountain some 14 or 15 miles to the west of the Mont Cenis Pass and under the Col de Fréjus would make the ideal connection between the two railroad systems.

The bold and revolutionary idea of boring an 8-mile tunnel without intermediate shafts through an Alpine mountain seemed foolish and frightening to many sensible people in the late 1850s. The summit tunnel on the Semmering railway in Austria, only ⅞ mile long, had been completed several years earlier, using nine shafts, of which five had been left permanently open for ventilation. No one was really certain that enough air to prevent asphyxiation of workmen would penetrate horizontally into a tunnel like the Mont Cenis; what would happen if and when a steam locomotive passed through was still another question. Nor did anyone actually know what the temperature of the rock would be a mile below the surface; geologists could only guess that it would be higher than at the surface. The geologists could not even guarantee that after months of blasting, the overlying strata might not weaken at a number of points and cause the tunnel roof to collapse without warning and bury the workmen. There was even some apprehension that the reputedly bottomless little lake high up in the Mont Cenis Pass might pour down into the tunnel at such a rate as to flood out the works and drown the workmen.

The Mont Cenis Tunnel was an Italian project throughout. Preliminary investigations, under the direction of three young, enthusiastic Italian engineers, Germain Sommeiller (1815–1871) and his colleagues Grandis and Grattoni, date from the 1850s. On August 15, 1857, the Italian parliament passed a bill authorizing the boring of the tunnel between Modane and Bardonecchia. Only two weeks later King Victor Emmanuel fired the first blast at the Italian end. Not much was accomplished in 1857, because working seasons in the high Alps are short. In the spring of 1858, however, work on both ends of the tunnel began in earnest although the surveyors had only barely started their triangulation and leveling. In the course of the summer of 1858 the surveyors spread a geodetic network of triangulation over the Mont Fréjus area above the tunnel site; this network enabled them to lay down its axis with certainty and to calculate its length. To this end they established 21 geodetic stations, or survey points or observatories, on nearby heights, and with a high-powered theodolite measured 86 angles, each of them at least 10 times, some 60 times, to ensure accuracy. They prolonged the axis on each end, thus giving ranging points to use as the boring progressed in order to make sure that the headings would meet. The engineers also carried a line of levels high over the summit of Mont Fréjus, in itself no mean feat considering the difficult terrain and the prevalence of storms. They were thus able to establish the tunnel grades from portal to portal. All of the surveying was completed by the fall of 1858. Never before had tunnel contractors been prepared to proceed with such confidence in the accuracy of the preliminary surveys.

Much of the time during the first two or three years was necessarily taken up with the establishment of extensive machine shops, power plants, and workmen's living quarters, all necessarily in duplicate, for the only connection between the Italian and French ends was over the mountain. By the end of 1860, the tunnel excavation had proceeded about ⅓ mile at the north end and nearly ½ mile at the south end, all by hand drilling. There were by this time many who, while not exactly scoffing, predicted that the tunnel could not be bored in less than thirty years. Thus for several years work on the tunnel progressed slowly. The promoters were not discouraged, but in 1863 they were glad to give John Barraclough Fell, of England, permission to build one of his patent railways along the edge of the Mont Cenis road from St. Michel up and over the pass and down to Susa, about 48 miles in all. This was a temporary permission, for everyone knew that if and when the tunnel should be completed the Fell road would forthwith become practically worthless. Fell began his extraordinary railway in 1866, finishing it in two years at a cost exceeding 2 million dollars. It was in operation only three years instead of the expected ten and was later sold to Brazilian interests for reinstallation in Brazil. The road had a 3-foot 7⅞-inches, or 1.10-meters, gauge. It climbed gradients as steep as 8 per cent, and some of its curves had radii as short as 132 feet. Instead of rack and pinion the track had a central third rail, laid on its side and 7½ inches higher than the running rails. Two smooth wheels on the locomotive, revolving horizontally, gripped this central rail between them to obtain sufficient traction.

The boring of the tunnel had been begun in 1857, using hand drills only. Compressed-air drills, or boring machines of the Sommeiller type, were introduced in 1861 on the Italian end and two years later on the north or French end. The total progress during 1858 by hand drilling was 1,506 feet; during 1870 with machines it was 5,364 feet. Machine drills thus proved to be 3½ times as rapid as hand drilling. In both cases the rock was schist; in fact, except for a vein of quartz and one of limestone, the tunnel passed entirely through two kinds of schist, with the calcareous type predominating. The boring machine consisted of a heavy carriage mounted on wheels, movable along a track, and carrying a battery of usually seven to nine drills or perforators, each of which could deliver 200 powerful strokes a minute against the rock at a spot selected. The holes were laid out according to a pattern so that each blast would loosen the greatest quantity of rock in pieces that could be easily handled. Compressed air operated the drills.

Two hydraulically operated compressor plants, one near each portal, furnished compressed air for the drills. At Bardonecchia, on the Italian side, a 2-mile canal was built to bring the water power under a head of 85 feet to the compressors. The compressor system furnished air at 6 atmospheres, 5 above normal. The compressor was a form of hydraulic ram (*compresseur à coup de bélier*), very wasteful of energy. At Modane a similar system was set up at first, but using Arc River water which two water wheels, each working two pumps, raised to the 85-foot level. A more direct and less wasteful system was finally installed at Modane. The compressed air throughout the tunnel was furnished to the machines through 8-inch pipes laid along one of the walls of the tunnel. Another pipe carried water to the drill points. The inaugural blast that King Victor Emmanuel fired in 1857 was the first of some three million blasts, extending over fourteen years. During this time nearly 1 million cubic yards of rock were excavated and disposed of; the average haul for the broken rock approximated 2 miles. Ordinary black powder was the only explosive used at Mont Cenis despite the fact that nitroglycerin dates back to 1847 and was used on the Hoosac Tunnel as early as 1866. Moreover, Alfred Nobel had invented dynamite in 1867, four years before the last blast at Mont Cenis.

The difficulties to which pessimists had directed attention before the work began were met and solved as they arose by the resourceful engineers. Ventilation was a troublesome question throughout although exhaust air from the drills freshened the air somewhat. Dividing the tunnel by a slightly arched horizontal brattice, or temporary low roof, helped to induce a circulation of air. The temperature of the rock seldom exceeded 75°, in part because of the welcome seepage of cooling water. A few cases of weakened rock strata were promptly and properly dealt with. The ceiling arch was varied in form and thickness to withstand varying pressures. Trouble from the "bottomless" little lake, which had been feared, did not materialize; the lake is 14 long miles from the tunnel.

The headings met on Christmas Day, 1870. The French side was a bit higher than the Italian—perhaps 2 feet; the error in alignment was somewhat less. The tunnel proved to be about 45 feet longer than calculated. The point of meeting was not midway on the tunnel; the Italian end had been excavated about 4⅔ miles, the French end about 3⅓ miles. The tunnel was finally completed during the summer of 1871 (Figure 14.8); France and Italy shared the total cost of about 15 million dollars.

Starting from St. Michel, France, in the valley of the Arc, the Mont Cenis Railway continues up the valley, close to the old Mont Cenis road, 10 miles to Modane, passing through 14 short tunnels. Here it takes a hairpin turn and winds up the hill for 3 more miles, reaching the tunnel entrance, Fourneaux, after having climbed 1,460 feet from St. Michel. It then continues on a 2.1 per cent gradient into and halfway through the tunnel, which is straight for more than 7½ miles. At the highest point the rails are 4,393 feet above sea level, and there is nearly a mile of solid rock over the tunnel roof up to the Col de Fréjus. From here the rest of the tunnel is nearly level to Bardonecchia, the southern portal. The railway on the Italian side for the next 25 miles winds down the slope at a rather steeper rate than on the French side. It passes over 8 viaducts and through 26 tunnels, descending to just less than 1,600 feet above the sea at Bussoleno, which is on the old Mont Cenis road just below Susa, where it connects with the Susa-Turin line. The Mont Cenis Tunnel led the way for other great Alpine tunnels even longer and of bolder design, the St.

Figure 14.8 Placing the last stone in the Bardonecchia portal arch of the Mont Cenis Tunnel, Aug. 18, 1871 (From *L'Illustration*, Sept. 9, 1871)

Gotthard, the Simplon, the Loetschberg, and several others. The construction of the 12¼-mile Simplon at the end of the century involved overcoming technical difficulties such as had not been met earlier.

The Mont Cenis Railway and Tunnel were being built in the years between 1857 and 1871. During nearly half of this period the Civil War was being waged in the United States. A decade before this war began, the country's first important railway tunnel, the Hoosac, had been planned on the Troy and Greenfield, later called the Fitchburg, now the Boston and Maine Railroad, in western Massachusetts. The Hoosac Tunnel, somewhat more than half as long as the Mont Cenis, was not, however, actually started until 1854. Unlike the Mont Cenis, the Hoosac pierced a relatively low mountain, which made it possible to drive two vertical shafts in the 4¾ miles, which vastly simplified the work. As at Mont Cenis, hand-rock drills were at first used on the Hoosac. Then came some not too successful experimenting with steam drills. These drills were originally an American invention, dating back to J. J. Couch in 1849, but they proved not to be well adapted to tunnel work, largely on account of the difficulty of piping steam a considerable distance and taking care of the exhaust. In the 1860s, Charles Burleigh improved and perfected a pneu-

Figure 14.9 Brunel's Thames Tunnel, completed 1843 (From Charles Knight (ed.), *London*, vol. 3, 1851)

matic drill. In England Thomas Bartlett patented a steam drill, much used in coal mines, that he insisted could operate also with compressed air. Sommeiller improved the Bartlett machine, which he began to use in 1861 on the Italian end of the Mont Cenis Tunnel.

Subaqueous Tunnels

The Thames Tunnel (Figure 14.9) at London is usually considered to be the first subaqueous tunnel. Actually, however, it had been preceded for many years by others of small bore in the coal mines of western and northeastern England. The Thames Tunnel is a monument to the courage, resourcefulness, and perseverance of one man, the French-born British engineer Marc Isambard Brunel (1769–1849), father of Isambard Kingdom Brunel. His 1,200-foot tunnel, begun in 1825, was built for highway traffic. It is about 37 feet wide and 23 feet high outside with two arched traffic lanes. The tunnel is lined with brick laid up in what was then called Roman cement.

Brunel designed for his tunnel what later came to be called a shield, a curious rectangular end of cast iron, faced with 3-inch planks and weighing 120 tons, for driving through the pressure-exerting clayey material under the Thames. The shield fitted over the tunnel proper and contained 36 compartments, or cells, each about 3 feet by 7, which could be opened one by one to allow workmen to excavate a small amount of the stiff blue clay of the bed of the river ahead of the shield. The entire shield was divided vertically into 12 sections called frames, some of which could be forced forward a few inches at a time by means of screw jacks. The roof of the tunnel is in some places barely 13 feet below the river bed. With such a crude device the river broke through five times and drowned seven workmen. After each such irruption, the tunnel had to be pumped out, and the overlying layer of soil strengthened by material dropped from boats. Once the work was suspended for seven long years. Nevertheless, the tunnel was finally finished in 1843 when Brunel was 74 years old and had suffered a paralytic stroke. Planned and used for highway traffic, it was not a financial success, and a railway later acquired it. Now, as part of London's underground system, it carries rapid-transit traffic never contemplated by its builder. It crosses under the Thames from Wapping Station, nearly 1½ miles east of London Bridge, to Rotherhithe.

The earliest American underwater tunnel, mentioned in Chapter 13, was bored for 2 miles at a depth of 30 feet below the bed of Lake Michigan in 1864 to 1867 to supply Chicago with water. Passing as it did through a dense blue clay, the tunnel was excavated without a shield. There was a circular vertical shaft at each end which kept the water out and through which the excavated material was hoisted to the surface. The tunnel was a cast-iron tube lined with brick and had an inside diameter of 5 feet.

Brunel's Thames Tunnel showed both the advantages and the disadvantages of his type of shield construction. It remained for a young British engineer of the next generation, the South African-born James Henry Greathead (1844–1896), to develop a circular, one-piece, close-fitting shield originally suggested by Peter W. Barlow (1809–1885) and which became the prototype of those used all over the world ever since. Greathead had been a pupil-assistant to Barlow and at his suggestion used such a shield in tunneling under the Thames in 1869. This tunnel, because it began close to the Tower of London, was called the Tower Subway (Figure 14.10). It was of much smaller cross section than Brunel's tunnel, indeed only 7 feet in diameter, and was used mainly by pedestrians.

Figure 14.10
The Tower Subway
(From G. W. Thornbury,
Old and New London,
vol. 2, 1873)

The tunnel was lined with heavy cast iron, built up in narrow cylindrical rings, each composed of a number of segments, bolted together as the shield advanced. The shield was forced forward, as Brunel's had been, by screw jacks. The Tower Subway, 1,350 feet long, was bored deeper than Brunel's tunnel, so that it passed through non-water-bearing clay. Entirely accomplished within the year 1869, it was for a time equipped with tiny cable cars, each carrying 12 passengers. The successful construction of the little Tower Subway suggested the use of the Greathead shield for a larger tunnel project, called at first the London and Southwark Subway, later the City and South London Railway. Greathead began work on this tunnel in 1886, nearly a generation after the inauguration of the comparatively shallow London underground system which was operated by steam locomotives. This railway started from a point near the Monument and, piercing the bed of the Thames somewhat obliquely, extended more than 3 miles to Stockwell in South London. It was completed and in use late in 1890.

Almost at once it was dubbed the "tuppenny tube." It was altogether different from the shallow underground. Modeled rather after the Tower Subway, it was of twin cast-iron tubes, each just over 10 feet in diameter. Greathead used a larger version of the shield he had devised and used on the Tower Subway twenty years earlier. Hydraulic jacks, pumped by hand power, advanced the shield. In addition, wherever he encountered difficult water-bearing soils, which incidentally were not under the river,

Figure 14.11 Electric locomotive and passenger cars on City and South London Railway (From *Scientific American*, Nov. 29, 1890)

he used compressed air to keep the water out. This method, patented in 1830, had been used in bridge-foundation construction and had indeed been employed on a few short tunnel projects but with only varying success. Greathead was thus the first to use these two techniques in combination, and he set the pattern for most subaqueous tunneling down to the present day. The maximum air pressure used by Greathead on this tunnel did not exceed 15 pounds per square inch above atmosphere.

The motive power for the City and South London Railway was electricity from the first, although the original plans contemplated cable cars. It was the first underground railway anywhere to be operated by electricity (Figure 14.11). The little articulated three-car trains, which fitted tightly in the tubes, were drawn by Siemens Brothers electric locomotives that weighed only 13½ gross tons. The cars had very small windows. They seated 32 not too comfortable passengers, and the average speed was 12 to 13 miles per hour. After about thirty years' service the tubes were rebuilt on a slightly more generous scale; traffic was kept running meanwhile—a difficult feat.

In the year 1879 two epoch-making attempts were made to apply compressed air without a shield to tunneling. The first, at Antwerp, was for a short tunnel, 4 feet high and 5 feet wide, in connection with some dock work at the river Scheldt, under engineer-contractor Hersent. The project seems to have accomplished its purpose. The second attempt was on a vastly larger scale, namely, the tunneling of the Hudson River at New York City. De Witt Clinton Haskin, an optimistic Californian of means, inspired perhaps by the success of Eads in 1869–1870 with compressed air, or of William Sooy Smith on his Missouri River Bridge, came to New York in 1873 with a most ambitious scheme. He hoped to bore a river tunnel that would enable the several railroads terminating in Jersey City and Hoboken to bring their lines together through it into a large union-terminal station to be built in Washington Square, New York City. The Hudson here is nearly a mile wide and in places about 60 feet deep. The underlying soil is in general almost liquid silt. Haskin may have known of Greathead's little Tower Subway. He seems, however, to have been serenely confident that compressed air would make a shield unnecessary for boring his proposed tunnel.

He started work on a shaft at 15th Street, Jersey City, in 1874, and after several years' delay due to litigation, began in 1879 to excavate for twin tunnels. They were of slightly elliptical form, about 16 feet wide and 18 feet high inside, lined with 3-foot-thick brickwork inside a shell of ¼-inch iron. In the summer of 1880 the compressed air failed to keep out the silt at one spot close to the New Jersey shaft, and 20 men were drowned. Work was resumed later, but ceased in 1882 when funds gave out. Haskin had completed about 1,600 feet of the northerly tunnel and 600 feet of the southerly, mostly from the New Jersey end. Seven years later (1889) a London firm, S. Pearson and Son, with much tunneling and subway experience, took up the project and kept at it for more than two years. Their work was to reconstruct the portion of the tunnel already built and to extend it, using a cast-iron lining which by that time had become standard practice. This firm had learned how to use a shield in connection with compressed air. The engineers encountered one unprecedented difficulty in the form of ledges of bedrock close to the New York shore, and no one yet knew how to excavate such a tunnel partly in silt and partly in rock. In 1892 the British company's funds were exhausted and work stopped. This company had built 1,800 feet of tunnel, all lined with cast iron.

At about this time a 10-foot-diameter tunnel, lined with cast iron and 2,500 feet long, was being bored under the East River at a depth of nearly 125 feet below low water. This was the East River Gas Company's tunnel which, passing under Welfare Island, connects 71st Street, Manhattan, with what was then Ravenswood, Long Island. When the construction contract was let it had been assumed that, deep as it was, the tunnel would pass almost entirely through rock, like a mining tunnel. No provision was therefore made for a shield, and there was little thought of compressed air. Before many months the top of the tunnel had advanced from each side into soft water-bearing material, which compressed air at safe pressures could not keep out. After prolonged effort and many discouragements, the contractor gave up in despair and the conduct of the work was put into the hands of an English-born engineer, Charles M. Jacobs (1850–1919), who immediately provided shields at each heading. The tunnel headings soon passed entirely into soft mud, and the work advanced more steadily. At the end of a year, late in 1894, the tunnel was completed and ready for its 36-inch gas main and narrow-gauge railway track. This pioneer tunnel had served as an experimental laboratory for subaqueous tunneling in the New York area. Shield and compressed-air problems were here successfully applied to a variety of conditions, even to blasting of rock just ahead of a tunnel whose top was in soft mud.

Shortly after Jacobs had completed the Ravenswood tunnel he was asked to report on the feasibility and probable cost of completing the two Hudson River tunnels which had lain embedded in the silt since the Pearson firm had stopped work in 1892. After a few years' delay an American company was incorporated in 1902, with William G. McAdoo as president and Charles M. Jacobs and J. Vipond Davies as engineers. The original plan was considerably altered, and the result, in 1904 and 1905, was the McAdoo Tunnel which forms part of the Hudson and Manhattan Railroad, a vastly more extensive and complicated system than Haskin had planned thirty years previously. It was decided not to let the tunnel work to a contractor; in fact, contractors were properly hesitant to attempt the completion of an undertaking which had already proved so hazardous. Instead, Jacobs organized and directed the constructional force for the entire tunneling operations. By this time the project had expanded to include two pairs of twin river tunnels somewhat over a mile apart, with elaborate connections in Jersey City and Hoboken and more than a mile of tunnel under Sixth Avenue, Manhattan, north to 33d Street.

Before the original twin river tunnels had been completed two unusual situations presented themselves. It was found that in the silt the aggregate thrust of 2,500 tons which the 11 hydraulic jacks could exert on a shield was often enough to force it forward without the necessity of admitting any of the material. Thereafter this blind driving, as it was called, became an accepted procedure. In fact as much as 72 feet of progress was made in twenty-four hours on one of the southerly tunnels, using this method. At one point on one of the northerly tunnels near the New York shore, ledge rock was encountered, and several blows had occurred. A blow is an irruption or escape of the compressed air in the tunnel, resulting in a huge bubble blowing up through the overlying silt and perhaps carrying men and equipment to the surface. Clay was dumped from barges in an attempt to reinforce the scanty covering of silt above the tunnel. The clay, as it filtered down, became almost fluid and failed to serve its purpose. It was, however, indurated, or hardened, by playing against it from within the tunnel blowpipe flames fed by kerosene tanks under compressed air. This hardened clay made a wall which protected the workmen and made it possible for them to drill and blast ahead of the shield under an overhanging steel apron. Jacobs later remarked that never before in history had man "made brick in the bottom of a river." The first of the northerly pair of tubes was completed in 1904, practically thirty years after Haskin had started the project.

As the McAdoo, or Hudson and Manhattan, tunnels were approaching completion the Pennsylvania Railroad's plan to tunnel the Hudson at 32d Street began to take shape. The Pennsylvania tunnels are about 1½ miles north of the northerly pair of the McAdoo tunnels; they are of slightly larger bore than the McAdoo tunnels, and the cast-iron shell is reinforced by a concrete lining at least 2 feet thick. The bed of the Hudson is not essentially different here from what it is farther south, a deep layer of thin silt with bedrock underlying it at a variable but considerable depth lower. At one point the rock is 300 feet below high water. It was realized that tunnels in soft silt might not be stable under such loads as heavy locomotives and cars, though there had been no appreciable settlement in the Hudson and Manhattan tunnels which, however, carried relatively light multiple-unit trains. It was at first thought that this objection could be met by providing long piers or screw piles extending down to harder material or even to rock. The work was let out by contract in 1904. While the original drawings showed screw piles or pipe supports, it was decided after elaborate tests that they were unnecessary. Instead, the roof and invert were strengthened by steel rods and the concrete lining was made somewhat thicker. The result was that the weight of each tunnel, foot by foot, including its maximum train load, was equal to that of the displaced silt; the tunnel was thus practically a buoyant cylinder hung at the ends and floating in the mud.

At Tappan Zee, 22 miles north of the Pennsylvania tunnels, the bed of the Hudson is still a layer of silt, in places 300 feet deep. Here the New York State Thruway will cross the river on a bridge to be completed in 1955, more than 3 miles long, whose main channel span of 2,400 feet is of cantilever truss design. The piers of this channel span are in large part supported by eight enormous semibuoyant closed caissons, or boxes of reinforced concrete, resting on steel piles which were driven down to bedrock. Pier foundations with such buoyant supports are a very recent engineering technique in the United States.

We have narrowed our discussion of the development of subaqueous tunneling to the point of including only London and New York projects and have omitted reference to the larger-bore highway tunnels more recently completed. If space permitted we might also have described such other early achievements as the long railway tunnels in England under the Severn and the Mersey, both completed in 1886, and the Sarnia tunnel under the St. Clair River between Michigan and Canada, completed in 1890. Historically, the importance of subaqueous tunnels is that they are a modern development beginning in the nineteenth century, whereas tunnels through rock have had a long technical evolution extending from antiquity.

Bibliography

Bossom, Alfred C.: *Building to the Skies,* The Studio Publications, Inc., New York, 1934.

Duluc, Albert: *Le Mont Cenis, sa route, son tunnel,* Hermann & Cie., Paris, 1952.

Freyssinet, E.: "The 600-ft. Concrete Arch Bridge at Brest, France," *Proceedings, American Concrete Institute,* vol. 25, pp. 83–97, 1929.

Jacobs, Charles M.: "The Hudson River Tunnels of the Hudson and Manhattan Railroad Company," *Minutes of Proceedings of the Institution of Civil Engineers,* vol. 181, pt. 3, pp. 169–257, 1909–1910.

Randall, Frank A.: *History of the Development of Building Construction in Chicago,* University of Illinois Press, Urbana, Ill., 1949.

Sopwith, Thomas: "The Actual State of the Works on the Mont Cenis Tunnel, and Description of the Machinery Employed," *Minutes of Proceedings of the Institution of Civil Engineers,* vol. 23, pp. 258–319, 1863–1864.

Sopwith, Thomas: "The Mont Cenis Tunnel," *Minutes of Proceedings of the Institution of Civil Engineers,* vol. 36, pp. 1–34, 1872–1873.

Starrett, William A.: *Skyscrapers and the Men Who Build Them,* Charles Scribner's Sons, New York, 1928.

Strauss, Joseph B.: *The Golden Gate Bridge,* Golden Gate Bridge and Highway District, San Francisco, 1938.

FIFTEEN

Reflections

Although this chapter is the last of *Engineering in History*, it certainly does not bring to a conclusion the history of engineering. Indeed it would appear from the accelerated rate of engineering development during the past hundred years that, midway in the twentieth century, society is only in the initial stages of an engineering advance which is dramatically changing the ways of human life. New sources of mechanical power, new prime movers, and automatic controls alone will lead to many transformations in human activity.

As previous chapters have shown, there is an orderly sequence in the history of technical progress, which Abbott Payson Usher has analyzed brilliantly.[1] However, rates of engineering advance have not been uniform. The most rapid pace has occurred since the wedding of engineering and science about the middle of the nineteenth century after a long flirtation. Of the many innovations discussed in the earlier chapters there can be no doubt that the development of sources of power other than man's own strength had the most extensive impact on Western civilization and facilitated far-reaching advances in engineering.

The substitution of mechanical for muscular effort began during the Middle Ages with the inventions of water wheels, windmills, the horse collar, and the principle of fore-and-aft rigs for ships. Ever since, mechanical power has made it possible to increase the welfare of man by relieving him of the drudgery of hard muscular effort such as pulling an oar, dragging a plow, sledge, or cart, and lifting water. With the invention of the steam engine in the eighteenth century, and turbines, internal-combustion

[1] Abbott P. Usher, *A History of Mechanical Inventions*, 2d ed., Harvard University Press, Cambridge, Mass., 1954, pp. 56–83.

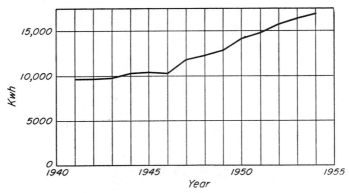

Figure 15.1 Kilowatthours per industrial worker per year in the United States (Based on data in Edison Electric Institute, *About the Electric Industry*, 1952–1953, 1954–1955)

engines and electric generators and motors in the nineteenth century, new applications of power to perform heavy tasks have recurred at an ever-increasing rate. The average power which a man can exert by muscular effort throughout the day has been estimated to be about 35 watts ($\frac{1}{20}$ horsepower).[2] If he worked 240 eight-hour days in a year, this would amount to 67 kilowatthours per year. Edison Electric Institute statistics indicate that the electrical energy used by each worker in industry in the United States amounted in 1954 to about 17,314 kilowatthours per worker (Figure 15.1). Electricity furnishes approximately 95 per cent of power used by industry. On this basis each individual engaged in manufacture had under his control an amount of energy equivalent to the muscular effort of 244 men! The average family in the United States used about 2,549 kilowatthours in 1954 (Figure 15.2). On the same basis of calculation the family employed the equivalent of about thirty-three laborers each day to help in the household duties! Each driver of a 100-horsepower automobile has under his control the equivalent in mechanical effort of more than two thousand slaves to speed him on the road!

The locomotive engineer and the airplane pilot control impressive amounts of power. The locomotive engineer often has under his full control an engine capable of exerting up to 5,000 horsepower. At cruising speed, the pilot of one of the newest British Comet passenger jet airplanes controls about 35,000 horsepower—the equivalent of 700,000 humans.

[2] C. M. Ripley, "A Kilowatthour," Edison Electric Institute *Statistical Bulletin*, vol. 7, p. 326, June, 1939.

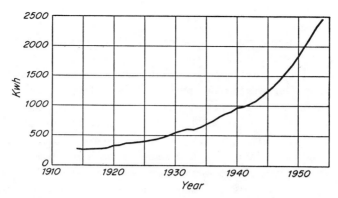

Figure 15.2 Kilowatthours per family per year in the United States (Based on data in National Electric Light Association *Bulletin,* 1931, and Edison Electric Institute *Statistical Bulletin,* 1954)

The switchboard operator of a large power plant may control by a twist of his fingers over 1,000,000 horsepower—more power than the entire population of the state of Connecticut could produce by muscular effort.

The largest source of power available to most Americans is the automobile engine producing upward of 100 horsepower. But the housewife, especially in the United States, has control of electric-motor appliances in her home, such as washing machines for dishes or clothes, vacuum cleaners, refrigerators, mixers, ventilating fans, and automatic furnaces, each of which does an amount of work equivalent to that of several human beings. Although it has been claimed that such electrical conveniences have solved the problems of the shortage of domestic labor, the authors have it on good authority that the solution is not entirely satisfactory. Nevertheless, as Philippe Le Corbeiller has stated: ". . . mechanical power has changed the status of woman in our society. It has brought it about that the motor which one worker in a plant has to control is operated by pushing a button or twisting a dial, and this can be done by a girl without expenditure of strength; she may well be more useful in that capacity than a big burly man with brawn and perhaps less brains. That means economic independence, the possibility of getting away from the family cell if she wants to and cares to. We have already said that easy transportation brought about social fluidity; this means for a woman the possibility of rebuilding in another town a life that for one reason or another has not been successful; it creates in her a

497

completely different psychology—one of self-reliance and independence. We could go on and on about the consequences of that!" [3]

It would be difficult to overemphasize the social importance of mechanical power as replacing muscular effort of human beings. Mechanical power has become something far more than a substitute for human effort. Obviously 700,000 people could not directly supply the power to fly the equivalent of a jet airplane carrying only fifty persons at over 500 miles per hour. Western societies, having adapted themselves to mechanical power, have used it for many purposes other than as a substitute for human muscles. The extension of power utilization has truly made the twentieth century the Golden Age of power.

The development of automatic controls in the twentieth century may have as great an effect on human welfare as has mechanical power. An automatic control is a device which performs automatic self-regulation by feeding back to an earlier stage information regarding the status of a process at a later stage to alter the action of the process and thereby to control the output. One of the most familiar of automatic controls is the household thermostat. It shuts off the furnace when the temperature of the room rises above a given setting and restarts the furnace when the temperature falls below that setting. The thermometer's information about the temperature of a room is fed back to the furnace, which in turn controls the room temperature. The most important link in this closed sequence is the feedback. The principle of control by feedback is the core of automatic controls, or servomechanisms as they are often called when they have a power-amplifying system in addition to feedback.

Although automatic controls were not an important factor in social history until the twentieth century, they originated in the eighteenth. One of the earliest automatic controls was on Newcomen's first steam engine of 1712. Since a full cylinder of steam produced a temporary exhaustion of Newcomen's small boiler, he ingeniously constructed an automatic control which regulated operation so that the engine would make one complete cycle only when a sufficient amount of steam was in the boiler to effect the cycle. After the cylinder had been filled with steam, thereby exhausting the boiler, the engine stopped. When the boiler had produced enough steam to refill the cylinder, a float arrange-

[3] Philippe Le Corbeiller, "Applications of Science and the Teaching of Science," *General Education in Science*, Harvard University Press, Cambridge, 1952, p. 136. (Quoted with the kind permission of the Harvard University Press.)

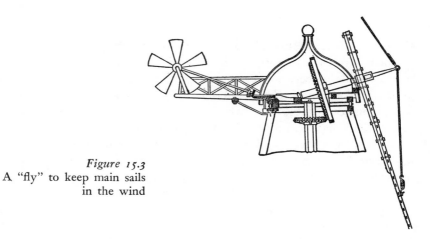

Figure 15.3
A "fly" to keep main sails
in the wind

ment opened the injection valve, admitting water to the cylinder to condense the steam. Atmospheric pressure forced down the piston, and a rod attached to the working beam first shut off the injection valve and then, when the piston reached the bottom of its stroke, opened the steam valve, thereby breaking the vacuum to allow the piston to rise to the top of the cylinder.

Later in the eighteenth century, several innovators added automatic controls to windmills. In 1750 Andrew Meikle (1719–1811) invented the fantail gear or fly. Meikle's device was a small windmill mounted at right angles to the main sails and diametrically opposite them on the turret (Figure 15.3). If the main sails were not correctly set, the wind drove the fly which rotated the turret until the main sails faced the wind, at which time the sails on the fly became stationary. Prior to 1787 when Thomas Mead patented it, a centrifugal flyball device was in use to regulate the distance between the grinding stones of windmills. Following the invention of the flyball regulator, various improvements in louvered sails led to William Cubitt patenting such a sail to control automatically the speed of the mill. As the force of the wind increased, the louvers opened to spill some of it, thereby keeping the mill going at a fairly constant speed.

These automatic controls, including Watt's adaptation of the windmill flyballs for regulating a throttle valve to stabilize the speeds of his engines, had a tendency to oscillate, or hunt, around a given setting. In the nineteenth century, engineers invented damping devices to prevent oscillation, or hunting, in the governor. During the same century, mathematicians and physicists, among them P. S. Laplace, J. Clerk Maxwell, Lord

Kelvin, and Oliver Heaviside, worked out the theory of control and developed differential equations representing oscillation and damping. It was not until the 1920s, however, that engineers began to apply this theory in the construction of reliable and quick-acting servomechanisms. Further innovations during the past two decades have minimized the tendency to hunt and thus have increased the applicability of automatic controls. The result has been a rapid growth of instrumentation in United States industries; for instance, industry purchased nearly twice as many instruments in 1951 as it did in 1946 to accomplish accurate control.

The social significance of the introduction of automatic controls is analogous to that of the power revolution of the Middle Ages. Whereas the development of mechanical power relieved men from being the major source of power for the operation of various processes, the application of automatic controls has tended to remove men from such operations. There is no indication, however, that men will be removed entirely from the productive system. However, as in the case of mechanical power, automatic controls have become something far more than just laborsaving devices. Many modern processes can operate only under automatic controls; human control would result in an expensive and inferior product. Eugene Ayres has given a striking example from the petroleum industry, illustrating this development.

"Last year [1951] one of the countries of Asia employed a U.S. contracting firm to design a modern oil refinery. The firm submitted a design for a $50 million plant, and it included the usual array of control instruments. After studying the plans, the officials of the country, which has an embarrassing surplus of manpower, asked the designers to eliminate all automatic controls from the plant. The country could provide any number of thousands of men to record measurements and to control processes, and it was prepared to pay the price of lower efficiency and poorer-quality products to create this opportunity for employment. The contracting firm gave sympathetic consideration to the request, but its engineers finally decided that under no circumstances could control instruments be eliminated from the design. It was not just a question of operating costs or efficiency; without suitable control instruments a modern refinery simply could not operate at all.

"If the 50,000 control devices in the oil refineries of the U.S. should go 'on strike,' we would be faced with social disaster. The refineries would become lifeless industrial monuments. If we undertook to replace them with old-fashioned, manually operated refineries to supply our present motor-fuel needs, we would have to build four or five times as much plant, cracking and some other modern chemical processes would have to be eliminated, yields of motor fuel from crude petroleum would drop to a quarter of those at present, costs would skyrocket, and quality would plummet. Automobile engines would have to be radically redesigned to function with inferior fuel. And because of lower motor-fuel yields, we would need to produce crude petroleum several times more rapidly as we produce it now. Technology in refining would be set back to the early 1920's." [4]

The control of combustion in the boiler of a modern power plant is another example of development in automatic control. The output of the generators is signaled or fed back to the boiler room, and any increase in power demand sets in motion the automatic mechanism controlling the feedwater pumps, the draft fans, and the fuel supply, so that no labor is needed in the boiler room except to maintain proper records of the operation and to stand by in case of emergency.

The history of automatic control reveals social implications which some view with alarm and others with equanimity. Automatic controls, like mechanizations, may be a possible threat to society in that they may produce technological unemployment, a potential problem which has existed since the initial stages of the Industrial Revolution. The widespread introduction of laborsaving steam power in the early nineteenth century led to rioting by workers who feared that the steam engine would permanently put them out of work. Actually, the construction and operation of the new engines and machines created new jobs which demanded greater skill and less drudgery. It is interesting to observe the reemployment of stagecoach drivers and canal boatmen who had been displaced by the introduction of railroads and to consider the numbers of those engaged in transportation now in comparison with those employed a century and a half ago. Nevertheless, at mid-century, one of the assumptions upon which Karl Marx constructed his theory of capital's exploitation of the worker was that of inevitable technological unemployment. Although it cannot be excluded as a possible event in a capitalistic economy, the history of the past century and a half has thoroughly discredited Marx's assumption of its inevitability. Moreover, the short history of automatic control indicates that it will not cause more than temporary technological un-

[4] Eugene Ayres, "An Automatic Chemical Plant," *Scientific American*, vol. 187, pp. 82-83, September, 1952. (Quoted with the kind permission of the *Scientific American*.)

employment, at least in an industrial society like that of the United States, because for each unit produced in the over-all economy in recent years there has been a decrease in capital investment as well as manpower required. Automatic controls will further decrease capital investment for each unit produced on account of a larger number of units being turned out. If capital's contribution were to increase while manpower requirements dropped, capital would receive an increasingly larger share of total income, and there would be technological unemployment. To date there is every indication that the reverse is the case.

There are some who feel that the mechanization of industry and automatic control will lead to a stagnation of human intelligence and enterprise by making it unnecessary for men even to think about problems. Actually, history shows that the significant advances in technology and science have often accompanied improvements in justice and expressions of beauty which are more apt to occur when men have leisure and material comfort and are not totally occupied with the business of keeping alive. The history of automatic control shows that it not only can produce things of finer quality in abundance, but it also can perform tasks that direct human control could not accomplish. In addition to furthering a reduction of working hours, automatic control, like mechanical power, has eliminated much drudgery. As a consequence many workers are now being trained to be the masters of their machines. This raising of the status of the worker contributes to the general welfare.

The Impact of Engineering on Society

In relating the history of engineering we have endeavored to show how human ingenuity and social circumstance have produced engineering progress. In turn, earlier chapters have also demonstrated the impact of engineering innovations on human activity and consequently upon history. It is the thesis of this book that engineering advance is vital as one of the interdependent variables on which the evolution of history depends. William Fielding Ogburn has elucidated a part of this thesis by showing how inventions influence history through affecting the size of populations.[5] However, the previous chapters have not analyzed the nature of the effect that engineering has had on human activity and values.

[5] William Fielding Ogburn, "Inventions, Population and History," *Studies in the History of Culture*, George Banta Publishing Company (The Collegiate Press), Menasha, Wis., 1942, pp. 232–245.

Certainly engineering has made major contributions to man's material welfare. Few in Western societies would elect to revert to the physical hardships of the early Middle Ages or even to those of the mid-nineteenth century, with its endemic intestinal diseases, its lack of widely available power, its seventy-hour working week, and its low average earned income. Improvement in prosperity and general health has been dramatic in recent decades. Power, machines, and automatic controls have made possible a reduction in the average number of hours worked per week. Not long ago a sixty-hour week was standard for most workers, and many worked twelve hours a day for six or seven days a week. In the United States, the normal present work week is five eight-hour days, which gives two days of leisure each week. Two or three paid vacation weeks are not uncommon. The individual worker's increased output capacity made possible by control of mechanical power has given him more hours of leisure to say nothing of the remarkable improvement in his standard of living. As for health, to cite but one example, deaths from typhoid fever in large cities in the United States dropped from 24.5 per 100,000 in 1910 to 0.2 per 100,000 in 1945—a 99 per cent reduction! Moreover, 1945 was before the discovery of a specific and effective curative treatment of the disease. In other words, this remarkable mortality reduction was due almost entirely to the advances in preventive methods consisting largely of the purification of water described in Chapter 13.

But this improvement in health was not due to engineering alone. Indeed, it is a typical illustration of the manner in which engineering must be integrated with other knowledge and human values to improve human welfare—even material welfare. Two important, nonengineering developments, one moral, the other scientific, were necessary before sanitary engineering could be effective. The first was the general rise in the humanitarian desire to improve well-being that began in the last half of the eighteenth century. There is little incentive to prevent deaths if a high value is not placed on human life. Almost incredible proof of this statement is the reduction of child mortality in London and elsewhere, between 1740 and 1800. About 1740 only one child out of every four born alive reached the age of five; children were not expected to live. By 1800 this terrible death rate had been cut nearly in half owing almost entirely to increased interest in children by the laity and the medical profession.[6] There was no new and effective treatment of disease; it was simply better

[6] Ernest Caulfield, *The Infant Welfare Movement in the Eighteenth Century*, Paul B. Hoeber, Inc., New York, 1931.

day-by-day care of well children that decreased the deaths. This higher value placed on life contributed to the successes of sanitary engineering in the nineteenth century.

The second factor was Pasteur's theory of the germ origin of infectious disease which has already been discussed in Chapter 13. Without the knowledge supplied by Pasteur's theory, earlier water-supply systems had fortuitously been either healthful or disease-ridden. Long aqueducts, from Roman times, brought water by gravity to cities from distant sources, relatively free from pollution. However, the introduction of water-wheel pumps, and later steam pumps, made it possible for a city to obtain water from a nearby river. The installation of pumps saved the expense of constructing aqueducts but at the same time brought death to many of the inhabitants from grossly polluted river waters. Engineering technique, in this instance, was in advance of relative scientific knowledge. This example of the effect of water-supply engineering on health is characteristic of the impact of engineering on general welfare. Engineering by itself is not enough; it must be integrated with other knowledge and directed by ethical principles. On the other hand, the most extensive knowledge combined with the highest human values would contribute little to society as a whole without benefit of engineering applications.

Religion, even before the general benefits of engineering application were available, occupied a major place in human thought and action. But while engineering has played no direct part in the formulation of any religion, technology has greatly facilitated the communication of religious teaching to many millions of people. The production of papyrus, paper, and inks has been of importance to the spread of religious beliefs as well as of general education. Likewise, the construction of churches has greatly enhanced the religious life, which would be difficult to imagine without them. The release of men from the burdensome task of food production and constant, unrelenting drudgery has made possible— but, of course, was not the stimulation of—the appearance of great religious leaders and a more extensive opportunity for the great masses of men to follow those leaders in faith and hope. It would be utterly ridiculous to contend that religion depends on engineering; nevertheless, to a degree not generally appreciated, engineering has enlarged religious experience.

In similar ways engineering has contributed to the increased production and enjoyment of beauty, the progress of knowledge, and the ele-

vation of social justice. Engineering has made possible many beautiful buildings, bridges, and other structures, though an advance in construction does not in itself guarantee a beautiful structure. As but one example of many, the introduction of the pointed arch, the unit-element bay, and the flying buttress made possible Gothic cathedrals of great beauty. Without these innovations it would not have been possible to express the aesthetic values in Gothic architecture, but it was necessary that engineering be combined with a reverence for beauty, which unfortunately was not always the case. Nevertheless, there are many examples where the first application of an engineering advance has been combined with the highest aesthetic values. Robert Maillart's Schwandbach Bridge is not only the first reinforced-concrete bridge to have a roadway platform with curved alignment, but it is also a structure of great beauty (Figure 15.4).

Progress in transportation and communication engineering has greatly extended the enjoyment of beauty and may well have contributed to the recent heightening of reverence for beauty. A. N. Whitehead has written: "Just when the urbanisation of the western world was entering upon its state of rapid development, and when the most delicate, anxious consideration of aesthetic qualities of the new material environment was requisite, the doctrine of the irrelevance of such ideas [of natural and artistic beauty] was at its height. In the most advanced industrial countries, art was treated as a frivolity. A striking example of this state of mind in the middle of the nineteenth century is to be seen in London where the marvellous beauty of the estuary of the Thames, as it curves through the city, is wantonly defaced by the Charing Cross railway bridge, constructed apart from any reference to aesthetic values." [7] The ugliness of many nineteenth-century industrial cities is too painfully familiar to make it necessary to belabor Whitehead's point. It is enough to say that if engineering ability is not combined with aesthetic values, only ugliness can result.

The railroad, steamship, airplane, and automobile have enabled millions to see the world's great works of art and natural beauty. Thousands of Americans visit European cathedrals each year, whereas it is doubtful that most medieval Europeans ever had the opportunity to see more than the one in their immediate neighborhood. The radio, despite its abundance of low-quality programs, gives opportunities, which would not otherwise be available for those interested, to hear great works of music. Engineering has played a vital role also in the advance of knowledge and has facilitated the diffusion of truth and learning. Earlier chapters have noted the scientist's dependence on engineering advances in prosecuting new investigations of the natural world. Beyond supplying instruments to the scientist, engineering has presented him with problems the solutions of which have proved to be important scientific advances.

Just as engineering accomplishments have contributed to faith, beauty, and truth, so have they added to the extension of social justice. The manner in which mechanical power as a substitute for human effort has permitted the increase of individual leisure, self-reliance, and independence has been discussed. By itself, this improvement in the equality of man has been a powerful force in the growth of the Western democracies. With the rise of equality there has been greater opportunity for the benevolent to be good. At the same time malevolent superstitions and the darkness of ignorance have been diminished. Many engineering devices make it more and more difficult for cruel and unjust men as well as criminals to remain undetected. One major type of eighteenth-century malefaction, piracy, has been practically eliminated, largely because of improved communications facilitating international cooperation. Ever since the Slough murder case mentioned on page 338, the telegraph, and subsequently the telephone and radio, have made it increasingly possible for the forces of law and order to apprehend criminals. Street illumination combined with a mobile police force have prevented some crime and wanton immorality.

[7] Alfred N. Whitehead, *Science and the Modern World*, The Macmillan Company, New York, 1931, p. 281. (Quoted with the kind permission of The Macmillan Company.)

Unfortunately, it has also been true that criminals have taken advantage of technological advances and have misused them.

Some of the values of social justice are improved over those of the eighteenth century, but there have been terrible regressions. Some modern governments have misused material power and machines for the lowering of social justice and have immorally employed the telegraph, telephone, and radio, to say nothing of gunpowder, dynamite, nitroglycerine, and internal-combustion engines, for the centralization of political control which, with a disregard for individual human liberties and values, has submitted some of their citizens to most evil indignities. These governments have clearly demonstrated that without a continuously evolving morality and a ready acceptance of new knowledge tempered with idealism the gains in social justice can be ruinously destroyed.

In addition to peacetime iniquities on the international scene, the so-called civilized countries have woefully misused mechanical power and machines for destruction in the two great wars of the twentieth century. For as long as world society is made up of antagonistic nations, the possibility of another moral collapse into annihilative war will continue to exist. The disregard of human values during the Fascist, Nazi, and Soviet regimes and the two World Wars, the development of fission and fusion bombs, plus the terror of a potential third war have led some to demand a moratorium on all engineering advances which may be made to work for evil as well as for good. Such a moratorium would be foolish, even were it possible to effect. The real problem is that other knowledge, particularly in the social sciences, must advance more rapidly, and the forces for good must be strengthened.

However, societies have always had power problems of one kind or another even before the invention of mechanical power. An early example of misuse of power is the story of Pharaoh's oppression of the Israelites in the Book of Exodus. The long series of oppressive acts culminated in the Pharaoh's deliberately vicious order to his foremen no longer to continue supplying the Israelite brickmakers with straw as theretofore. Nevertheless, the Israelites were required to produce the same quota of bricks as previously. Having to gather straw in addition to making bricks, the Israelites were unable to maintain their former quotas and were cruelly chastised. The Nazi regime misused its power to persecute the same race in the 1930s albeit for entirely different reasons.

Or to quote again from Professor Whitehead, who was speaking in

1925: "At the present moment a discussion is raging as to the future of civilisation in the novel circumstances of rapid scientific and technological advance. The evils of the future have been diagnosed in various ways, the loss of religious faith, the malignant use of material power, the degradation attending a differential birth rate favouring the lower types of humanity, the suppression of aesthetic creativeness. Without doubt, these are all evils, dangerous and threatening. But they are not new. From the dawn of history, mankind has always been losing its religious faith, has always suffered from the malignant use of material power, has always suffered from the infertility of its best intellectual types, has always witnessed the periodical decadence of art. In the reign of the Egyptian king, Tutankhamen, there was raging a desperate religious struggle between Modernists and Fundamentalists; the cave pictures exhibit a phase of delicate aesthetic achievement as superseded by a period of comparative vulgarity; the religious leaders, the great thinkers, the great poets and authors, the whole clerical caste in the Middle Ages, have been notably infertile; finally if we attend to what actually has happened in the past, and disregard romantic visions of democracies, aristocracies, kings, generals, armies, and merchants, material power has generally been wielded with blindness, obstinacy and selfishness, often with brutal malignancy. And yet, mankind has progressed." [8]

American Engineering

It is important for a nation of the world, as it is constituted today with its economic, political, and military rivalries, to know thoroughly its own engineering potential for security and welfare. The United States has made more extensive *applications* of engineering innovations than has any other country. It is also widely realized, as pointed out in the first chapter, that the great majority of *basic scientific advances* upon which recent engineering developments were based had been made by Europeans. This fact has influenced the United States government to direct its scientific policy toward stimulating training in the United States that would produce scientists capable of making the fundamental advances in scientific theory which heretofore have been made by Europeans. The National Science Foundation is an immediate result of the new policy.

The lack of interest of Americans in purely theoretical work has long

[8] *Ibid.*, pp. 293–294. (Quoted with the kind permission of The Macmillan Company.)

been the subject of discussion. Alexis de Tocqueville analyzed it in 1840 when he first published the second volume of *De la démocratie en Amérique*.[9] Having conceded that "it must be acknowledged that in few of the civilized nations of our time have the higher sciences made less progress than in the United States," Tocqueville disagreed with many Europeans who thought that this lack of science was "a natural and inevitable result" of democracy. Tocqueville felt that the Americans' "strictly Puritanical origin, their exclusively commercial habits, even the country they inhabit, which seems to divert their minds from the pursuit of science, literature, and the arts, the proximity of Europe, which allows them to neglect these pursuits without relapsing into barbarism, a thousand special causes, of which I have only been able to point out the most important, have singularly concurred to fix the mind of the American upon purely practical objects." However, he was convinced that "if the Americans had been alone in the world, with the freedom and the knowledge acquired by their forefathers and the passions which are their own, they would not have been slow to discover that progress cannot long be made in the application of the sciences without cultivating the theory of them." A century later there is evidence from the awards of Nobel Prizes to support Tocqueville's conviction that Americans living in equality in a democracy could do important scientific work. An analysis of percentages of Nobel Prizes in science awarded to Americans from 1901, the year of the first prizes, to 1953 indicates that scientific work in the United States has produced increasingly important results. In the first ten-year period, Americans received 3 per cent, in the second, 4 per cent, in the third, 5 per cent, but in the last two periods, the percentages were 28 and 39, respectively.

Because the United States has made such wide use of engineering, the impression apparently is general that Americans have been preeminent in engineering innovation and applied research as used by W. R. Maclaurin in classifying the five types of discovery and innovation typical of modern engineering. These five are (1) fundamental research, (2) applied research, (3) engineering development, (4) production engineering, and (5) service engineering.[10] Such groups as the President's Scientific Research Board and the Engineers' Joint Council have made

[9] Alexis de Tocqueville, *Democracy in America*, 2 vols., Alfred A. Knopf, Inc., New York, 1946, vol. 2, pp. 35–47, Book 1, Chaps. 9–10.
[10] William R. Maclaurin, *Invention and Innovation in the Radio Industry*, The Macmillan Company, New York, 1949, p. xvii.

statements giving the impression that many fundamental advances in basic applied research have been the work of Americans.

However, a review of the fundamental engineering innovations discussed in previous chapters covering the nineteenth and twentieth centuries shows that less than a third were made by Americans in the last hundred years, although Americans made more than half of the important advances between 1800 and 1850. Certainly during the latter half of the nineteenth century men who were European-born and -trained made the principal original contributions in a large majority of the salient developments. Such men originated commercial steel, reinforced concrete, radio, electric generators and motors, the internal-combustion engine, steam turbines, and automobiles. The importance of twentieth-century developments is more difficult to evaluate. For instance, atomic energy is not an important source of power, although it may be in the future. However, of the twentieth-century fundamental innovations which have already proved to be basic, at least two-thirds have been made by men trained in Europe.

The history of chemical and industrial engineering is not included in this book, and many phases of mechanical, mining, and metallurgical engineering history have either been touched on only lightly or altogether omitted. However, an examination of Samuel Lilley's tabulation of inventions in mechanical and mining and metallurgical engineering [11] reinforces the observations regarding European contributions. Moreover, since the chemical industry in the United States lagged far behind that of Europe, particularly Germany, until the First World War, it seems unlikely that the inclusion of advances in chemical engineering would materially alter the comparison. Even for the fields covered in this book, this analysis of the relative contributions of Americans and Europeans is far from absolute. Another observation is that even when allowances have been made for the difficulties in evaluation, the basic innovations of the twentieth century appear to be fewer than those in the last half of the nineteenth century.

The great genius of American engineering has been in what Maclaurin terms engineering development, production engineering, and service engineering. American engineers have clearly displayed their creative power in electrical engineering and automotive engineering. The United States generates over 40 per cent of the world's electric power, yet

[11] Samuel Lilley, *Men, Machines and History*, Cobbett Press, London, 1948, pp. 214–220.

Europeans made most of the primary inventions. The growth of American automotive engineering is even more outstanding. In this case Europeans made all of the basic inventions. Moreover, in 1894, only two years after the first automobile was built in the United States and two years before Henry Ford made his original car, the world's first automobile road race was run between Paris and Rouen. Twenty-one cars, all European, participated, and at that time there were five automobile manufacturers in France and Germany, with none in the United States. Sixty years later the United States had 55 million motor vehicles—approximately 75 per cent of the world's automobiles. At the present time, no other nation can come close in approximating the advances made by American development, production, and service engineering.

American industries have excelled in mass production and in making the tools necessary for it. Indeed, one type of engineering, industrial engineering, or scientific management, originated in the United States. More than any other one man, Frederick Winslow Taylor (1856–1915) was responsible for the introduction of scientific management. Taylor did his most important early work in a steel plant at Philadelphia in the 1890s.

Taylor's contribution to scientific management included development of tools and machines for manufacturing processes. He also introduced time-and-motion studies in the investigation of industrial plants and of handling material, together with the determination, based on the analyses of these studies, of methods which were most efficient and involved the least waste. These studies were both psychological and mechanical, and he had to sell the idea to both employees and employers. He introduced systems of incentive pay for labor that were more effective than the piecework system which had become discredited on account of the exploitation practiced by certain types of industrialists.

Taylor was also active in the development of a tool steel which would cut metal at a far higher speed than anything previously used and therefore increased the productivity of both labor and capital. This steel alloy containing chromium and tungsten had already been developed in England, but about 1895 Taylor introduced a technique for preparing the steel at what was then the exceedingly high temperature of 1890°F, just under the melting point. This heat treatment produced a cutting tool which became harder the faster it cut. Machine-tool practice was thus revolutionized, and speeds were doubled, tripled, and even quadrupled.

It is significant that, while Taylor's contributions to industrial effi-

ciency were at first unpopular in his native land, although he insisted
that the advantages of increased efficiency ought to be shared equitably
by capital and labor, they were seized upon enthusiastically in Europe.
Early in the Soviet regime in Russia, Lenin wrote in *Pravda* on April 28,
1918, "We must introduce in Russia the study and teaching of the new
Taylor System."

"The Past Is Prologue to the Future"

It is possible to discern many vital world problems that must be solved
in the relatively near future. From past experience it would seem that
engineering will contribute to their solution, although it would be impossible to predict precisely how. Of the major problems facing the
world in 1955, those resulting from the almost completely novel and rapid
increase in world population are among the foremost. The novelty of the
present situation is not merely the unprecedented speed of world increase
but also the increase in non-European populations. The conquests of preventive medicine have been largely among European peoples, but the
more recent triumphs of curative medicine over infectious disease seem
to be having a relatively greater effect on the size of non-European populations. It has been estimated that the increase in world population during
1950 to 1980 may be as much as the total world population in 1900.

As many observers, including Sir Harold Hartley in the first Fawley
Foundation Lecture (1954), have pointed out, huge exertion will be
necessary to supply the new population not only with food but also with
fuel, power, and raw materials, if malnutrition and a world-wide lowering
of standards of living are to be avoided. Despite the fact that time is short
and that there are not obvious many new developments in the production of foodstuffs, fuel, power, and transportation, it seems probable that
engineering will assist in making it possible, if indeed it turns out to be
possible, for the world's population of 1980 to eat adequately and to improve its living standards. Engineering has contributed in large part to
the solution of analogous problems accompanying the growth of great
cities. But such a historical analogy does not make possible the prediction
that engineering will inevitably be a prime element in such social transi-
tions of the future.

Nevertheless, an inescapable conclusion to be drawn from the story
of engineering in history is that engineering has become an increasingly
powerful factor in the development of civilization. Midway in the

twentieth century engineering is playing a predominant role in the greatest social evolution that the world has seen. Rapid developments in our ways of life are unsettling and confusing, but only primitive societies stand still and look backward to emulate the past. A civilization worthy of its name looks and moves forward. It may lag at times, but it remains dynamic. Associated with this dynamism is inevitably a degree of instability, and the inherent human desire for tranquillity magnifies the instability of the present. But this instability should not be confused with a lack of security. Some of the most productive periods in human history have also been the most tremulous. Witness Greece in the fifth and fourth centuries B.C. If knowledge of man and society can be increased and if ethical principles and human values can be bettered, there is no need to view the uncertain future with dark pessimism, although constant vigilance will be necessary if progress is to be achieved.

General Bibliography

Bibliographies of titles having particular pertinence or extensive sources for subjects treated in the individual chapters will be found at the ends of the chapters. The following list consists of helpful works that discuss topics treated in many of the previous chapters; they therefore have been brought together at the end of the book. Many of the following titles have bibliographies or bibliographical footnotes which will assist in the further pursuit of a subject of interest.

Black, Archibald: *The Story of Tunnels*, McGraw-Hill Book Company, Inc., New York, 1937.

Finch, James K.: *Engineering and Western Civilization*, McGraw-Hill Book Company, Inc., New York, 1951.

Forbes, Robert J.: *Man, the Maker*, Abelard-Schuman, Inc., Publishers, New York, 1950.

Gibson, Charles E.: *The Story of the Ship*, Abelard-Schuman, Inc., Publishers, New York, 1948.

Gregory, John W., and C. J. Gregory: *The Story of the Road*, 2d ed., A. & C. Black, Ltd., London, 1938.

Hodgins, Eric: *Behemoth: The Story of Power*, Doubleday & Company, Inc., New York, 1932.

Kiely, Edmond R.: *Surveying Instruments, Their History and Classroom Use*, Bureau of Publications, Teachers College, Columbia University, New York, 1947.

Kirby, Richard S., and Philip G. Laurson: *The Early Years of Modern Civil Engineering*, Yale University Press, New Haven, Conn., 1932.

Lilley, Samuel: *Men, Machines and History*, Cobbett Press, London, 1948.

Mumford, Lewis: *Technics and Civilization*, Harcourt, Brace and Company, Inc., New York, 1934.

Rasmussen, Steen E.: *Towns and Buildings Described in Drawings and Words*, Harvard University Press, Cambridge, Mass., 1951.

Rickard, Thomas A.: *Man and Metals*, 2 vols., McGraw-Hill Book Company, Inc., New York, 1932.

516 GENERAL BIBLIOGRAPHY

Robins, Frederick W.: *The Story of Water Supply*, Oxford University Press, London, 1946.

Steinman, David B., and Sarah R. Watson: *Bridges and Their Builders*, G. P. Putnam's Sons, New York, 1941.

Straub, Hans: *A History of Civil Engineering*, L. Hill, London, 1952.

Timoshenko, Stephen P.: *History of Strength of Materials*, McGraw-Hill Book Company, Inc., New York, 1953.

Usher, Abbott P.: *A History of Mechanical Inventions*, 2d ed., Harvard University Press, Cambridge, Mass., 1954.

Index

Aerodynamics, 415–416
Agricola, Georg, 153
Air pressure, 128
Air pump, 128
Airplane engines (*see* Internal-combustion engines, gasoline; Turbines, gas)
Airplanes, 415–418
  first, 417 (fig.)
  significance of, 414
Albert-Louppe Bridge, 477–479 (fig.)
Alcantara Bridge, 68
Alexander the Great, 50
Allegheny Portage Road, 288
Allen, Horatio, 285–286
Ammann, O. H., 470, 474
Ampère, André-Marie, 332–333
Analytic geometry, 128–129
Anio Novus, 66–67 (fig.)
Anthemios, 92
Apollodorus, 71
Appian Way, 73–75 (fig.)
Aqua Claudia, 66–67 (fig.)
Aqueducts, Anio Novus, 66–67 (fig.)
  Aqua Claudia, 66–67 (fig.)
  Catskill, 437–440
  Colorado River, 459–460 (fig.)
  Coutances, 117–118 (fig.)
  Croton, 434–435
  Delaware, 440–442
  Marseilles, 431–432 (fig.)
  Roman, 63–64
    Middle Ages, 117–118
  Sennacherib's, 15–16 (fig.)
  Zempola, 456
Arabic (Moslem) science, 96
Arboga Canal, 142
Arc lighting, 355–356
  Brush system, 356
  Jablochkoff candle, 355
Arch, corbeled, 13, 40 (fig.)
  Gothic, 104
  pointed, 104
  true, 13

Archimedes, 53
Architecture, Gothic, 103
  Romanesque, 102–103
Aristotle, 42
Armstrong, E. M., 347
Aspdin, Joseph, 197
Ausonius, 97
Automatic controls, of boiler combustion, 501
  flyball governors, 499–500
  on Newcomen engine, 498–499
  social significance of, 498–502
  windmills, 499 (fig.)
Automobiles, buses, first, 408–409 (fig.)
  effect on society, 404–405
  engines (*see* Internal-combustion engines, gasoline)
  first, 406–407 (fig.)
  trucks, early, 408
  in United States, 510–511
  (*See also* Locomotives, road)
Avignon Bridge, 109 (fig.)
Ayres, Eugene, 500–501

Babylon, 9–11 (fig.)
Bacon, Roger, 115
Bahr Yusuf Canal, 33
Baird, J. L., 349
Baker, Benjamin, 312–317
Baldwin, Loammi, 215
Ballasting of railroad tracks, 284
Balloon frame, 325
Balloons, 414–415
  dirigible, 415
Baltimore & Ohio Railroad, 287
Barattieri, Nicolò, 120
Barker's mill, 185–186 (fig.)
Barlow, P. W., 488
Barton Aqueduct, 208–209 (fig.)
Battery, electric, 332, 372
Baur, Julius, 296
Beau de Rochas, 405
Belisarius, 88

517

Bell, Alexander Graham, 343–344
Bellows Falls Bridge, 223
Bénezet, 108–109
Benz, K. F., 406–409
Berliner, A. E., 421
Berliner, H. A., 421
Bessemer, Henry, 291–294
Bessemer converter, 294 (fig.)
Besson, Jacques, 151
Bicycles, 404
Biringuccio, Vannoccio, 151–152
Blenkinsop, John, 275
Blondel, François, 139
Blücher, 276
Bogardus, James, 320, 322
Bogardus cast-iron building, 322–323
   (fig.)
Boilers, early, 177–179
   fusible plug in, 176
   modern, 370–371 (fig.)
Boulder Dam, 457–459
Boulton, Matthew, 171–172
Bow, R. H., 231
Boyle, Robert, 129
Brakes, compressed-air, 387–389
   dynamic braking, 384
   vacuum, 388
Branca, Giovanni, 151
Brayton, G. B., 406
Briare Canal, 141
Bridges, Albert-Louppe, 477–479 (fig.)
   Alcantara, 68
   Avignon, 109 (fig.)
   Barton Aqueduct, 208–209 (fig.)
   Bellows Falls, 223
   Britannia, 238–243 (figs.)
   Brooklyn, 305–311 (figs.)
   Cabin John, 475
   cantilever, first, 224–225
   cast-iron, 227–228
   cellular bracing, 239
   "Chain Bridge," 233–234 (fig.)
   Charlemagne, Mainz, 107
   Coalbrookdale, 227–228 (fig.)
   Conway, 243–244
   Forth, 312–317 (figs.)
   foundations, 139–140
      Roman, 69–70
   Garabit Viaduct, 317–318 (fig.)
   George Washington, 470–471
   Golden Gate, 471–473 (fig.)
   Hell Gate Arch, 473–474 (fig.)
   iron, 228–229
      first, 227–228 (fig.)
   Julius Caesar's Rhine, 70–71 (fig.)
   Kill van Kull, 473–474
   London, 110–112 (fig.)

Bridges, McCall's Ferry, 224
   Martorell, 69 (fig.)
   medieval, 107–112
      drawbridges, 108
   Menai Straits, 235–236 (fig.)
   Mesopotamian, 19
   mule-back, 68–69 (fig.), 108
   Narni, 68
   New Hope, 225
   "Permanent Bridge," Philadelphia, 223
      (fig.)
   Plougastel, 477–479 (fig.)
   Pons Fabricius, 70 (fig.)
   Pont de la Concorde, 221 (fig.)
   Pont du Gard, 63 (fig.)
   Pont Neuf, 149 (fig.)
   reinforced-concrete, 476–479
   Risorgimento, 477
   Roman, 68–72
   Roquefavour, 431–432 (fig.)
   Royal Albert, 231–232 (fig.)
   Saint Esprit, 110
   St. Louis, 298–305 (figs.)
   Saltash, 231–232 (fig.)
   Schwandbach, 505 (fig.)
   steel, 298, 317, 473–475
   stone, 220–222, 475–476
   suspension, 305–306, 470–473
      first, 232–233
      Fribourg, 237
      Veranzio's plan for, 139 (fig.)
   Sydney Harbor, 473–475 (fig.)
   Syra River, 475–476
   Tappan Zee, 493–494
   Trajan's Danube, 71–72 (fig.)
   Trezzo, 110
   truss, 222–232
      Ithiel Town's, 226 (fig.)
      Palladio's, 137–138 (fig.)
      stresses in, 229
      various early American, 227 (fig.)
   Wandipore, 225 (fig.)
   Wettingen, 223
   wooden, 222–227
   wrought-iron, 317–318
Bridgewater, Duke of, 208
Bridgewater Canal, 208
Briggs, Henry, 131
Brindley, James, 208, 210
Britannia, 262
Britannia Bridge, 238–243 (figs.)
Brooklyn Bridge, 305–311 (figs.)
   design, 307–308
   financing of, 311
Brothers of the Bridge, 108
Bruges canal lock, 116–117
Brunel, Isambard Kingdom, 231, 263–266

Brunel, Marc Isambard, 487–488
Brush, C. F., 356
Brustlein, H. A., 296
Building (see Bridges; Construction; Skyscrapers; Tunneling)
Buildings (see Skyscrapers)
Burr, Theodore, 223–224
Buses (see Automobiles)

Cabin John Bridge, 475
Cables, suspension-bridge, 308–310, 472
  first wire, 237 (fig.)
  transatlantic, 340–342
  underwater, 340–342
Caisson disease, 301–303
Caissons, 301–302 (fig.), 306, 315–316
  incandescent lamps in, 315
  semibuoyant supporting bridge, 493–494
Calculus, differential, 130
Campanile, Venice, 121–122 (fig.)
Canal du Midi, 141
Canals, Arboga, 142
  Bahr Yusuf, 33
  Briare, 141
  Bridgewater, 208
  Canal du Midi, 141
  Chain of Rocks, 453
  Charlemagne's Danube-Rhine, 113
  Chesapeake and Ohio, 212
  Durance River, 113–114
  eighteenth- and nineteenth-century, 207–220
  Erie, 216–219
  Farmington, 219–220
  Gotha, 211
  Grand, China, 112
  Grand Trunk, 210
  inclined planes, 214 (fig.)
  Languedoc, 141
  locks (see Locks)
  Manchester Ship, 210
  medieval, 112–115
  Mesopotamian, 13–14
  Middlesex, 212–213
  Morris, 214 (fig.)
  Naviglio Grande, 114–115
  Panama, 453–456
  Renaissance, 140–143
  Santee, 212
  Schuylkill and Susquehanna, 215
  ship, 450–456
  South Hadley, 213–214
  Suez, 266, 450–451
  Willebroeck, 140
Capacitor, 330–331

Carnegie, Andrew, 297
Caselli, Giovanni, 349
Cathode rays, 335
Catskill Aqueduct, 437–440
Cellular bracing, 239, 264
Cement, natural, 196
  first in United States, 217–218
  portland, 197
  pozzuolana, 62
Cement concrete, 476
Chadwick, Edwin, 442
"Chain Bridge," 233–234 (fig.)
Chain of Rocks Canal, 450
Chaley, Joseph, 237
Chanute, Octave, 415–416
Chapman, William, 285
Chappe, Claude, 336
Charlemagne, bridge at Mainz, 107
  Danube-Rhine canal, 113
Charles, J. A. C., 415
Charlotte Dundas, 250
Chauvenet, William, 303
Chesapeake and Ohio Canal, 212
Chesbrough, E. S., 428
"Chicago construction," 325
Cholera, 428
Chorobates, 82–83
Christianity, effect of, on engineering, 57
  on medieval engineering, 96
Chryses of Alexandria, 92
Churches, Cluny Abbey, 103
  Gothic cathedrals, 103–104
  Old St. Peter's, Rome, 102
  St. Mark's, 120–121
  Santa Sophia, 93 (fig.)
Cities, rise of, 6
City planning (see Town and city planning)
Clark, Edwin, 240–241
Clermont, 252–253 (figs.)
Clinton, DeWitt, 218
Cluny Abbey, 103
Coalbrookdale Bridge, 227–228 (fig.)
Cochlea, 86 (fig.)
Colorado River, 457–458
Colorado River Aqueduct, 459–460 (fig.)
Colossus, bridge at Philadelphia, 225
Concrete, cement, 476
  reinforced, 476–477, 479
Condenser, electric, 330–331
Construction, balloon frame, 325
  Byzantine, 91–93
  "Chicago," 325
  Egyptian, 24–32
    materials and methods, 23–24, 31–32
  Gothic, 103

Construction, Greek, materials and methods, 44–46
  specifications, 44
  medieval, 102–106
    drawings, 105
    materials, 105–106
    plans, 105
    specifications, 105
    templates, 105
    Venice, 120–121
  Mesopotamian, 8–13
  materials and methods, 11
  Minoan, materials and methods, 37–39
  Mycenaean, 39–40
  Renaissance, 133–137
  Roman, influence of, Middle Ages, 102–103
    materials and methods, 60–62, 84
  skyscrapers (see Skyscrapers)
  (See also Bridges; Tunneling)
Conway Bridge, 243–244
Cooke, William Fothergill, 338–339
Cornu, Paul, 421
Corps des Ponts et Chaussées, 200
Cort, Henry, 194–195
Couch, J. J., 486
Coulomb, C. A. de, 331
Coutances Aqueduct, 117–118 (fig.)
Crank, earliest, 97 (fig.)
  for steam engine, 169–170
Crapponne, Adam de, 140
Croton Aqueduct, 434–435
Croton Dam, new, 436–437
  old, 434
Crozet, Claude, 206
Crystal Palace, 229–230 (fig.)
Ctesibius, 87
Cugnot, N.J., 267–268
Cumberland Road, 204–206
Curtis, C. G., 369

D'Acres, R., 155
Daimler, Gottlieb, 406–408
Damme dike, 116–117
Dams, Croton, new, 436–437
  old, 434
  Daras, 92
  Furens, 431
  Holland, 462
  Hoover, 457–459
  Keokuk, 448
  Marib, 15
  Mesopotamian, 14
  Olive Bridge, 437
  Parker, 459
  Yemen, 15

Daras Dam, 92
Darby, Abraham (1677–1717), 191
Darby, Abraham (1711–1763), 193
Darby, Abraham (1750–1791), 193, 227
Dark Ages (Middle Ages), 95–123
Davenport, Thomas, 353
Davy, Humphry, 355
Decimal fractions, 131
Decompression sickness, 301–303
Defoe, Daniel, 199–200
De Forest, Lee, 346
Delaware Aqueduct, 440–442
Derrick, Roman, 84–85 (fig.)
Descartes, René, 128–129
Detector, radio, 345–346
Dickens, Charles, 289
Diesel, Rudolph, 383
Diesel-electric locomotives, 382–385
Diesel engines, 383–384, 405
Differential calculus, 130
Dikes, Damme, 116–117
  medieval, 116–117
Dinocrates, 51
Dioptra, 82 (fig.)
Dirigibles, 415
Disease, infectious, germ origin of, 430
Domesday Book, 98
Douglass, D. B., 214
Doyne, W. T., 231
Drake, E. L., 422
Drawbridges, medieval, 108
Dredging, Mississippi River, 449
  Panama Canal, 455 (fig.)
  sixteenth-century, 144 (fig.)
  Suez Canal, 451–452 (fig.)
Drilling (see Tunneling)
Dry docks, Portsmouth, England, 116
Dudley, Dud, 192
Dunlop, J. B., 404
Durance River, canal, 113–114
Duryea, C. E., 407–408
Duryea, Frank, 407–408
Dynamo (see Electric generators)

Eads, James B., 298–305, 446
École des Ponts et Chaussées, 327
Eddystone Lighthouse, 196
Edison, Thomas Alva, 335, 356–357, 359
"Edison effect," 335
Egerton, Francis, 208
Eiffel, Alexandre G., 317–318
Eiffel Tower, 318–319 (fig.)
Electric battery, 332
  solar, 372

Electric generators, 352–355
  field excitation, 352–353
  Gramme type, 355
  Pacinotti type, 353 (fig.)
Electric locomotives (see Locomotives)
Electric motors, 353–354
  Gramme type, 355
Electric power, alternating-current systems, 360–362
  central stations, 356–359
  direct-current systems, 357–359
  flexibility of, 351–352
  grids, 364–365
  hydroelectric, 363–365
    Niagara Falls, 364
  kilowatt-hours, per United States family, 496–497 (fig.)
    per United States worker, 496 (fig.)
  polyphase systems, 361
  rotary converter, 362
  static rectifier, 362
  transmission of, 360, 363–366 (fig.)
  United States production of, 373 (fig.)
Electric street railways (see Street railways)
Electric valve (radio tube), 335, 346–347
Electricity, animal, 331–332
  history of, 329–336
  induction of, 333–334
  one-fluid theory of, 331
  producing motion, 333
  two-fluid theory of, 330
Electrodynamics, 333
Electromagnetic waves, 334–335
Electromagnetism, discovery of, 332–333
Elevated railways (see Rapid-transit systems)
Elevators, 319–321
  Eiffel Tower, 321
  electric, 465–466
  passenger, first, 320
Empire State Building, 468–469 (fig.)
Engineering, and beauty, 504–506
  definition of, 2–3
  effect of, on health, 503–504
    on science, 126
    on society, 502–508
  effect on, of printing, 132
    of rise of cities, 7–8
    of writing, 6–7
  fundamental innovations, United States and European, 510
  history of, principal events in, 4
  values, 1–2
  and religious activity, 504
  and social justice, 506–508
  in United States, 508–512

Engineering schools, 327–328
Engineering science, 328
Engineers, activities of, 3
  Egyptian, 32
  Greek, 43–44
  medieval, 105
  professional, 327
  Renaissance, 136–137
  Roman, 57–60
  training of, 59
Engines (see Internal-combustion engines; Steam engines; Turbines)
Enterprise, 254
Ericsson, John, 262
Erie Canal, 216–219
  effect of, 218–219
  locks, 216–217 (fig.)
Euclid, 52–53
Eupalinus of Megara, 48
Evans, Oliver, 172–175, 269–270

Fairbairn, William, 179, 238–239, 243
Faiyûm, 33
Faraday, Michael, 333–334, 352
Farmington Canal, 219–220
Fay, C. F. de C. du, 330
Fermat, Pierre, 128
Fessenden, R. A., 347
Filtration of water, 429, 431
Finley, James, 232
Fitch, John, 250
Flaminian Way, 73, 75
Flying buttresses, 104 (fig.)
Flywheel, 169
Folding paddle wheel, 258 (fig.)
Fontana, Domenico, 133–136
Ford, Henry, 408
Forth Bridge, 312–317 (figs.)
  compared with Britannia Bridge, 316
  design of, 313–315
Fourneyron, Benoît, 185–186
Fowler, John, 312–316
Franklin, Benjamin, 331
French, Daniel, 254
Frères Pontifes, 108
Freyssinet, E., 477
Fribourg suspension bridge, 237
Frontinus, 66
Fteley, Alphonse, 436
Fucino, Lake, 76
Fuller, G. W., 430
Fulton, Robert, 249–254
Furens Dam, 431

Galileo Galilei, 125
Gallatin, Albert, 204–207, 213–214, 233

Galvani, Luigi, 331-332
Garabit Viaduct, 317-318 (fig.)
Garnier, Tony, 410
Gas engines, 405
Gas turbines, 418-421 (fig.)
Gasoline engines (see Internal-combustion engines)
Gaulard, Lucien, 360
Geometry, analytic, 128-129
George Washington Bridge, 470-471
Germ origin of infectious disease, 430
Ghega, Carl von, 481
Gibbs, J. D., 360
Gliders, 415-416
Goethals, G. W., 453-454
Golden Gate Bridge, 471-473 (fig.)
Gorgas, W. C., 453
Gotha Canal, 211
Gothic cathedrals, 103-104
Gramme, Z. T., 355
Grand Canal, China, 112, 141
Grand Trunk Canal, 210
Gray, Elisha, 343
Gray, Stephen, 329-330
Great Britain, 262-263 (fig.)
Great Eastern, 263-266 (fig.), 341
  significance of, 266
Great Western, 260-261 (fig.)
Greathead, J. H., 488
Greeks, 41-42
Groma, 81-82 (fig.)
Grubenmann, Hans Ulrich, 223
Grubenmann, Johannes, 223
Gudea, 10-11 (fig.)
Guericke, Otto von, 128
Gunpowder, medieval, 115
Gunter's chain, 130
Gurney, Goldsworthy, 271-272

Hale, Enoch, 223
Hall, Samuel, 259
Harbors, medieval, 115-116
  Ostia, 78 (fig.)
  sixteenth-century, Le Havre, 143-145
Harecastle Tunnels, 210-211 (fig.)
Hartley, Sir Harold, 512
Haskin, DeWitt Clinton, 490-491
Haupt, Herman, 231
Haussmann, G.-E., 409-410
Hazen, Allen, 430
Health, 503-504
Helicopters, 421-422 (fig.)
Hell Gate Arch Bridge, 473-474 (fig.)
Henry, Joseph, 334
Hero, 54, 154

Hertz, H. R., 334-335
Hertzian waves (radio waves), 334-335, 345
Highways (see Roads)
Hippodamus of Miletus, 48-49
Hornblower, J. C., 177
Horse collar, 101
Horsepower as unit of power, 171-172
Horseshoe, 101
Holland (see Reclamation)
Holley, A. L., 297
Home Insurance Company Building, 323
Hooke, Robert, 129
Hoosac Tunnel, 480
Hoover Dam, 457-459
Howe, William, 227
Hulls, Jonathan, 247
Huntsman, Benjamin, 293
Huygens, Christian, 129-130
Hydraulic engineering (see Canals; Dams; Dikes; Harbors; Irrigation; Reclamation; River improvement)
Hydraulic jacks, 241-242 (fig.)
Hydraulics, 127-128, 190

Illuminating gas, 354
Imhotep, 32
Inclined planes, canal, 214 (fig.), 288
Industrial engineering, 511-512
Industrial Revolution, effect of, 159-162, 501
Ineni, 32
Internal-combustion engines, diesel, 383-384
  in ships, 403
  gas, 405
  gasoline, 406
    airplane, early, 416
    electric ignition, 406
    radial, 418-419 (fig.)
    radiator, 406
Iron, cast, definition of, 292
  puddling process, 194-195
  restrictions on manufacturing, British colonies, 193-194
  smelting, with charcoal, 190-192
    with coal, 192-195
  Roman, 90
  wrought, definition of, 292
Iron Act of 1750, 193
Irrigation, Egypt, 33-34
  Mesopotamia, 17
  Utah, 457
Isidoros, 92
Ives, H. E., 349

Jablochkoff, Paul, 355
Jackson, C. T., 339
Jacobi, Hermann de, 353
Jacobs, C. M., 492
Jefferson, Thomas, 204-205
Jenkins, C. F., 349
Jenney, William LeBaron, 323-324
Jerusalem tunnel, 16-17
Jervis, John Bloomfield, 285-286, 434
Jet engines, 418-421 (fig.)
Johnson, Edward, 203
Julius Caesar's Rhine Bridge, 70-71 (fig.)

Karnak, temple, 28-30 (fig.)
Kelly, William, 294
Kelvin, Lord, 341, 499
Keokuk Dam, 448
Kill van Kull Bridge, 473-474
Kirkwood, J. P., 428-429
Knight, Jonathan, 287
Koch, Robert, 430
Krupp steel works, 297

Lamps, efficiencies of, 362-363
  fluorescent, 363
  incandescent, 356-357
    GEM, 362
    tungsten, 362-363
Lancaster Turnpike, 203-204
Langen, Eugene, 405
Langley, S. P., 416
Languedoc Canal, 141
Lardner, Dionysius, 259, 261
Lateen sail, 101
Latrobe, B. H. (1764-1820), 174
Latrobe, B. H. (1806-1878), 287
Laval, C. G. P. de, 367-368
Lawrence Experiment Station, 430
Laxey water wheel, 183-184 (fig.)
Learned societies, early, 132
Lebon, Philippe, 354
Le Corbeiller, Philippe, 497
Leeghwater, J. A., 461
Leibniz, G. W., 130
Lenin, Nicolai, 512
Lenoir, J. E., 405
Leonardo da Vinci, 124, 126-127, 138
Lesseps, F. M. de, 451, 453
Leyden jar, 330-331
Lilienthal, Otto, 415
Lilley, Samuel, 510
Limelight, 354
Lindenthal, Gustav, 473
Lintlaer, Jean, 149
Little Juliana, 251

Liverpool and Manchester Railway, 277-278
Livingston, Robert R., 251-254
Locks, canal, Bruges, 116-117
  Erie, 216-217 (fig.)
  Panama, 455-456
  seventeenth-century, 114 (fig.)
Locomotion No. 1, 277
Locomotives, diesel-electric, 382-385
    (fig.)
  advantages of, 384-385
  electric, first freight, 380-381 (fig.)
    first United States main-line, 381-382
    (fig.)
  road, Cugnot gun tractor, 267-268
    (fig.)
    early, 267-268
    English restrictions on, 272
    Gurney's steam carriage, 271-272
      (fig.)
    Orukter Amphibolos, 269-270 (fig.)
  steam, 377
    American type, 377-378 (fig.)
    Blücher, 276
    compound, 379
    coupled drivers, 275, 377
    early, 272-276
    efficiency of, 379
    first, 273-274
    Locomotion No. 1, 277
    oil-fired, 380-381
    Rainhill trials, 278
    Rocket, 278-279 (fig.)
    steam pressures, 380
    stoking, 380
    Whyte system designating wheel arrangement, 378
Lodge, Oliver Joseph, 345-346
Logarithms, 131
London Bridge, 110-112 (fig.)
London water-supply systems, 147-149
    (fig.)
Lumen, definition of, 362
Lusitania, 403

McAdam, J. L., 202
McAdoo Tunnel, 491-493
McCall's Ferry Bridge, 224
Maclaurin, W. R., 509
Magna Charta, rights on rivers, 99-100
Maillart, Robert, 479, 505
Mallet, J.-T.-A., 379
Manchester Ship Canal, 210
Manly, C. M., 416
Marconi, Guglielmo, 346
Marcus, A. S., 406

Marib Dam, 15
Mariotte, Edmé, 129
Marly pumps, 150 (fig.)
Marseilles Aqueduct, 431–432 (fig.)
Martin, Émile, 295
Martin, Pierre, 295
Martorell Bridge, 69 (fig.)
Marx, Karl, 501
Masonry (see Concrete; Construction)
Mauretania, 403
Maxwell, James Clerk, 334, 499
Mechanics, 127
Menai Straits Bridge, 235–236 (fig.)
Menier chocolate works, 322–323 (fig.)
Mercantile system, 160
Metallurgy (see Iron; Steel)
Middle Ages, 95–123
Middlesex Canal, 212–213
Miller, Patrick, 250
Mills, H. F., 430
Mining, Renaissance, 151–153
    Roman, 89–90
    (See also Pumps)
Minot, Charles, 391
Mississippi River, 445–450
Mitis, Ignaz von, 236
Modulus of elasticity, 188
Monier, Joseph, 476
Monopolies on Hudson and Mississippi
    Rivers, 254, 256
Mont Cenis Railway, 485–486
Mont Cenis Tunnel, 480–486
    surveying, 483
Montgolfier brothers, 414
Morice, Peter, 147
Morland, Samuel, 156
Morrill Land Grant Act of 1862, 328
Morris, William, 316
Morris Canal, 214 (fig.)
Morse, S. F. B., 339–340
Morse Code, 340
Moslem science, 96
Motion, laws of, 130
Motor ships, 403
Municipal planning (see Town and city
    planning)
Murdock, William, 269, 354
Mushet, Robert F., 295
Musschenbroek, Pieter van, 188, 330–331
Myddelton, Hugh, 148–149

Napier, John, 131
Narni Bridge, 68
Nasmyth, James, 263
National Road, 204–206
Natural gas, 354, 424

Naviglio Grande Canal, 114–115
Neolithic Age, 4
New Hope Bridge, 225
New Orleans, 254
New Stone Age, 4
New York City water supply (see Water-
    supply systems)
Newcomen, Thomas, 162–167
Newton, Isaac, 130
Nipkow, P. G., 349
Norse Mill, 97

Obelisks, Caligula's, moving of, 133–136
    (fig.)
    handling of, 30–31
    quarrying of, 23
Oersted, Hans Christian, 332–333
Ogburn, W. F., 502
Ohio River, 446–447
Ohm, Georg Simon, 333
Oil, 422–424
Old St. Peter's, Rome, 102
Old Stone Age, 4
Olive Bridge Dam, 437
Olynthus, 49 (fig.)
Opus incertum, 61
Opus quadratum, 60
Opus reticulatum, 61
Orukter Amphibolos, 269–270 (fig.)
Ostia harbor, 78 (fig.)
Otis, Elisha G., 320–321
Otto, N. A., 405
Oughtred, William, 131

Pacinotti, Antonio, 352
Paddle wheels, early suggestions for,
    247–248 (figs.)
    folding, 258 (fig.)
Page, C. G., 342
Paleolithic Age, 4
Palladio, Andrea, 137–138
Palmer, Timothy, 223
Panama Canal, French work on, 452–453
    locks, 455–456
    map of, 454
    traffic through, 456
    United States construction of, 453–455
Panama Railroad, 452
Papin, Denis, 156, 267
Parent-Duchâtelet, A.-J.-B., 427
Parker, James, 196
Parker Dam, 459
Parsons, C. A., 367, 369
Pascal, Blaise, 128, 393
Pasteur, Louis, 430
Pauger, Adrien de, 445–446

Pavements (see Roads)
Paxton, Joseph, 229
Pennsylvania Railroad Hudson tunnels, 493
"Permanent Bridge," Philadelphia, 223 (fig.)
Perronet, J. R., 221
Perry, John, 142–143
Peter of Colechurch, 110
Peter the Great, 142
Petroleum, 422–424
Pharos, Alexandria, 51 (fig.)
Phoenix, 251
Pipelines, 422–424 (fig.)
    gas, 424
    pumps, 424
Pipes, ancient, 52, 63–64
    cast-iron, 150–151
    lead, 63–64
Pitard, Jean, 118
Pitot, Henri, 190
Pixii, Hippolyte, 352
Plougastel Bridge, 477–479 (fig.)
Poncelet, J. V., 184
Pons Fabricius, 70 (fig.)
Pont de la Concorde, 221 (fig.)
Pont du Gard, 63 (fig.)
Pont Neuf, 149 (fig.)
Pope, Thomas, 234
Population, world, 512
Port of New York Authority, 470
Porta, G. B. della, 154
Portland cement, 197
Power, human, 5
    Egypt, 22 (fig.)
    mechanical, social importance of, 495–498
    Middle Ages, 96–102
    misuse of, 507–508
    nonhuman, development of, 96
    (See also Electric power)
Pozzuolana, 62
Pratt, Caleb, 227
Pratt, T. W., 227
President's Scientific Research Board, 1
Prime movers (see Electric battery; Internal-combustion engines; Ships, sailing; Steam engines; Turbines; Water wheels; Windmills
Printing, effect on engineering, 132
Propellers, ship, 262–263
Pulitzer Building, 321–322 (fig.)
Pulleys, Greek, 46
    Roman, 84 (fig.)
Pumps, air, 128
    Ctesibian, 87
    London Bridge, 147–148 (fig.)

Pumps, Marly, 150 (fig.)
    mine, seventeenth-century, 154 (fig.)
        sixteenth-century, 98 (fig.), 153
    Roman, 87
    Samaritaine, 149 (fig.)
    steam vacuum, 154–158
    (See also Water-raising machines)
Pyramids, 24–28
Pytheos, 50

Radar, 350
Radio, 344–348
    amplitude modulation, 347
    applications, 350
    broadcasting, 347
    detector, 345–346
    feed-back circuit, 347
    frequency modulation, 347–348
    heterodyne circuit, 347
    superheterodyne circuit, 347
Radio astronomy, 351
Radio detector, 345–346
Radio tube, 335, 346–347
Radio waves, 334–335, 345
Railroad cars, couplers, 389
    early, 385–387 (fig.)
    first all-steel, 389–390
    flanged wheels, 281–282
    illumination of, 389
    swivel trucks, 285
    tank cars, 423
Railroad locomotives (see Locomotives)
Railroad signals, 390–393
    block, 391–392
    interlocking, 391
    telegraphic, 391
    train-control systems, 392–393
Railroad tracks, ballasting, 284
    "battle of the gauges," 282–283
    crossties, 282
    curve easement, 285
    early, 279–282 (figs.)
    rails, 283–284
Railroads, centralized traffic control, 393
    early, European, 286–287
        United States, 287–288
    effect on United States Civil War, 289–290
    electrification, 381–383 (figs.)
    nineteenth-century, last half, 376
    resistance of trains, 276
    (See also Rapid-transit systems; Street railways)
Rails, 283–284
Rainhill trials, 278
Rapid-transit systems, 397–400

Rapid-transit systems, City and South London Railway, 490 (fig.)
electrification, 398
London, 398–399
New York, 400
Paris, 399
tunneling, 398–400
Ravenswood Tunnel, 491–492
Reclamation, Holland, 460–462
dams, 462
pumps, 461–462
Zuider Zee, 462
Roman, 76
Regenerative furnace, 295
Reinforced concrete, 476–477, 479
Reinforced-concrete bridges, 476–479
Reis, Philipp, 343
Relays, 340–341
Renwick, James, 214
Riquet, R.-P. de, 141
Risorgimento Bridge, 477
Rittenhouse, David, 204, 215
River improvement, Mississippi delta, 445–446
Mississippi River, 448–450
Ohio River, 446–447
Roman, 77
Riveting, 243–244
Road locomotives (see Locomotives)
Roads, bituminous-concrete, 411
British, 202–203
cement-concrete, 411–412
construction machinery, 412
Cumberland Road, 204–206
design of, twentieth century, 412–413
effect of automobiles on, 411
eighteenth- and nineteenth-century, 199–207
financing of, 412
French, 201
good-roads movement, 411
intersections with grade separation, 414 (fig.)
Lancaster Turnpike, 203–204
medieval, Paris, 117
Mesopotamian, 19–20
Minoan, 39
National Road, 204–206
penetration macadam, 411
Roman, 72–76
England, 74
Middle Ages, 107
superhighways, 413–414
United States, colonial, 203
early policy on, 204
influence of French and British on, 206

Roads, United States, nineteenth-century, 203–207
various cross sections, 201 (fig.)
Via Appia, 73–75 (fig.)
Via Flaminia, 73, 75
Rocket, 278–279 (fig.)
Roebling, John A., 305–306
Roebling, Washington A., 305–307
Roebuck, John, 171
Roman traditions in Middle Ages, 95–96
Rome water-supply system, 67–68
Roosevelt, Nicholas J., 254
Roosevelt, Theodore, 453
Roquefavour Bridge, 431–432 (fig.)
Rosing, Boris, 349
Royal Albert Bridge, 231–232 (fig.)
Royal William, 259–260
Rozier, J.-F. P. de, 414
Rumsey, James, 249

Sailing ships, 101, 401
Saint Esprit bridge, 110
St. Louis Bridge, 298–305 (figs.)
objections to, 300
St. Mark's, Venice, 120–121
Saltash Bridge, 231–232 (fig.)
Sanitary engineering (see Sewage-disposal systems; Water-supply systems)
Santa Sophia, 93 (fig.)
Santee Canal, 212
Saulnier, Jules, 322–323
Savage, John L., 458
Savannah, 258–259 (fig.)
Savery, Thomas, 157–158
Schuylkill and Susquehanna Canal, 215
Schwandbach Bridge, 505 (fig.)
Science, applied, 327–329
effect of engineering on, 126
Greek, 42–43
applications of, 43
interrelations with engineering, 351
Moslem, 96
Roman, 56
seventeenth-century, 126
in United States, 508–509
Scientific management, 511–512
Seguin, Marc, 179, 237
Selden, G. B., 406
Semmeringbahn, 481
Senmut, 32
Sennacherib's aqueduct, 15–16 (fig.)
Servomechanisms (see Automatic controls)
Sewage-disposal systems, 442–444
Chicago, 429
purification, 443–444

Sewage farms, 443
Sewers, storm water, 442
  Harappa, 18–19 (fig.)
  (*See also* Sewage-disposal systems)
Shandaken tunnel, 440
Sheffield, Lord, 194
Ships, earliest, 20–21
  Egyptian, 20–21 (fig.)
  Mediterranean merchantman, 48
  motor, 403
  paddle wheels, early suggestions for, 247–248 (figs.)
  Roman merchantman, 79
  sailing, clippers, 401
    medieval rigs, 101
  steamships, 401–403
    Britannia, 262
    cellular bracing, 264
    Charlotte Dundas, 250
    Clermont, 252–253 (figs.)
    compound condensing engines in, 401–402
    early, development of, 246
      European, 256–257
      transatlantic, 258–260
      United States, 250–251
    Enterprise, 254
    Great Britain, 262–263 (fig.)
    Great Eastern, 263–266 (fig.)
    Great Western, 260–261 (fig.)
    Little Juliana, 251
    Lusitania, 403
    Mauretania, 403
    New Orleans, 254
    Phoenix, 251
    propellers, 262–263
    Royal William, 259–260
    Savannah, 258–259 (fig.)
    Sirius, 260 (fig.)
    steam turbines in, 402–403
    Turbinia, 402
    Washington, 255–256
Shreve, Henry M., 254–256
Siemens, E. W. von, 395
Siemens, William, 292, 295
Signals (*see* Railroad signals)
Sikorsky, I. V., 418, 421–422
Sirius, 260 (fig.)
Skyscrapers, Chicago, 466–467
  Empire State Building, 468–469 (fig.)
  Home Insurance Company, 465–466 (fig.)
  New York, 467
  steel framework, 466
Slavery, 5, 22
Slide rule, 131
Slipway, Corinth, 47

Slough, Matthias, 204
Slough murder case, 338–339
Smeaton, John, 166–167, 183, 196
Smith, Charles S., 312
Smith, F. P., 262
Smith, William S., 317
Snow, George Washington, 325
Snow, John, 428
Somerset, Edward, 156
Sommeiller, Germain, 482
Sostratus, 51
South Hadley Canal, 213–214
Spiegeleisen, 295
Stanley, William, 360
Statics, 127
Steam automobiles (*see* Automobiles; Locomotives, road)
Steam engines, automatic controls on, 498–499
  compound, 176–177 (figs.)
  compound condensing, 401–402
  Corliss, 180–181 (fig.)
  crank and flywheel for, 169
  grasshopper type, 173 (fig.)
  high-pressure, 172–175
  Newcomen, 162–167 (fig.)
    efficiency of, 180
  noncondensing, 175
  reciprocating, 366–367 (fig.)
  Savery's, 157–158 (fig.)
  surface condenser, 402
  United States, early, 174
  Watt, double-acting, 168 (fig.)
    efficiency of, 180
    first, 167
    flyball governor, 169
    limitations of, 172
    sun-and-planet gears for, 170
    types of service in 1800, 172
Steam hammer, 263
Steam locomotives (*see* Locomotives)
Steam turbines (*see* Turbines)
Steamships (*see* Ships)
Steel, alloys, 296–297
  Bessemer process, 293–294
  cast, 293
  cementation process, 293
  chrome, 296–297
  crucible process, 293
  definition of, 292
  heat treatment of, 297
  nickel, 297
  production of, nineteenth century, 297–298
  Siemens-Martin process, 295–296
  in skyscrapers, 323–324, 466
  stimulus for invention of, 291

Steel, structural, 317
  Thomas process, 296
Stephenson, George, 179, 275–279
Stephenson, Robert, 238, 275
Stevens, John, 179, 250–251
Stevens, John F., 453
Stevens, Robert L., 284
Stevenson, David, 288
Stevin, Simon, 127, 131
Stockton and Darlington Railway, 277
Stourbridge Lion, 285
Strada, Jacob de, 151
Street railways, 393–397
  cable, 394–395 (fig.)
  electric, circuits and motors, 396
    early United States, 395–396
    first, at Lichterfelde, 395 (fig.)
    interurban, 396
    trackless, 397
    trolley, 396–397 (fig.)
  horse-drawn, 393–394
  (*See also* Rapid-transit systems)
Streets (*see* Roads)
Strength of materials, 188–190
Sualem, Rennequin, 150
Subways (*see* Rapid-transit systems;
    Tunnels)
Suez Canal, 266, 450–451
Sun-and-planet gears, 170
Surrey Iron Road, 281
Surveying, Mont Cenis Tunnel, 483
  Roman, 79–84
  seventeenth-century, 130–131
Suspension bridges (*see* Bridges)
Swan, J. W., 356
Swivel trucks, 285
Sydney Harbor Bridge, 473–475 (fig.)
Symington, William, 250
Syra River Bridge, 475–476

Tain-Tournon Bridge, 237
Tappan Zee bridge, 493–494
Taylor, F. W., 511–512
Telecommunications, need for, 336
Telegraph, Cooke and Wheatstone sys-
    tem, 338–339
  electromagnetic, 337–340
  importance for communications, 342
  incentives for invention of, 342
  Morse code, 340
  relays, 340
  semaphore, 336–337
  Slough murder case, 338–339
Telephone, 342–343
  first central station, 344–345 (fig.)
  Reis system, 343

Television, 348–350
  applications, 350
  Baird system, 349
Telford, Thomas, 142, 202, 211, 235
Tesla, Nikola, 361
Thales, 42
Thames Tunnel, 487–488 (fig.)
Thomas, Sidney Gilchrist, 296
Thomson, J. J., 335
Thomson, Sir William, 341, 499
Tocqueville, Alexis de, 509
Torricelli, Evangelista, 127–128
Tower Subway, 488–489 (fig.)
Town, Ithiel, 226
Town and city planning, Greek, 48–49
    Alexandria, 51
    Olynthus, 49 (fig.)
  medieval, 119
  nineteenth-century, 409–410
    Paris, 409–410
  twentieth-century, 410
Town planning (*see* Town and city plan-
    ning)
Tracks (*see* Railroad tracks)
Trajan's Danube bridge, 71–72 (fig.)
Transportation, distribution of traffic
    among various types in United
    States, 375–376
  effect of Industrial Revolution on, 199–
    200
  effect on society, 374–375
  (*See also* Airplanes; Automobiles; Ca-
    nals; Railroads; Rapid-transit sys-
    tems; River improvement; Ships;
    Street railways)
Trésaguet, P. M. J., 200
Trevithick, Richard, 175–176, 269, 273–275
Trezzo bridge, 110
Trispastos, 84–85 (fig.)
Trucks, 408
Truss (*see* Bridges)
Tunneling, blind driving, 492
  Catskill Aqueduct, 438
  compressed air, 490
  Croton Aqueduct, 436
  cut-and-cover, 398–399
  drilling, 480–481 (fig.)
    compressed-air, 484
    steam, 486
  railroad mountain, 479–487
  shields, 487–489
Tunnels, Harecastle, 210–211 (fig.)
  Hoosac, 480
  Jerusalem, 16–17
  McAdoo, 491–493
  medieval, 115
  Mont Cenis, 480–486

Tunnels, Pennsylvania Railroad Hudson
    River, 493
  Ravenswood, 491–492
  Shandaken, 440
  Thames, 487–488 (fig.)
  Tower Subway, 488–489 (fig.)
Turbines, gas, 418–421 (fig.)
  stationary, 419
  steam, 367–370 (figs.)
    Branca's design, 152 (fig.)
    de Laval type, 368 (fig.)
    Hero's, 53–54 (fig.)
    modern, 369–370 (fig.)
    Parsons's first, 369 (fig.)
    in ships, 402–403
  water, 185–188
Turbinia, 402
Turner, C. A. P., 479
Turnpikes (see Roads)
Tympanum, 86
Typhoid fever, 430
  Poughkeepsie, New York, 429
  United States urban death rate from,
    1886–1925, 430
    1910–1945, 503

United States Waterways Experiment
    Station, 450
Usher, A. P., 495
Utrecht Psalter, 97 (fig.)

Vacuum tube (radio tube), 335, 446–447
Vault, groined, 103
  ribbed, 103
Venice, medieval, 120–122
  water-supply system, 121
Veranzio, Fausto, 139
Verbiest, Ferdinand, 267
Via Appia, 73–75 (fig.)
Via Flaminia, 73, 75
Vicat, L. J., 196
Villard de Honnecourt, 105
  Sketch-Book, 106 (fig.)
Villes neuves, 119
Vitruvius, 60, 65–66, 87, 96–97
Volta, Alessandro, 332
Voltaic pile, 332

Walls, nonstructural, in Gothic cathe-
    drals, 103–104
  in skyscrapers, 466
  structural, 322
Walter, Raymond F., 458

Wandipore Bridge, 225 (fig.)
Washington, 255–256
Water, importance of, 426
Water pipes (see Pipes)
Water-raising machines, cochlea, 86
    (fig.)
  Toledo water works, 146 (fig.)
  tympanum, 86
Water-supply systems, Chicago, 428
  Constantinople, 91–92
  filters, 429, 431
  Greek, 48
  healthful or disease-ridden, 504
  medieval, 117–118
    Venice, 121
  New York City, Catskill Aqueduct,
    437–440
    Croton Aqueduct, 434–435
    Croton Dam, new, 436–437
      old, 434
    Delaware Aqueduct, 440–442
    first reservoir, 433 (fig.)
    Murray Hill Reservoir, 434–435
      (fig.)
    Olive Bridge Dam, 437
    purification, 438
    Shandaken Tunnel, 440
  Pergamon, 52
  purification, 427–431
  Rome, 67–68
  seventeenth-century, London, 147–149
    Marly, 150–151 (fig.)
    Paris, 149 (fig.)
  sixteenth-century, Augsburg, 145–146
    Le Havre, 145
    London, 147–149 (fig.)
    Toledo, 146–147 (fig.)
Water wheels, medieval, 97–100
  nineteenth-century, 182–188
  overshot, 99–100 (fig.)
  pitchback, 183–184
  Poncelet type, 184–185 (fig.)
  Roman, 87–88, 97
  undershot, 98 (fig.)
Watkins, Francis, 353
Watt, James, 167–172
Wedgwood, Josiah, 210
Wernwag, Lewis, 224
Westinghouse, George, 360–361, 387–388
Weston, E. B., 430
Weston, William, 204, 213, 215, 223
Wettingen Bridge, 223
Wheatstone, Charles, 338–339
Wheels, flanged, 281–282
Whipple, G. C., 430
Whipple, Squire, 229

White, Canvass, 215, 217
White, Lynn, 95–96
Whittle, Frank, 419
Wilkinson, John, 171, 193, 227
Willebroeck Canal, 140
Windmills, bonnet, 100 (fig.)
  medieval, 100
Wireless (*see* Radio)
Wood, Nicholas, 276–277, 282, 286
Woolf, Arthur, 177
Worcester, Marquis of, 156
Wright, Benjamin, 217
Wright, Orville, 416–417

Wright, Wilbur, 416–417
Writing, significance for engineering, 6–7

Xerxes, 18

Yemen Dam, 15
Young, Thomas, 188

Zempola Aqueduct, 456
Ziggurat, Ur, 12 (fig.)
Zworykin, V. K., 349–350

A CATALOG OF SELECTED
DOVER BOOKS
IN ALL FIELDS OF INTEREST

A CATALOG OF SELECTED DOVER
BOOKS IN ALL FIELDS OF INTEREST

CONCERNING THE SPIRITUAL IN ART, Wassily Kandinsky. Pioneering work by father of abstract art. Thoughts on color theory, nature of art. Analysis of earlier masters. 12 illustrations. 80pp. of text. 5⅜ × 8½. 23411-8 Pa. $3.95

ANIMALS: 1,419 Copyright-Free Illustrations of Mammals, Birds, Fish, Insects, etc., Jim Harter (ed.). Clear wood engravings present, in extremely lifelike poses, over 1,000 species of animals. One of the most extensive pictorial sourcebooks of its kind. Captions. Index. 284pp. 9 × 12. 23766-4 Pa. $12.95

CELTIC ART: The Methods of Construction, George Bain. Simple geometric techniques for making Celtic interlacements, spirals, Kells-type initials, animals, humans, etc. Over 500 illustrations. 160pp. 9 × 12. (USO) 22923-8 Pa. $9.95

AN ATLAS OF ANATOMY FOR ARTISTS, Fritz Schider. Most thorough reference work on art anatomy in the world. Hundreds of illustrations, including selections from works by Vesalius, Leonardo, Goya, Ingres, Michelangelo, others. 593 illustrations. 192pp. 7⅛ × 10¼. 20241-0 Pa. $9.95

CELTIC HAND STROKE-BY-STROKE (Irish Half-Uncial from "The Book of Kells"): An Arthur Baker Calligraphy Manual, Arthur Baker. Complete guide to creating each letter of the alphabet in distinctive Celtic manner. Covers hand position, strokes, pens, inks, paper, more. Illustrated. 48pp. 8¼ × 11.
24336-2 Pa. $3.95

EASY ORIGAMI, John Montroll. Charming collection of 32 projects (hat, cup, pelican, piano, swan, many more) specially designed for the novice origami hobbyist. Clearly illustrated easy-to-follow instructions insure that even beginning papercrafters will achieve successful results. 48pp. 8¼ × 11. 27298-2 Pa. $2.95

THE COMPLETE BOOK OF BIRDHOUSE CONSTRUCTION FOR WOOD-WORKERS, Scott D. Campbell. Detailed instructions, illustrations, tables. Also data on bird habitat and instinct patterns. Bibliography. 3 tables. 63 illustrations in 15 figures. 48pp. 5¼ × 8½. 24407-5 Pa. $1.95

BLOOMINGDALE'S ILLUSTRATED 1886 CATALOG: Fashions, Dry Goods and Housewares, Bloomingdale Brothers. Famed merchants' extremely rare catalog depicting about 1,700 products: clothing, housewares, firearms, dry goods, jewelry, more. Invaluable for dating, identifying vintage items. Also, copyright-free graphics for artists, designers. Co-published with Henry Ford Museum & Green-field Village. 160pp. 8¼ × 11. 25780-0 Pa. $9.95

HISTORIC COSTUME IN PICTURES, Braun & Schneider. Over 1,450 costumed figures in clearly detailed engravings—from dawn of civilization to end of 19th century. Captions. Many folk costumes. 256pp. 8⅜ × 11¾. 23150-X Pa. $11.95

AUTOBIOGRAPHY: The Story of My Experiments with Truth, Mohandas K. Gandhi. Boyhood, legal studies, purification, the growth of the Satyagraha (nonviolent protest) movement. Critical, inspiring work of the man responsible for the freedom of India. 480pp. 5⅜ × 8½. (USO) 24593-4 Pa. $8.95

CELTIC MYTHS AND LEGENDS, T. W. Rolleston. Masterful retelling of Irish and Welsh stories and tales. Cuchulain, King Arthur, Deirdre, the Grail, many more. First paperback edition. 58 full-page illustrations. 512pp. 5⅜ × 8½. 26507-2 Pa. $9.95

THE PRINCIPLES OF PSYCHOLOGY, William James. Famous long course complete, unabridged. Stream of thought, time perception, memory, experimental methods; great work decades ahead of its time. 94 figures. 1,391pp. 5⅜ × 8½. 2-vol. set.
Vol. I: 20381-6 Pa. $12.95
Vol. II: 20382-4 Pa. $12.95

THE WORLD AS WILL AND REPRESENTATION, Arthur Schopenhauer. Definitive English translation of Schopenhauer's life work, correcting more than 1,000 errors, omissions in earlier translations. Translated by E. F. J. Payne. Total of 1,269pp. 5⅜ × 8½. 2-vol. set.
Vol. 1: 21761-2 Pa. $11.95
Vol. 2: 21762-0 Pa. $11.95

MAGIC AND MYSTERY IN TIBET, Madame Alexandra David-Neel. Experiences among lamas, magicians, sages, sorcerers, Bonpa wizards. A true psychic discovery. 32 illustrations. 321pp. 5⅜ × 8½. (USO) 22682-4 Pa. $8.95

THE EGYPTIAN BOOK OF THE DEAD, E. A. Wallis Budge. Complete reproduction of Ani's papyrus, finest ever found. Full hieroglyphic text, interlinear transliteration, word-for-word translation, smooth translation. 533pp. 6½ × 9¼. 21866-X Pa. $9.95

MATHEMATICS FOR THE NONMATHEMATICIAN, Morris Kline. Detailed, college-level treatment of mathematics in cultural and historical context, with numerous exercises. Recommended Reading Lists. Tables. Numerous figures. 641pp. 5⅜ × 8½. 24823-2 Pa. $11.95

THEORY OF WING SECTIONS: Including a Summary of Airfoil Data, Ira H. Abbott and A. E. von Doenhoff. Concise compilation of subsonic aerodynamic characteristics of NACA wing sections, plus description of theory. 350pp. of tables. 693pp. 5⅜ × 8½. 60586-8 Pa. $14.95

THE RIME OF THE ANCIENT MARINER, Gustave Doré, S. T. Coleridge. Doré's finest work; 34 plates capture moods, subtleties of poem. Flawless full-size reproductions printed on facing pages with authoritative text of poem. "Beautiful. Simply beautiful."—*Publisher's Weekly*. 77pp. 9¼ × 12. 22305-1 Pa. $6.95

NORTH AMERICAN INDIAN DESIGNS FOR ARTISTS AND CRAFTSPEOPLE, Eva Wilson. Over 360 authentic copyright-free designs adapted from Navajo blankets, Hopi pottery, Sioux buffalo hides, more. Geometrics, symbolic figures, plant and animal motifs, etc. 128pp. 8⅜ × 11. (EUK) 25341-4 Pa. $7.95

SCULPTURE: Principles and Practice, Louis Slobodkin. Step-by-step approach to clay, plaster, metals, stone; classical and modern. 253 drawings, photos. 255pp. 8⅛ × 11. 22960-2 Pa. $10.95

CATALOG OF DOVER BOOKS

ANATOMY: A Complete Guide for Artists, Joseph Sheppard. A master of figure drawing shows artists how to render human anatomy convincingly. Over 460 illustrations. 224pp. 8⅜ × 11¼. 27279-6 Pa. $10.95

MEDIEVAL CALLIGRAPHY: Its History and Technique, Marc Drogin. Spirited history, comprehensive instruction manual covers 13 styles (ca. 4th century thru 15th). Excellent photographs; directions for duplicating medieval techniques with modern tools. 224pp. 8⅜ × 11¼. 26142-5 Pa. $11.95

DRIED FLOWERS: How to Prepare Them, Sarah Whitlock and Martha Rankin. Complete instructions on how to use silica gel, meal and borax, perlite aggregate, sand and borax, glycerine and water to create attractive permanent flower arrangements. 12 illustrations. 32pp. 5⅜ × 8½. 21802-3 Pa. $1.00

EASY-TO-MAKE BIRD FEEDERS FOR WOODWORKERS, Scott D. Campbell. Detailed, simple-to-use guide for designing, constructing, caring for and using feeders. Text, illustrations for 12 classic and contemporary designs. 96pp. 5⅜ × 8½. 25847-5 Pa. $2.95

OLD-TIME CRAFTS AND TRADES, Peter Stockham. An 1807 book created to teach children about crafts and trades open to them as future careers. It describes in detailed, nontechnical terms 24 different occupations, among them coachmaker, gardener, hairdresser, lacemaker, shoemaker, wheelwright, copper-plate printer, milliner, trunkmaker, merchant and brewer. Finely detailed engravings illustrate each occupation. 192pp. 4⅝ × 6. 27398-9 Pa. $4.95

THE HISTORY OF UNDERCLOTHES, C. Willett Cunnington and Phyllis Cunnington. Fascinating, well-documented survey covering six centuries of English undergarments, enhanced with over 100 illustrations: 12th-century laced-up bodice, footed long drawers (1795), 19th-century bustles, 19th-century corsets for men, Victorian "bust improvers," much more. 272pp. 5⅜ × 8¼. 27124-2 Pa. $9.95

ARTS AND CRAFTS FURNITURE: The Complete Brooks Catalog of 1912, Brooks Manufacturing Co. Photos and detailed descriptions of more than 150 now very collectible furniture designs from the Arts and Crafts movement depict davenports, settees, buffets, desks, tables, chairs, bedsteads, dressers and more, all built of solid, quarter-sawed oak. Invaluable for students and enthusiasts of antiques, Americana and the decorative arts. 80pp. 6½ × 9¼. 27471-3 Pa. $7.95

HOW WE INVENTED THE AIRPLANE: An Illustrated History, Orville Wright. Fascinating firsthand account covers early experiments, construction of planes and motors, first flights, much more. Introduction and commentary by Fred C. Kelly. 76 photographs. 96pp. 8¼ × 11. 25662-6 Pa. $8.95

THE ARTS OF THE SAILOR: Knotting, Splicing and Ropework, Hervey Garrett Smith. Indispensable shipboard reference covers tools, basic knots and useful hitches; handsewing and canvas work, more. Over 100 illustrations. Delightful reading for sea lovers. 256pp. 5⅜ × 8½. 26440-8 Pa. $7.95

FRANK LLOYD WRIGHT'S FALLINGWATER: The House and Its History, Second, Revised Edition, Donald Hoffmann. A total revision—both in text and illustrations—of the standard document on Fallingwater, the boldest, most personal architectural statement of Wright's mature years, updated with valuable new material from the recently opened Frank Lloyd Wright Archives. "Fascinating"—*The New York Times*. 116 illustrations. 128pp. 9¼ × 10¾. 27430-6 Pa. $10.95

THE INFLUENCE OF SEA POWER UPON HISTORY, 1660–1783, A. T. Mahan. Influential classic of naval history and tactics still used as text in war colleges. First paperback edition. 4 maps. 24 battle plans. 640pp. 5⅜ × 8½.
25509-3 Pa. $12.95

THE STORY OF THE TITANIC AS TOLD BY ITS SURVIVORS, Jack Winocour (ed.). What it was really like. Panic, despair, shocking inefficiency, and a little heroism. More thrilling than any fictional account. 26 illustrations. 320pp. 5⅜ × 8½.
20610-6 Pa. $8.95

FAIRY AND FOLK TALES OF THE IRISH PEASANTRY, William Butler Yeats (ed.). Treasury of 64 tales from the twilight world of Celtic myth and legend: "The Soul Cages," "The Kildare Pooka," "King O'Toole and his Goose," many more. Introduction and Notes by W. B. Yeats. 352pp. 5⅜ × 8½.
26941-8 Pa. $8.95

BUDDHIST MAHAYANA TEXTS, E. B. Cowell and Others (eds.). Superb, accurate translations of basic documents in Mahayana Buddhism, highly important in history of religions. The Buddha-karita of Asvaghosha, Larger Sukhavativyuha, more. 448pp. 5⅜ × 8½.
25552-2 Pa. $9.95

ONE TWO THREE . . . INFINITY: Facts and Speculations of Science, George Gamow. Great physicist's fascinating, readable overview of contemporary science: number theory, relativity, fourth dimension, entropy, genes, atomic structure, much more. 128 illustrations. Index. 352pp. 5⅜ × 8½.
25664-2 Pa. $8.95

ENGINEERING IN HISTORY, Richard Shelton Kirby, et al. Broad, nontechnical survey of history's major technological advances: birth of Greek science, industrial revolution, electricity and applied science, 20th-century automation, much more. 181 illustrations. ". . . excellent . . ."—Isis. Bibliography. vii + 530pp. 5⅜ × 8¼.
26412-2 Pa. $14.95

Prices subject to change without notice.

Available at your book dealer or write for free catalog to Dept. GI, Dover Publications, Inc., 31 East 2nd St., Mineola, N.Y. 11501. Dover publishes more than 500 books each year on science, elementary and advanced mathematics, biology, music, art, literary history, social sciences and other areas.